Docker+Kubernetes
容器 实战派

赵渝强◎著

电子工业出版社
Publishing House of Electronics Industry
北京·BEIJING

内 容 简 介

本书基于作者多年的教学与实践经验编写，分为上下两篇，共 20 章。

上篇（第 1～11 章）介绍 Docoker，包含：Docker 入门、Docker 的镜像、Docker 的容器、Docker 的网络通信、使用 Docker Compose 进行服务编排、使用 Docker Machine 进行远程管理、使用 Docker Swarm 构建集群、在 Docker 中实现持续集成与持续部署、基于 Consul 实现 Docker 的服务注册与发现、利用图形工具管理 Docker 及 Docker 应用实战。

下篇（第 12～20 章）介绍 Kubernetes，包含：Kubernetes 体系架构、部署 Kubernetes 集群、Kubernetes 中的最小可部署对象 Pod、使用控制器管理 Pod、通过 Service 访问 Pod、持久化存储、Kubernetes 的安全认证、Kubernetes 中的日志收集与监控、Kubernetes 集成与运维管理。

本书适合对虚拟化容器技术感兴趣的平台架构师、运维管理人员和项目开发人员等阅读。

未经许可，不得以任何方式复制或抄袭本书之部分或全部内容。
版权所有，侵权必究。

图书在版编目（CIP）数据

Docker+Kubernetes 容器实战派 / 赵渝强著. —北京：电子工业出版社，2022.5
ISBN 978-7-121-43313-9

Ⅰ.①D… Ⅱ.①赵… Ⅲ.①Linux 操作系统—程序设计 Ⅳ.①TP316.85

中国版本图书馆 CIP 数据核字（2022）第 071499 号

责任编辑：吴宏伟
印　　刷：三河市良远印务有限公司
装　　订：三河市良远印务有限公司
出版发行：电子工业出版社
　　　　　北京市海淀区万寿路 173 信箱　邮编：100036
开　　本：787×980　1/16　印张：28.75　字数：690 千字
版　　次：2022 年 5 月第 1 版
印　　次：2023 年 6 月第 4 次印刷
定　　价：118.00 元

凡所购买电子工业出版社图书有缺损问题，请向购买书店调换。若书店售缺，请与本社发行部联系，联系及邮购电话：(010) 88254888，88258888。
质量投诉请发邮件至 zlts@phei.com.cn，盗版侵权举报请发邮件至 dbqq@phei.com.cn。
本书咨询联系方式：(010) 51260888-819，faq@phei.com.cn。

前言

随着云计算的不断发展，为了实现复杂应用的高效部署、管理及持续集成，出现了 Docker 和 Kubernetes 等虚拟化容器技术。虚拟化容器技术的发展，使得整个云原生领域更加成熟与健壮。本书正是在这样的背景下编写的。

笔者拥有多年的教学与实践经验，因此想系统地编写一本虚拟化容器技术方面的书，力求能够系统地介绍 Docker 与 Kubernetes。通过本书，笔者一方面希望总结自己在虚拟化容器技术方面的经验，另一方面希望对相关从业者有所帮助，从而为云原生体系在国内的发展贡献一份力量。

1. 本书特色

（1）一次讲解两种技术。

对于单一的应用，我们可以利用 Docker 将应用及其依赖打包到镜像中，从而很好地解决应用部署与集成的问题。但是，当我们需要将应用进行大规模部署时，则应使用 Kubernetes。因为，Docker 本质上是一种单一的容器技术（或者说是一种工具），并不能很好地将应用组织起来，难以独立地支撑起生产环境中应用的大规模容器化部署。Kubernetes 的功能包括应用的服务编排、容器集群的部署和集群的管理。

为了降低读者的学习成本，本书一次讲解了上述两种技术。书中覆盖了这两种技术的核心内容，可以帮助读者快速入门并应对工作中的大部分需求。

（2）主线清晰，循序渐进。

笔者在长期的教学过程中反复修订了自己的讲解主线，发现这样的主线更利于读者顺利地从"入门小白"成为"开发高手"。

（3）突出实战，注重效果。

全书采用"理论+实操"的方式进行讲解，在读者了解了概念、原理、方法后会进行实操，这样真正做到"知行合一"，而不只是停留在"知道""了解"的层面。

书中还提供了大量来源于真实项目的技术解决方案，这些解决方案可以在实际的生产环境中给技术人员提供相应的指导。

（4）言简意赅，阅读性强。

全书采用通俗易懂的文字编写，避免了复杂的语法和生僻的辞藻。书稿经过多次打磨，力求做到表达精确、前后内容衔接顺畅，以期让从未接触过容器技术的读者也可以读得懂、学得会。

（5）深入原理，关注难点和易错点。

本书不只停留在操作层，还深入介绍了容器技术的底层原理和机制。笔者在教学过程中发现有些内容对很多学员来说是难点、易错点，所以，书中对这些知识进行了详细讲解，希望读者可以很轻松地绕开这些"漩涡"。

2. 读者对象

本书读者对象如下：

- 初学容器技术的自学者；
- 培训机构的老师和学员；
- 中高级技术人员；
- 相关专业的大学毕业学生；
- 开发工程师；
- 测试工程师；
- 容器技术爱好者；
- 技术运维人员；
- 高等院校的教师和学生；
- 技术管理人员。

尽管笔者在本书写作过程中尽可能地追求严谨，但仍难免有纰漏之处，欢迎读者加入本书读者交流群批评与指正。

赵渝强

2022 年 2 月

读者服务

微信扫码回复：43313

- 获取本书配套代码
- 加入本书读者交流群，与更多读者互动
- 获取【百场业界大咖直播合集】（持续更新），仅需 1 元

目录

上篇　Docker 从原理到实战

第 1 章　Docker 入门 2
- 1.1　为什么需要容器技术 2
- 1.2　Docker 简介 4
- 1.3　Docker 的体系架构与基本概念 4
- 1.4　安装 Docker 5
  - 1.4.1　安装 Linux 操作系统 6
  - 1.4.2　使用 YUM 方式安装 Docker 16
  - 1.4.3　使用二进制包方式安装 Docker 18
  - 1.4.4　验证 Docker 环境 21
- 1.5　【实战】在 Docker 中部署第一个应用 22

第 2 章　Docker 的镜像 26
- 2.1　什么是 Docker 的镜像 26
  - 2.1.1　使用 Docker 默认的镜像存储路径 27
  - 2.1.2　自定义 Docker 的镜像存储路径 28
- 2.2　使用 Docker 的公有镜像仓库 28
  - 2.2.1　【实战】访问 Docker 官方的公有镜像仓库 29
  - 2.2.2　【实战】配置和使用阿里云 Docker 镜像加速仓库 32
- 2.3　使用命令行工具管理 Docker 的镜像 34
- 2.4　构建自己的镜像 36
  - 2.4.1　【实战】使用 "docker commit" 命令构建镜像 36
  - 2.4.2　【实战】使用 Dockerfile 文件构建镜像 39
  - 2.4.3　Dockerfile 文件详解 40
  - 2.4.4　【实战】使用 Dockerfile 文件的综合案例 42

2.5 搭建私有镜像仓库 Harbor 44
2.5.1 安装 Docker 和 Docker Compose 45
2.5.2 安装与配置 Harbor 46
2.5.3 【实战】在 Docker 中使用 Harbor 48

第 3 章 Docker 的容器 50
3.1 Docker 容器的基本概念与操作 50
3.2 Docker 的日志 53
3.2.1 【实战】访问 Docker 引擎的日志 53
3.2.2 【实战】访问 Docker 应用的日志 54
3.3 管理容器的资源 56
3.3.1 什么是 Linux CGroup 56
3.3.2 【实战】Docker 对 CPU 的使用 62
3.3.3 【实战】Docker 对内存的使用 64
3.3.4 【实战】Docker 对 I/O 带宽的使用 65
3.4 管理 Docker 容器中的数据 67
3.4.1 在 Docker 容器中实现数据管理的两种方式 67
3.4.2 【实战】使用数据卷管理 Docker 容器中的数据 68
3.4.3 【实战】使用数据卷容器管理 Docker 容器中的数据 71

第 4 章 Docker 的网络通信 75
4.1 Docker 容器网络通信的基本原理 75
4.2 使用命令查看 Docker 的网络配置信息 78
4.3 Docker 的 4 种网络通信模式 80
4.3.1 bridge 模式 80
4.3.2 host 模式 82
4.3.3 container 模式 83
4.3.4 none 模式 85
4.4 容器间的通信 86
4.4.1 【实战】通过 IP 地址进行通信 86
4.4.2 【实战】通过 Docker DNS Server 进行通信 87
4.4.3 【实战】通过 Joined 方式进行通信 88
4.4.4 容器间的跨节点通信 89
4.5 容器的网络访问控制 95
4.5.1 容器内的应用访问外部网络 95
4.5.2 从外部网络访问容器内的应用 96

第 5 章 使用 Docker Compose 进行服务编排 97
5.1 配置 Docker Compose 97
5.2 进行服务编排 98
5.2.1 【实战】使用手动方式部署应用 99
5.2.2 【实战】使用 Docker Compose 部署应用 101
5.2.3 【实战】使用 Docker Compose 进行服务的在线扩容/缩容 102
5.2.4 【实战】在 Docker Compose 中控制模块启动和停止的顺序 104
5.3 Docker Compose 中的网络 108
5.3.1 Docker Compose 中的默认网络环境 108
5.3.2 在 Docker Compose 中自定义模块的网络环境 109

第 6 章 使用 Docker Machine 进行远程管理 111
6.1 使用 Docker Machine 112
6.1.1 安装 Docker Machine 112
6.1.2 在远端宿主机上安装 Docker 112
6.2 Docker Machine 的基本用法 115
6.2.1 【实战】使用 Docker Machine 的命令 115
6.2.2 【实战】管理远端的 Docker 节点 116
6.3 Docker Machine 的高级用法 118
6.3.1 【实战】使用 Docker Machine 创建基于 VirtualBox 的虚拟主机 118
6.3.2 【实战】使用 Docker Machine 创建基于 vSphere 的虚拟主机 120

第 7 章 使用 Docker Swarm 构建集群 123
7.1 Docker Swarm 集群的体系架构 123
7.2 构建 Docker Swarm 集群 124
7.3 在 Docker Swarm 集群中部署应用与 HAProxy 126
7.3.1 【实战】在集群中部署应用 126
7.3.2 【实战】测试集群的高可用性 128
7.3.3 【实战】使用 HAProxy 为集群添加外部负载均衡功能 130
7.3.4 【实战】实现服务的滚动更新 131
7.4 Docker Swarm 集群的数据持久化 132
7.4.1 【实战】通过 volume 实现集群的数据持久化 133
7.4.2 【实战】通过 NFS 实现集群的数据持久化 134
7.5 Docker Swarm 集群的负载均衡 138
7.5.1 【实战】测试 Docker Swarm 集群的负载均衡 138

第 8 章　在 Docker 中实现持续集成与持续部署 142

- 8.1　什么是持续集成与持续部署（CI/CD） 142
- 8.2　Jenkins 简介与部署 143
  - 8.2.1　Docker 与 Jenkins 集成的体系架构 143
  - 8.2.2　【实战】部署 Jenkins 144
  - 8.2.3　【实战】使用 Jenkins 部署第一个应用 146
- 8.3　基于 Jenkins 实现 Docker 应用的持续集成与持续部署 149
  - 8.3.1　【实战】准备私有代码仓库 SVN 150
  - 8.3.2　开发 Dockerfile 文件 152
  - 8.3.3　【实战】集成 Jenkins 和 Docker 152

第 9 章　基于 Consul 实现 Docker 的服务注册与发现 155

- 9.1　服务的注册与发现 155
  - 9.1.1　什么是服务的注册与发现 155
  - 9.1.2　为什么需要服务的注册与发现 156
  - 9.1.3　常见的服务注册中心 159
- 9.2　注册中心 Consul 的基本使用 159
  - 9.2.1　Consul 的安装与启动 160
  - 9.2.2　【实战】使用 JSON 文件在 Consul 中注册服务 161
  - 9.2.3　【实战】使用 API 在 Consul 中注册服务 163
- 9.3　集成 Consul 与 Docker 166
  - 9.3.1　Docker 服务注册与发现的体系架构 166
  - 9.3.2　【实战】使用 Registrator 镜像实现 Docker 服务的注册 167
  - 9.3.3　【实战】使用 Consul-Template 实现 Docker 服务的发现ш.... 170

第 10 章　利用图形工具管理 Docker 173

- 10.1　单机环境中的 Docker 图形工具：Docker UI 173
  - 10.1.1　部署 Docker UI 173
  - 10.1.2　【实战】使用 Docker UI 管理镜像与容器 174
- 10.2　轻量级的 Docker 图形工具：Portainer 176
  - 10.2.1　在单机环境中部署 Portainer 177
  - 10.2.2　【实战】使用 Portainer 管理 Docker 的镜像与容器 179
  - 10.2.3　【实战】使用 Portainer 管理远端主机上的 Docker 180
  - 10.2.4　在 Docker Swarm 集群中部署 Portainer 182

7.5.2　选择 Docker Swarm 集群的负载均衡模式 139

10.3 开源的 Docker 图形工具——Shipyard 184
　　10.3.1 Shipyard 的组件 184
　　10.3.2 部署 Shipyard 184
　　10.3.3 【实战】使用 Shipyard 创建容器 187

第 11 章 Docker 应用实战 188

11.1 Docker 与数据库 188
　　11.1.1 在 Docker 容器中部署 MySQL 188
　　11.1.2 数据库不适合 Docker 容器化的原因 189
11.2 【实战】Docker 与 Python 190
11.3 【实战】Docker 与 PHP 192

下篇　Kubernetes 从原理到实战

第 12 章 Kubernetes 体系架构 196

12.1 什么是 Kubernetes 196
12.2 Kubernetes 集群 197
　　12.2.1 集群的架构体系 198
　　12.2.2 Kubernetes 的核心组件 198
　　12.2.3 Kubernetes 的常用附加组件 199
12.3 Kubernetes 的对象 200
　　12.3.1 对象的管理 200
　　12.3.2 对象与命名空间 201
　　12.3.3 对象的标签 202

第 13 章 部署 Kubernetes 集群 204

13.1 Kubernetes 的部署方式 204
　　13.1.1 使用 kubeadmin 部署 Kubernetes 集群 204
　　13.1.2 使用 YUM 方式部署 Kubernetes 集群 208
　　13.1.3 使用二进制包部署 Kubernetes 集群 212
　　13.1.4 使用 minikube 工具部署 Kubernetes 单机版集群 231
　　13.1.5 Kubernetes 集群的高可用 236
13.2 Kubernetes 的客户端工具 237
　　13.2.1 Kubernetes 图形管理工具——DashBoard UI 237
　　13.2.2 Kubernetes 命令行管理工具——kubectl 240

13.3 【实战】使用 Kubectl 在 Kubernetes 中部署第一个应用 242

第 14 章 Kubernetes 中的最小可部署对象 Pod 246

14.1 什么是 Pod 246
14.2 【实战】Pod 的基本使用方法 247
14.3 Pod 中的容器 251
  14.3.1 基础容器 251
  14.3.2 初始化容器 252
  14.3.3 临时容器 253
  14.3.4 业务容器 254
14.4 Pod 的生命周期 255
  14.4.1 Pod 的阶段与容器的状态 255
  14.4.2 Pod 中容器的重启策略 256
  14.4.3 【实战】Pod 的健康检查 257
14.5 Pod 的调度策略 261
  14.5.1 Pod 的创建过程 262
  14.5.2 【实战】自定义 Pod 调度的约束策略 262
14.6 Pod 资源的使用限制 264
14.7 Pod 的镜像拉取策略 267
14.8 Pod 的配置管理 268
  14.8.1 为什么需要配置管理 268
  14.8.2 【实战】使用 ConfigMap 管理 Pod 的配置信息 268
  14.8.3 【实战】使用 Secret 管理 Pod 的配置信息 275

第 15 章 使用控制器管理 Pod 280

15.1 为什么需要控制器 280
15.2 Deployment 控制器 281
  15.2.1 【实战】创建和使用 Deployment 控制器 282
  15.2.2 【实战】验证 Deployment 控制器的不同状态 286
  15.2.3 【实战】Deployment 控制器的清理策略 289
  15.2.4 应用的部署 292
  15.2.5 编写 Deployment 控制器的规则 304
15.3 DaemonSet 控制器 305
  15.3.1 DaemonSet 控制器的创建 305
  15.3.2 DaemonSet 控制器的调度 307
15.4 Job 控制器 307

15.4.1	【实战】单工作队列的 Job 串行方式	307
15.4.2	【实战】多工作队列的 Job 并行方式	308
15.4.3	Job 的终止与清理	310

15.5 CronJob 控制器 ... 312

15.5.1	【实战】运行第一个 CronJob 控制器	313
15.5.2	CronJob 控制器中的时间表示	314
15.5.3	CronJob 控制器的限制	315

15.6 StatefulSets 控制器 ... 316

15.6.1	【实战】创建 StatefulSets 控制器	316
15.6.2	StatefulSets 控制器的扩容/缩容	317
15.6.3	StatefulSets 控制器的更新/回滚	318

第 16 章 通过 Service 访问 Pod ... 320

16.1 Service 的概念与使用 ... 320

16.1.1	【实战】通过 Service 向外部暴露 Pod	321
16.1.2	Service 的多端口设置	323
16.1.3	集群内部的 DNS 服务	324
16.1.4	【实战】无头 Service	325

16.2 Service 的发布类型 ... 328

16.2.1	NodePort	328
16.2.2	ClusterIP	329
16.2.3	LoadBalance	331
16.2.4	ExternalName	332

16.3 虚拟 IP 与 Service 的代理模式 ... 333

16.3.1	userspace 代理模式	333
16.3.2	iptables 代理模式	334
16.3.3	IPVS 代理模式	336

16.4 集群外部的请求访问集群内应用的最佳方式——Ingress ... 339

16.4.1	Ingress 是什么	339
16.4.2	【实战】使用 Ingress Controller 创建 Ingress	340
16.4.3	【实战】使用 Ingress 的注解	344
16.4.4	基于 Ingress 的高可用架构	348

第 17 章 持久化存储 ... 349

17.1 Kubernetes 持久化存储方式 ... 349

17.1.1	【实战】使用节点数据卷	350

17.1.2 【实战】使用网络数据卷 351
17.1.3 【实战】使用临时数据卷 353
17.2 持久卷 355
17.2.1 持久卷是什么 355
17.2.2 【实战】第一个持久卷示例 355
17.2.3 持久卷的访问模式 357
17.2.4 【实战】持久卷的回收策略 359
17.3 持久卷声明 361
17.3.1 持久卷和持久卷声明的区别 361
17.3.2 【实战】在 Pod 中使用持久卷声明 362
17.3.3 storageClass 详解 364
17.4 【实战】实现持久卷的动态供给 365

第 18 章 Kubernetes 的安全认证 372

18.1 Kubernetes 的安全框架 372
18.2 Kubernetes 的用户认证 374
18.3 Kubernetes 的鉴权管理 378
18.3.1 基于角色的访问控制（RBAC 鉴权）....................... 379
18.3.2 基于属性的访问控制（ABAC 鉴权）....................... 384
18.3.3 基于节点的访问控制（node 鉴权）......................... 386
18.3.4 基于 Webhook 的访问控制 387
18.4 管理服务账号（Service Account）................................... 391
18.4.1 服务账号与用户账号 391
18.4.2 【实战】创建和使用服务账号 391
18.4.3 服务账号的工作机制 396

第 19 章 Kubernetes 中的日志收集与监控 398

19.1 收集哪些日志 398
19.2 日志收集方案 399
19.2.1 初识 ELK 399
19.2.2 日志收集的架构 399
19.2.3 日志收集方案详解 400
19.3 实现 Kubernetes 集群的日志收集 401
19.3.1 安装 ELK 402
19.3.2 【实战】采集 Kubernetes 系统组件的日志 405
19.3.3 【实战】采集 Nginx 应用的日志 410

19.3.4 【实战】采集 Tomcat 应用的日志 414
19.4 监控 Kubernetes 418
19.4.1 Kubernetes 监控方案 418
19.4.2 【实战】部署 Kubernetes 监控系统 419

第 20 章 Kubernetes 集成与运维管理 427

20.1 Jenkins 与 Kubernetes 的持续集成与持续部署 427
20.1.1 基于 Kubernetes 的 Jenkins 集群架构 427
20.1.2 【实战】Jenkins 与 Kubernetes 的集成 428
20.2 使用 Helm 简化 Kubernetes 应用的部署和管理 435
20.2.1 什么是 Helm 435
20.2.2 部署 Helm 435
20.2.3 【实战】使用 Helm 管理 Kubernetes 437

上篇
Docker 从原理到实战

第 1 章
Docker 入门

Docker 是虚拟化容器引擎中非常重要的一员。现在让我们开始学习 Docker。

1.1 为什么需要容器技术

1. 面临的问题

在项目迭代开发和部署过程中，运维人员不可避免会遇到这些情况：同样的代码需要被部署到不同运行环境中；由于运行环境的不同导致了项目部署失败，无法正常运行。

导致上述问题的原因可能不同。下面列举了几种比较典型的原因。

- 运行环境发生了变化。例如，在开发阶段，代码是运行在程序员的本地环境或测试环境中的；而在生产环境中，代码则被切换到集群环境或云平台上了。
- 代码的依赖发生了变化。例如，在开发阶段使用的是 JDK 1.7，而在生产环境中使用的是 JDK 1.8。
- 操作系统发生了变化。例如：在开发阶段使用的是 Redhat Linux，而在生产环境中使用的是 CentOS。
- 随着系统架构不断变复杂，微服务架构得到了广泛的应用，这就要求每个功能模块需要单独进行部署。图 1-1 展示了一个典型的微服务架构。

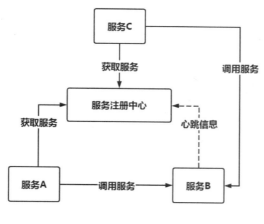

图 1-1

这些问题都会对参与项目的人员造成以下影响。

- 对于开发人员而言，除需要关注业务代码本身的实现外，还需要额外花费精力去处理这种底层执行环境的问题。
- 对于运维人员而言，需要考虑如何将开发人员新开发的业务代码不断集成与发布。

有没有一种办法能使得开发人员和运维人员只要关注他们需要关注的问题，而不需要关注其他的问题呢？答案是有的——使用虚拟化容器技术。

有了虚拟化容器技术，开发人员在业务开发过程中只需要关注业务代码的实现；而运维人员也可以很方便地实现项目的持续集成与持续发布（CI/CD）。Docker 是虚拟化容器技术中的典型代表，也是云计算中的重点。

2. Docker 虚拟化容器的价值

Docker 虚拟化容器有以下两方面的价值。

- 从系统的架构层面上看：Docker 可以方便地支持并实现微服务架构，从而更方便灵活地实现架构的变化和系统的扩展。同时，Docker 虚拟化容器有助于 DevOps 的落地，可以大大提升开发效率，加速迭代。
- 从底层基础层面上看：利用 Docker 虚拟化容器技术可以方便地实现系统的移植，帮助实现企业应用上云，让应用在自有数据中心和云端之间实现动态迁移。

随着云计算技术的不断发展，在产生 Docker 虚拟化容器技术后，一批相关的虚拟化容器管理技术也随之诞生，例如 Kubernetes（K8s）等。这样的工具极大地推动了技术的分工，也极大地促进了技术和业务的创新。

1.2 Docker 简介

Docker 是一个开源的虚拟化容器引擎，让开发者可以打包他们的应用及依赖到一个可移植的容器中，然后发布到 Linux 环境中以实现虚拟化的管理。这些 Linux 环境包括 CentOS、Redhat、Ubuntu 等。在 Windows 上也可以部署 Docker，但不推荐。

Docker 中的虚拟化容器完全使用"沙箱"机制，相互之间不会有任何接口。可以把这些容器理解为是逻辑隔离的。

一个完整的 Docker 由以下几部分组成：

- Docker 客户端。
- Docker 守护进程（Daemon）。
- Docker 镜像（Image）。
- Docker 容器（Container）。
- 镜像仓库（Repository）。

Docker 实现了应用代码与底层运行环境之间的耦合。它可以将一个复杂系统中的各个模块进行容器化，同时提供了负载均衡和失败迁移功能。应用的容器化，满足了敏捷开发、动态迁移、标准化的要求，从而大大提高了效率。

1.3 Docker 的体系架构与基本概念

Docker 是一个客户端服务器（Client-Server）架构。Docker 客户端和 Docker 守护进程交流，而 Docker 守护进程是运作 Docker 的核心，起着非常重要的作用（如构建、运行和分发 Docker 容器等）。

> Docker 客户端和 Docker 守护进程可以运行在同样的系统上，也可以运行在不同的系统上。用户可以将一个 Docker 客户端连接到一个远程 Docker 守护进程中。Docker 客户端和守护进程，通过 sockets 或 RESTFul API 进行沟通。

我们在使用 Docker 创建容器时需要有镜像。镜像是一个只读的模板。而存放镜像的地方叫作"镜像仓库"。镜像仓库，可以是公有镜像仓库（例如官方提供的公有镜像仓库 Docker Hub），也可以是私有镜像仓库（例如 Harbor）。

图 1-2 展示了 Docker 的体系架构。

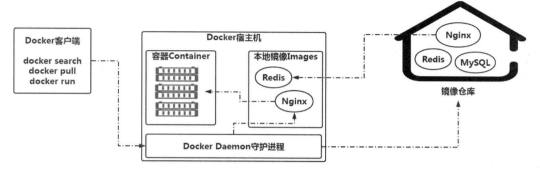

图 1-2

表 1-1 列出了 Docker 体系架构中的组成部分及其功能特性。

表 1-1

组成部分	功能特性
Docker 客户端	通常指 Docker 提供的命令行工具，是 Docker 最基本的用户接口。用户通过 Docker 客户端提交 Docker 指令，Docker 守护进程接收并执行该指令。Docker 也有图形化的客户端工具
Docker 守护进程	在 Docker 宿主机上运行 Docker，实际上运行的是 Docker 守护进程。用户并不直接和 Docker 守护进程交互，而是通过 Docker 客户端的命令来和它进行交互
Docker 镜像	一个 Docker 镜像是一个只读的模板。例如，一个 Docker 镜像可以包含一个 CentOS 的操作系统、一个 MySQL 的数据库和一个 Tomcat 的应用服务器。Docker 镜像被用来创建 Docker 容器。Docker 提供了一个简单的方式来构建一个新的镜像或更新一个已经存在的镜像。用户也可以从镜像仓库下载其他人已经创建好的 Docker 镜像
Docker 容器	通过 Docker 镜像可以创建 Docker 容器。Docker 的容器可以保存任何东西，而这些东西是运行一个应用所必需的。可以把 Docker 容器看成是一个虚拟机。Docker 容器可以被运行、开启、停止、移动和删除。每一个容器都是一个分离的、安全的应用平台。Docker 容器是 Docker 的运行组件
镜像仓库	Docker 的镜像仓库用于保存 Docker 镜像，它可以是公共的存储地方，也可以是私有的存储地方。 • 公共的镜像仓库由 Docker Hub 提供，它提供了一个用户可以使用的已有镜像的集合。这些集合中的镜像可以是你自己创建的，也可以是别人创建的。 • 私有的镜像仓库需要自己在私有环境中搭建，例如在企业内网中自行搭建。Harbor 是一个典型的私有的镜像仓库

1.4 安装 Docker

　　Docker 实现虚拟化的本质是：在已经运行的 Linux 中创建了一个逻辑隔离的运行环境。因此，其执行效率几乎等同于宿主机的 Linux 主机。

Docker 必须部署在 Linux 系统上。如果想在其他系统（如 Windows）上部署 Docker，则需要先安装一个虚拟 Linux 环境。表 1-2 列举了本书所使用的实验环境。

表 1-2

所需软件	软件说明
CentOS-7-x86_64-Everything-1708.iso	CentOS Linux 7 64 位安装文件，将作为运行 Docker 的宿主机
VMware® Workstation 12 Pro	虚拟机管理器，用于部署 CentOS 环境
Xshell 7	远程 SSH 登录工具，用于使用命令行方式登录 CentOS

1.4.1 安装 Linux 操作系统

由于 Docker 必须运行在 Linux 上，因此，在安装 Docker 之前需要先安装 Linux。在安装 Linux 时，需要选择 NAT 网络模式，这主要是让 Docker 宿主机能够访问外部的互联网。下面是安装 Linux 的步骤。

（1）检查 VMWare Workstations NAT 网络模式的配置。在图 1-3 中选择"虚拟网络编辑器"命令，然后在如图 1-4 所示的"虚拟网络编辑器"对话框中确定 NAT 模式的子网地址是 192.168.79.0。在设置 CentOS Linux 的 IP 地址时，需要将 IP 地址设置在这个网段内。

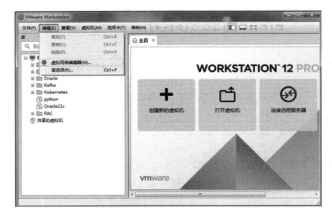

图 1-3

图 1-4

（2）在"虚拟网络编辑器"对话框中单击"NAT 设置"，可以查看 NAT 模式的网关地址和详细信息，如图 1-5 所示。之后就可以创建虚拟机了。

（3）在 VMWare Workstations 的主页上单击"创建新的虚拟机"，如图 1-6 所示。

图 1-5

图 1-6

（4）在"欢迎使用新建虚拟机向导"界面中选择"自定义（高级）（C）"单选框，如图 1-7 所示。然后在"选择虚拟机硬件兼容性"界面中单击"下一步"按钮，如图 1-8 所示。

图 1-7

图 1-8

（5）在"安装客户机操作系统"界面中，选择"安装程序光盘镜像文件(iso)(M)："单选按钮，并单击"浏览"按钮找到 CentOS-7-x86_64-Everything-1708.iso 文件，然后单击"下一步"按钮，如图 1-9 所示。

（6）在"命名虚拟机"界面中，输入虚拟机名称和保存的位置。这是第 1 个 Docker 的宿主机，使用"master"作为它的名字，如图 1-10 所示。这主要是为了后续在搭建 Docker 集群时可以方便地进行集群的规划。

图 1-9　　　　　　　　　　　　图 1-10

（7）在"处理器配置"界面中直接单击"下一步"按钮，如图 1-11 所示。然后在"此虚拟机的内存"界面中为虚拟机分配适当的内存（如 2048MB），如图 1-12 所示。

图 1-11　　　　　　　　　　　　图 1-12

（8）在"网络类型"界面中，选择"使用网络地址转换（NAT）（E）"单选框，然后单击"下一步"按钮，如图 1-13 所示。

（9）在"I/O 控制器类型"和"磁盘类型"界面中，直接单击"下一步"按钮。

（10）在"选择磁盘"界面中，选择"创建新虚拟磁盘（V）"单选框，单击"下一步"按钮，如图 1-14 所示。

图 1-13

图 1-14

（11）在"指定磁盘容量"界面中设定磁盘的大小。这里可以根据需要进行设置（如 50GB），然后单击"下一步"按钮，如图 1-15 所示。

（12）在"指定磁盘文件"界面中直接单击"下一步"按钮。

（13）在"已准备好创建虚拟机"界面中，勾选"创建后开启此虚拟机（P）"，然后单击"完成"按钮，如图 1-16 所示。

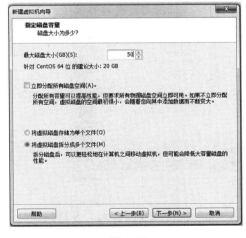
图 1-15

图 1-16

（14）在创建完虚拟机后，会自动运行 CentOS 的启动界面。选择"Install CentOS 7"，如图 1-17 所示。

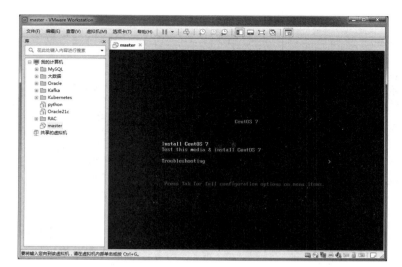

图 1-17

（15）进入"CentOS 7 欢迎界面"，如图 1-18 所示，单击"Continue"按钮。

图 1-18

（16）进入"INSTALLATION SUMMARY"界面，如图 1-19 所示，在这里将完成对 CentOS 的配置。

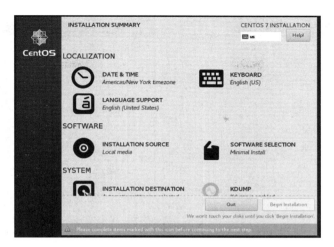

图 1-19

（17）在"INSTALLATION SUMMARY"界面中，选择"DATE & TIME"进入时区和时间的设置。这里可以单击地图上的"Shanghai"，然后单击界面左上角的"Done"按钮完成设置。

（18）在"INSTALLATION SUMMARY"界面中，选择"SOFTWARE SELECTION"，进入软件配置界面。这里勾选"Development Tools"复选框，在安装过程中会自动安装必需的开发工具，如 gcc 的编译器等。然后单击界面左上角的"Done"按钮完成设置，如图 1-20 所示。

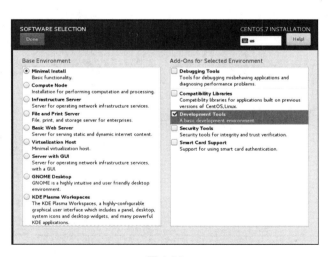

图 1-20

（19）在"INSTALLATION SUMMARY"界面中，选择"KDUMP"和"SECURITY POLICY"，进入它们的配置界面，分别禁用 KDUMP 和 SECURITY POLICY，如图 1-21 和图 1-22 所示。

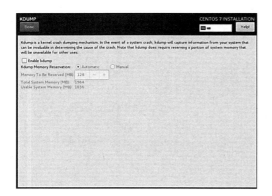
　　　　　　图 1-21

　　　　　　图 1-22

（20）在"INSTALLATION SUMMARY"界面中，选择"NETWORK & HOSTNAME"进入网络和主机名的配置界面。

- 将 Host name 设置为"master"，单击"Apply"按钮生效。
- 将界面右上角的网络开关设置为"ON"状态，这时会自动为虚拟机分配一个 IP 地址。由于使用的是 NAT 网络模式，我们可以看到分配的 IP 地址是：192.169.79.226。
- 为了方便后期集群的规划，可以对 IP 地址进行适当修改，然后单击"Configure"按钮，如图 1-23 所示。

（21）在"Editing ens33"对话框中选择"General"选项卡，勾选"Automatically connect to this network when it is available"复选框，如图 1-24 所示。

　　　　　　图 1-23
　　　　　　图 1-24

（22）在"Editing ens33"对话框中选择"IPv4 Setttings"选项卡，并在"Method"下拉列表中选择"Manual"，单击"Add"按钮添加一个 IP 地址，如图 1-25 所示。参数设置如下，然后单击"Save"按钮。

- IP 地址：192.168.79.11；
- 子网掩码：255.255.255.0；
- 网关 Gateway：192.168.79.2；
- DNS Server：192.168.79.2。

（23）完成网络和主机名设置后的界面如图 1-26 所示。

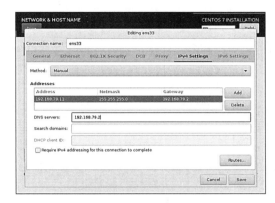

图 1-25

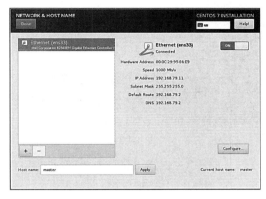

图 1-26

（24）在"INSTALLATION SUMMARY"界面中选择"INSTALLATION DESTINATION"，然后选择之前创建的 50GB 磁盘，如图 1-27 所示。

（25）图 1-28 展示了完成 CentOS 设置后的界面，单击"Begin Installation"按钮开始安装。

图 1-27

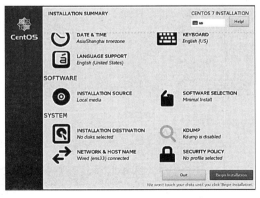

图 1-28

（26）图 1-29 展示了 CentOS 的安装过程，可以看到"Root password is not set"（提示需要设置 root 用户的密码），单击"ROOT PASSWORD"按钮。

（27）在"ROOT PASSWORD"界面中自行设置 Root 的用户密码。完成后单击界面左上角的"Done"按钮，如图 1-30 所示。

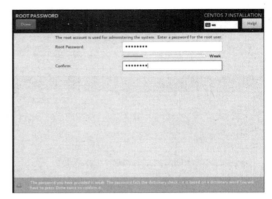

图 1-29　　　　　　　　　　　　　　图 1-30

（28）图 1-31 展示了 CentOS 的安装进度条。安装完成后的界面如图 1-32 所示，单击"Reboot"按钮重启虚拟机。

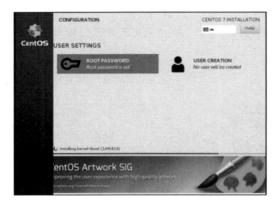
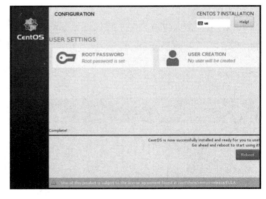

图 1-31　　　　　　　　　　　　　　图 1-32

（29）在虚拟机重启完成后，会自动进入登录界面，如图 1-33 所示。

图 1-33

（30）为了方便操作，可以通过命令行工具 Xshell 进行远程登录。图 1-34 展示了配置好的 Xshell 会话，可以看到这里配置了一个 master 的会话，它的主机地址就是 master 虚拟主机的 IP 地址，单击"连接"按钮。

图 1-34

（31）由于是第一次通过 Xshell 进行远程登录，所以需要在如图 1-35 所示的界面中单击"接收并保存"按钮以保存主机的密钥信息。

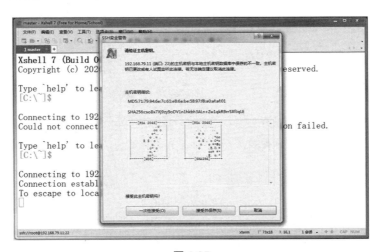

图 1-35

（32）图 1-36 展示了通过 Xshell 登录成功的界面。在 Xshell 命令行中输入"ifconfig"命令来确定虚拟机的 IP 地址。

```
[root@master ~]# ifconfig
ens33: flags=4163<UP,BROADCAST,RUNNING,MULTICAST>  mtu 1500
        inet 192.168.79.11  netmask 255.255.255.0  broadcast 192.168.79.255
        inet6 fe80::9927:4703:e66d:3c67  prefixlen 64  scopeid 0x20<link>
        ether 00:0c:29:95:e8:e9  txqueuelen 1000  (Ethernet)
        RX packets 60  bytes 7263 (7.0 KiB)
        RX errors 0  dropped 0  overruns 0  frame 0
        TX packets 63  bytes 9729 (9.5 KiB)
        TX errors 0  dropped 0 overruns 0  carrier 0  collisions 0

lo: flags=73<UP,LOOPBACK,RUNNING>  mtu 65536
        inet 127.0.0.1  netmask 255.0.0.0
        inet6 ::1  prefixlen 128  scopeid 0x10<host>
        loop  txqueuelen 1  (Local Loopback)
        RX packets 0  bytes 0 (0.0 B)
        RX errors 0  dropped 0  overruns 0  frame 0
        TX packets 0  bytes 0 (0.0 B)
        TX errors 0  dropped 0 overruns 0  carrier 0  collisions 0

[root@master ~]#
```

图 1-36

（33）为了便于进行 Docker 的相关操作，需要关闭 Linux 防火墙和 SELinux。输入以下的命令即可，如图 1-37 所示。

```
systemctl stop firewalld.service
systemctl disable firewalld.service

setenforce 0
```

```
[root@master ~]# systemctl stop firewalld.service
[root@master ~]# systemctl disable firewalld.service
Removed symlink /etc/systemd/system/multi-user.target.wants/firewalld.service.
Removed symlink /etc/systemd/system/dbus-org.fedoraproject.FirewallD1.service.
[root@master ~]# setenforce 0
[root@master ~]#
```

图 1-37

至此，CentOS 虚拟机已经成功安装，它将作为运行 Docker 的宿主机环境。在安装 Docker 的过程中，可以使用两种方式进行安装：YUM 方式和二进制包的方式。下面分别介绍这两种方式。

1.4.2　使用 YUM 方式安装 Docker

YUM 的全称是 Yellow dogUpdater Modified。利用 YUM 方式可以很方便地添加、删除和更新 Linux 系统的程序包，并且能够自动解决包的依赖性问题。使用 YUM 也能够方便地管理大量的系统更新问题。一般使用 YUM 方式需要连接外部的网络。

（1）为了验证虚拟机是否可以访问外部的网络，这里输入以下的命令访问百度主页，如图 1-38 所示。

```
ping www.baidu.com
```

```
[root@master ~]# ping www.baidu.com
PING www.a.shifen.com (110.242.68.3) 56(84) bytes of data.
64 bytes from 110.242.68.3 (110.242.68.3): icmp_seq=1 ttl=128 time=90.0 ms
64 bytes from 110.242.68.3 (110.242.68.3): icmp_seq=2 ttl=128 time=20.9 ms
64 bytes from 110.242.68.3 (110.242.68.3): icmp_seq=3 ttl=128 time=11.0 ms
64 bytes from 110.242.68.3 (110.242.68.3): icmp_seq=4 ttl=128 time=11.2 ms
^C
--- www.a.shifen.com ping statistics ---
4 packets transmitted, 4 received, 0% packet loss, time 3006ms
rtt min/avg/max/mdev = 11.035/33.318/90.043/32.995 ms
[root@master ~]#
```

图 1-38

（2）执行以下命令使用 YUM 方式安装 Docker。

```
yum -y install docker
```

> 这里没有指定 Docker 的版本信息，默认安装最新版本的 Docker。也可以指定具体的版本信息，例如执行以下命令将安装 Docker 1.13.1。
>
> yum -y install docker-1.13.1

（3）安装完成后的界面图 1-39 所示。

```
        usermode.x86_64 0:1.111-6.el7
        yajl.x86_64 0:2.0.4-4.el7

Dependency Updated:
        audit.x86_64 0:2.8.5-4.el7
        audit-libs.x86_64 0:2.8.5-4.el7
        libselinux.x86_64 0:2.5-15.el7
        libselinux-python.x86_64 0:2.5-15.el7
        libselinux-utils.x86_64 0:2.5-15.el7
        libsemanage.x86_64 0:2.5-14.el7
        libsepol.x86_64 0:2.5-10.el7
        libxml2.x86_64 0:2.9.1-6.el7_9.6
        policycoreutils.x86_64 0:2.5-34.el7
        selinux-policy.noarch 0:3.13.1-268.el7_9.2
        selinux-policy-targeted.noarch 0:3.13.1-268.el7_9.2

Complete!
[root@master ~]#
```

图 1-39

（4）执行以下命令启动 Docker 的服务。

```
systemctl start docker.service
systemctl enable docker.service
```

（5）执行以下命令确定 Docker 的版本，如图 1-40 所示。

```
docker version
```

图 1-40

从图 1-40 可以看到,Docker 分为 Client 端和 Server 端,当前安装的 Docker 是 1.13.1 版本。

1.4.3 使用二进制包方式安装 Docker

使用 YUM 方式安装 Docker 非常简单,但需要连接外部的网络。而在实际的企业生产环境中,通常不能直接访问外部的网络。这时可以使用 Docker 官方提供的二进制包进行 Docker 的离线安装。图 1-41 展示的是 Docker 官方提供的二进制包下载网页。

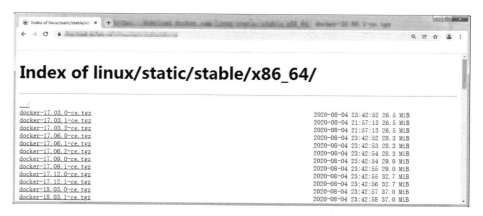

图 1-41

(1)安装 wget 下载工具。

```
yum -y install wget
```

(2)下载 Docker 安装包。

```
wget https://download.docker.com/linux/static/stable/x86_64/docker-20.10.9.tgz
```

（3）使用 tar 命令解压缩 Docker 二进制安装包。

```
tar -zxvf docker-20.10.9.tgz
```

（4）查看 Docker 二进制包提供的执行命令，如图 1-42 所示。

```
ls docker
```

图 1-42

（5）将 Docker 的可执行命令复制到"/usr/bin/"目录下。

```
cp docker/* /usr/bin/
```

（6）执行以下命令启动 Docker 服务。启动成功后输出的日志信息如图 1-43 所示。

```
/usr/bin/dockerd
```

图 1-43

（7）新开启一个命令窗口，执行以下语句查看 Docker 的版本信息（包括 Docker 客户端与 Docker 服务器端的版本），可以看到 Docker 的版本为 20.10.9，如图 1-44 所示。

```
/usr/bin/docker version
```

> 如果想将 Docker 运行在后台，则可以通过以下方式启动 Docker。
> nohup /usr/bin/dockerd >/tmp/docker.log 2>&1 &

以上命令除将 Docker 运行在后台外，还会保存 Docker 的日志到"/tmp/docker.log"文件中。

```
[root@master ~]# /usr/bin/docker version
Client:
 Version:           20.10.9
 API version:       1.41
 Go version:        go1.16.8
 Git commit:        c2ea9bc
 Built:             Mon Oct  4 16:03:22 2021
 OS/Arch:           linux/amd64
 Context:           default
 Experimental:      true

Server: Docker Engine - Community
 Engine:
  Version:          20.10.9
  API version:      1.41 (minimum version 1.12)
  Go version:       go1.16.8
  Git commit:       79ea9d3
  Built:            Mon Oct  4 16:07:30 2021
  OS/Arch:          linux/amd64
  Experimental:     false
 containerd:
  Version:          v1.4.11
  GitCommit:        5b46e404f6b9f661a205e28d59c982d3634148f8
 runc:
  Version:          1.0.2
  GitCommit:        v1.0.2-0-g52b36a2d
 docker-init:
  Version:          0.19.0
  GitCommit:        de40ad0
[root@master ~]#
```

图 1-44

（8）在"/etc/systemd/system/docker.service"文件中输入以下内容：

```
[Unit]
Description=Docker Application Container Engine
Documentation=https://docs.docker.com
After=network-online.target firewalld.service
Wants=network-online.target
[Service]
Type=notify
ExecStart=/usr/bin/dockerd
ExecReload=/bin/kill -s HUP $MAINPID
LimitNOFILE=infinity
LimitNPROC=infinity
TimeoutStartSec=0
Delegate=yes
KillMode=process
Restart=on-failure
StartLimitBurst=3
StartLimitInterval=60s
[Install]
WantedBy=multi-user.target
```

（9）在创建"/etc/systemd/system/docker.service"文件后，需要给该文件添加可执行权限：

```
chmod +x /etc/systemd/system/docker.service
```

（10）启动 Docker，并设置 Docker 为开机自启模式，如图 1-45 所示。

```
systemctl start docker
systemctl enable docker
```

```
[root@master ~]# systemctl start docker
[root@master ~]# systemctl enable docker
Created symlink from /etc/systemd/system/multi-user.target.wants/docker.service to /etc/systemd/system/docker.service.
[root@master ~]#
```

图 1-45

至此，通过使用二进制包完成了 Docker 安装。不管使用 YUM 方式还是二进制包方式，在安装完成后，Docker 的使用方式是完全一样的。但在使用之前，最好验证一下 Docker 的环境。

1.4.4 验证 Docker 环境

在命令行工具 Xshell 中，使用"docker info"命令可以查看 Docker 运行状态的详细信息，如图 1-46 所示。

```
docker info
```

```
[root@master ~]# docker info
Client:
 Context:    default
 Debug Mode: false

Server:
 Containers: 0
  Running: 0
  Paused: 0
  Stopped: 0
 Images: 1
 Server Version: 20.10.9
 Storage Driver: overlay2
  Backing Filesystem: xfs
  Supports d_type: true
  Native Overlay Diff: true
  userxattr: false
 Logging Driver: json-file
 Cgroup Driver: cgroupfs
 Cgroup Version: 1
 Plugins:
  Volume: local
  Network: bridge host ipvlan macvlan null overlay
  Log: awslogs fluentd gcplogs gelf journald json-file local logentries splunk syslog
 Swarm: inactive
 Runtimes: io.containerd.runc.v2 io.containerd.runtime.v1.linux runc
 Default Runtime: runc
 Init Binary: docker-init
 containerd version: 5b46e404f6b9f661a205e28d59c982d3634148f8
 runc version: v1.0.2-0-g52b36a2d
 init version: de40ad0
 Security Options:
  seccomp
   Profile: default
 Kernel Version: 3.10.0-693.el7.x86_64
 Operating System: CentOS Linux 7 (Core)
 OSType: linux
 Architecture: x86_64
 CPUs: 1
 Total Memory: 1.938GiB
 Name: master
 ID: JVCF:I2WN:LJMN:HGQ6:WZ5U:QHQE:DMFY:DWKJ:SDJM:KKAV:XAOT:247D
 Docker Root Dir: /var/lib/docker
 Debug Mode: false
 Registry: https://index.docker.io/v1/
 Labels:
 Experimental: false
 Insecure Registries:
  127.0.0.0/8
 Live Restore Enabled: false
 Product License: Community Engine

[root@master ~]#
```

图 1-46

在图 1-46 的 Server 端配置参数中，Registry 表示镜像仓库的地址。可以看出这里使用的是官方提供的 Docker Hub 镜像仓库。

通过使用系统服务命令可以查看 Docker 的运行状态。执行以下命令，可以查看 Docker 服务的状态，如图 1-47 所示。

```
systemctl status docker
```

图 1-47

1.5 【实战】在 Docker 中部署第一个应用

在成功安装 Docker 后，就可以通过镜像来创建容器，从而运行应用。下面将演示如何在 Docker 中，通过使用 Nginx 镜像来部署第一个应用，并在浏览器中访问它。图 1-48 展示了最终在浏览器中访问的效果。

图 1-48

下面是具体的步骤。

（1）在镜像仓库中搜索 Nginx 的镜像，如图 1-49 所示。其中，OFFICAL 列中标有[OK]的镜像是 Docker 官方提供的镜像。

```
docker search nginx
```

```
[root@master ~]# docker search nginx
INDEX       NAME                                              DESCRIPTION                                       STARS     OFFICIAL   AUTOMATED
docker.io   docker.io/nginx                                   Official build of Nginx.                          16043     [OK]
docker.io   docker.io/jwilder/nginx-proxy                     Automated Nginx reverse proxy for docker c...     2104                 [OK]
docker.io   docker.io/richarvey/nginx-php-fpm                 Container running Nginx + PHP-FPM capable ...     819                  [OK]
docker.io   docker.io/jc21/nginx-proxy-manager                Docker container for managing Nginx proxy ...     301
docker.io   docker.io/linuxserver/nginx                       An Nginx container, brought to you by Linu...     161
docker.io   docker.io/tiangolo/nginx-rtmp                     Docker image with Nginx using the nginx-rt...     148                  [OK]
docker.io   docker.io/jlesage/nginx-proxy-manager             Docker container for Nginx Proxy Manager          147                  [OK]
docker.io   docker.io/alfg/nginx-rtmp                         NGINX, nginx-rtmp-module and FFmpeg from s...     112                  [OK]
docker.io   docker.io/nginxdemos/hello                        NGINX webserver that serves a simple page ...     81                   [OK]
docker.io   docker.io/privatebin/nginx-fpm-alpine             PrivateBin running on an Nginx, php-fpm & ...     61                   [OK]
docker.io   docker.io/nginx/nginx-ingress                     NGINX and  NGINX Plus Ingress Controllers         59
docker.io   docker.io/nginxinc/nginx-unprivileged             Unprivileged NGINX Dockerfiles                    56
docker.io   docker.io/nginxproxy/nginx-proxy                  Automated Nginx reverse proxy for docker c...     33
docker.io   docker.io/staticfloat/nginx-certbot               Opinionated setup for automatic TLS certs ...     25                   [OK]
docker.io   docker.io/nginx/nginx-prometheus-exporter         NGINX Prometheus Exporter for NGINX and NG...     22
docker.io   docker.io/schmunk42/nginx-redirect                A very simple container to redirect HTTP t...     19                   [OK]
docker.io   docker.io/centos/nginx-112-centos7                Platform for running nginx 1.12 or buildin...     16
docker.io   docker.io/centos/nginx-18-centos7                 Platform for running nginx 1.8 or building...     13
docker.io   docker.io/bitwarden/nginx                         The Bitwarden nginx web server acting as a...     12
docker.io   docker.io/flashspys/nginx-static                  Super Lightweight Nginx Image                     11                   [OK]
docker.io   docker.io/mailu/nginx                             Mailu nginx frontend                              10                   [OK]
docker.io   docker.io/webdevops/nginx                         Nginx container                                   9                    [OK]
docker.io   docker.io/sophos/nginx-vts-exporter               Simple server that scrapes Nginx vts stats...     7                    [OK]
docker.io   docker.io/ansibleplaybookbundle/nginx-apb         An APB to deploy NGINX                            3                    [OK]
docker.io   docker.io/wodby/nginx                             Generic nginx                                     1                    [OK]
[root@master ~]#
```

图 1-49

（2）通过以下命令从镜像仓库拉取 Nginx 的镜像到本地，这里拉取了 docker.io/library/nginx 镜像，即官方提供的 Nginx 镜像，如图 1-50 所示。

```
docker pull nginx
```

```
[root@master ~]# docker pull nginx
Using default tag: latest
Trying to pull repository docker.io/library/nginx ...
latest: Pulling from docker.io/library/nginx
a2abf6c4d29d: Downloading [================>                ] 9.792 MB/31.36 MB
a9edb18cadd1: Downloading [================>                ] 8.895 MB/25.35 MB
589b7251471a: Download complete
186b1aaa4aa6: Download complete
b4df32aa5a72: Download complete
a0bcbecc962e: Download complete
```

图 1-50

（3）使用"docker images"命令查看本地的镜像信息，如图 1-51 所示。

```
docker images
```

```
[root@master ~]#
[root@master ~]# docker images
REPOSITORY          TAG         IMAGE ID         CREATED          SIZE
docker.io/nginx     latest      605c77e624dd     38 hours ago     141 MB
[root@master ~]#
```

图 1-51

（4）执行以下的命令将使用镜像来创建 Nginx 的容器。

```
docker run -d -p 1234:80 nginx
```

其中的参数说明如下。

- -d：启动容器的守护进程。
- -p：将容器内的 80 端口映射到宿主机的 1234 端口。这样就可以通过宿主机访问容器内部了。

（5）在容器创建并启动成功后，可以查看 Docker 的容器信息，如图 1-52 所示。

```
docker ps
```

```
[root@master ~]# docker run -d -p 1234:80 nginx
f40755edf7d62f1cb8c49160ef0a3a7d51b00404b18aac45e4bf69d017604117
[root@master ~]# docker ps
CONTAINER ID    IMAGE          COMMAND                CREATED         STATUS
                PORTS                          NAMES
f40755edf7d6    nginx          "/docker-entrypoin..." 3 seconds ago   Up 2 sec
onds            0.0.0.0:1234->80/tcp           brave_torvalds
[root@master ~]#
```

图 1-52

> 如果直接使用"docker images"和"docker ps"命令显示的信息不便于查看，则可以使用 Linux 的 alias 别名进行定制格式的显示。
>
> 在"~/.bash_profile"文件中输入以下的内容：
>
> alias myps='docker ps --format "table {{.ID}}\t{{.Image}}\t{{.Names}}\t{{.Ports}}"'
>
> alias myimages='docker images --format "table {{.ID}}\t{{.Repository}}\t{{.Tag}}"'
>
> 执行以下 source 命令生效"~/.bash_profile"文件：
>
> source ~/.bash_profile
>
> 这时就可以使用定制的 myps 和 myimages 命令进行格式化显示了。在图 1-53 中，对于镜像，只显示镜像的 ID、镜像仓库的信息和镜像的 TAG 标签；而对于容器，只显示容器的 ID、容器中运行的镜像、容器的名称和端口的映射。

```
[root@master ~]# myimages
IMAGE ID              REPOSITORY              TAG
605c77e624dd          docker.io/nginx         latest
[root@master ~]# myps
CONTAINER ID     IMAGE          NAMES              PORTS
f40755edf7d6     nginx          brave_torvalds     0.0.0.0:1234->80/tcp
[root@master ~]#
```

图 1-53

（6）打开浏览器访问"http://192.168.79.11:1234/"，将打开 Nginx 的首页。

（7）如果要终止和销毁容器，则执行以下命令。

```
docker rm -f f40755edf7d6
```

（8）再次查看容器信息，可以看到没有任何运行的容器信息，如图 1-54 所示。

图 1-54

（9）可以通过"docker help"命令查看命令的帮助信息，在后续章节中会陆续介绍这些命令的使用方法。

第 2 章 Docker 的镜像

本章将介绍什么是 Docker 的镜像、如何管理 Docker 的镜像,以及如何构建自己的私有镜像仓库。

2.1 什么是 Docker 的镜像

Docker 的镜像是一个模板,或者说其是一个只读文件。在该模板中包含应用和应用运行时所需要的依赖环境。Docker 的镜像采用的是分层的文件系统,每一次对镜像的修改将以"读写层"的形式增加到原来的只读文件的模板上。

图 2-1 展示了 Docker 镜像的分层结构。下面进行介绍。

- 内核(bootfs):用来加载 Linux 的内核以启动 Linux 环境。Docker 的用户不会与这一层直接打交道。
- 根镜像:可以将其理解成操作系统,图 2-1 中的根镜像使用的是 CentOS。
- 在根镜像之上就是叠加的每一层应用,图 2-1 中的 MySQL、Tomcat 等。

另外,在物理存储上,镜像的本质其实是磁盘上一系列文件的集合,如图 2-2 所示。

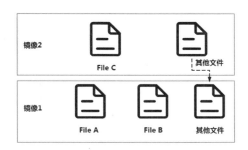

图 2-1　　　　　　　　　　　图 2-2

Docker 默认的镜像存储路径是"/var/lib/docker",也可以自定义其他路径。

2.1.1　使用 Docker 默认的镜像存储路径

通过执行以下语句可以得知 Docker 镜像的存储路径,可以看出在默认情况下,Docker 将拉取的镜像存储在"/var/lib/docker"目录下,如图 2-3 所示。

```
docker info|grep "Docker Root Dir"
```

```
[root@master ~]# docker info|grep "Docker Root Dir"
  WARNING: You're not using the default seccomp profile
Docker Root Dir: /var/lib/docker
[root@master ~]#
```

图 2-3

以 1.5 节拉取的 Nginx 镜像为例,通过以下步骤可以得知该 Nginx 镜像的存储路径。

(1)确定镜像的 ID,这里是 605c77e624dd,如图 2-4 所示。

```
myimages
```

```
[root@master ~]# myimages
IMAGE ID          REPOSITORY            TAG
605c77e624dd      docker.io/nginx       latest
[root@master ~]#
```

图 2-4

(2)切换到"/var/lib/docker"目录下。

```
cd /var/lib/docker
```

(3)执行"find"命令查找镜像文件,如图 2-5 所示。

```
find . -name 605c77e624dd*
```

```
[root@master ~]# cd /var/lib/docker
[root@master docker]# find . -name 605c77e624dd*
./image/overlay2/imagedb/content/sha256/605c77e624ddb75e6110f997c5
8876baa13f8754486b461117934b24a9dc3a85
[root@master docker]#
```

图 2-5

2.1.2 自定义 Docker 的镜像存储路径

在实际生产环境中，Docker 默认的镜像存储路径往往不能满足磁盘空间大小的要求。可以根据以下步骤来修改这个存储路径。

（1）创建新的镜像存储路径"/data/docker"，该路径用于保存 Docker 的镜像文件。

```
mkdir -p /data/docker
```

（2）创建 Docker 守护进程的配置文件。

```
cd /etc/docker/
vi daemon.json
```

（3）在 daemon.json 文件中输入以下内容。

```
{
    "graph": "/data/docker"
}
```

（4）重新加载 Docker 的服务，并重启 Docker。

```
systemctl daemon-reload
systemctl restart docker
```

（5）重新查看 Docker 的镜像存储路径，会发现路径变成了"/data/docker"，如图 2-6 所示。

```
[root@master docker]# docker info|grep "Docker Root Dir"
 Docker Root Dir: /data/docker
[root@master docker]#
```

图 2-6

2.2 使用 Docker 的公有镜像仓库

存储 Docker 镜像的地方叫作镜像仓库。镜像仓库分为公有镜像仓库和私有镜像仓库。下面通过实战来介绍如何使用 Docker 公有镜像仓库。

2.2.1 【实战】访问 Docker 官方的公有镜像仓库

Docker 官方提供了一个公有镜像仓库 Docker Hub。在通过 Docker 客户端操作 Docker 时，如果没有指定镜像仓库的地址，则默认使用该镜像仓库的地址。图 2-7 展示了 Docker Hub 的主页。

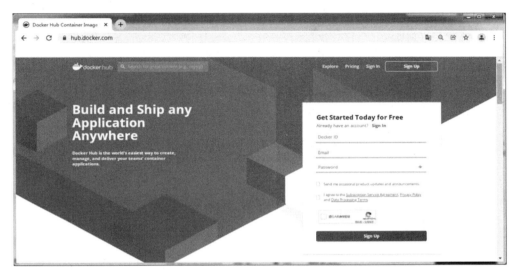

图 2-7

下面通过具体步骤演示来介绍如何使用 Docker Hub。

（1）使用 "docker search" 命令搜索镜像仓库中的一个镜像，如 MySQL。

```
docker search mysql
```

（2）使用 "docker search" 命令拉取 MySQL 的镜像。如果在拉取镜像时没有指定版本，则默认拉取最新版本的镜像。也可以在拉取时指定版本信息，例如拉取 mysql:5.7.19。

```
docker pull mysql:5.7.19
```

> 这时会发现拉取的速度很慢。这是因为，拉取命令默认从国外的 Docker Hub 上拉取镜像。因此，在国内访问 Docker Hub 时，一般需要配置国内的镜像加速器。
> 在 2.2.2 节中会介绍如何使用国内的 Docker 镜像加速器。

（3）用拉取的 MySQL 镜像创建一个容器，用来运行 MySQL 数据库。

```
docker run --name my_mysql -p 2206:3306 \
-e MYSQL_ROOT_PASSWORD=Welcome_1 \
-d mysql:5.7.19
```

其中的参数说明如下。

- -p：2206 表示宿主机的端口号；3306 表示容器的端口号。这里是将容器的 3306 端口映射到了宿主机的 2206 端口。
- -e：传递一个变量给容器。这里通过变量 MYSQL_ROOT_PASSWORD 设置了 MySQL 的 root 用户的密码为 Welcome_1。

（4）查看容器的信息，如图 2-8 所示。这里的容器 ID 是 6ee77955894e。

```
[root@master ~]# docker run --name my_mysql -p 2206:3306 \
> -e MYSQL_ROOT_PASSWORD=Welcome_1 \
> -d mysql:5.7.19
6ee77955894e4e17f8a41e01957f0141793254ad4ea5a542d799a019a7eab854
[root@master ~]# myps
CONTAINER ID        IMAGE               NAMES               PORTS
6ee77955894e        mysql:5.7.19        my_mysql            0.0.0.0:2206->3306/tcp
[root@master ~]#
```

图 2-8

（5）使用一个 MySQL 的客户端工具连接 MySQL，会出现不能连接的错误，如图 2-9 所示。

图 2-9

（6）使用 Docker 命令行进入容器。

```
docker exec -it 6ee77955894e /bin/bash
```

其中，-it 表示启动一个虚拟的标准输入终端。

（7）登录 MySQL 数据库时输入密码"Welcome_1"，如图 2-10 所示。

```
[root@master ~]# docker exec -it 6ee77955894e /bin/bash
root@6ee77955894e:/# mysql -uroot -p
Enter password:
Welcome to the MySQL monitor.  Commands end with ; or \g.
Your MySQL connection id is 3
Server version: 5.7.19 MySQL Community Server (GPL)

Copyright (c) 2000, 2017, Oracle and/or its affiliates. All rights reserved.

Oracle is a registered trademark of Oracle Corporation and/or its
affiliates. Other names may be trademarks of their respective
owners.

Type 'help;' or '\h' for help. Type '\c' to clear the current input statement.

mysql>
```

图 2-10

（8）在 MySQL 的命令行中执行以下命令则允许 root 用户远程登录，如图 2-11 所示。

```
grant all privileges on *.* to root@"%" identified by "Welcome_1" with grant option;
```

```
mysql> grant all privileges on *.* to root@"%"
    -> identified by "Welcome_1" with grant option;
Query OK, 0 rows affected, 1 warning (0.00 sec)

mysql>
```

图 2-11

（9）重新使用 MySQL 客户端登录，这时就可以成功访问 Docker 容器中的 MySQL 数据库了，如图 2-12 所示。

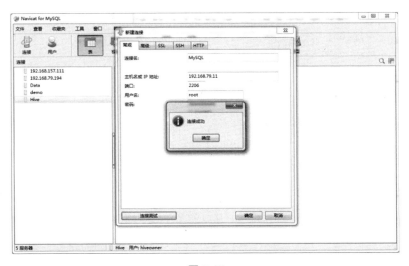

图 2-12

2.2.2 【实战】配置和使用阿里云 Docker 镜像加速仓库

在 2.2.1 节提到，在国内访问 Docker Hub 是非常缓慢的。为了提高效率，可以在 Docker 引擎中配置国内的镜像加速器。Docker 可以配置的国内镜像加速器有很多，比如阿里云、网易蜂巢、DaoCloud、Docker 中国区官方镜像加速器等。

下面以阿里云的镜像加速器进行演示。

首先确定默认的镜像仓库地址，可以看到默认使用的是官方的 Docker Hub，如图 2-13 所示。

```
[root@master ~]# docker info |grep registry
  WARNING: You're not using the default seccomp profile
[root@master ~]# docker info |grep Registry
  WARNING: You're not using the default seccomp profile
Registry: https://index.docker.io/v1/
[root@master ~]#
```

图 2-13

以下步骤演示了如何使用阿里云的镜像加速器。

（1）注册一个阿里云账号。

（2）访问阿里云的容器镜像服务网站。选择左边导航条中的"镜像工具"→"镜像加速器"，在右边的"操作文档"中选择"CentOS"，如图 2-14 所示。

图 2-14

（3）根据操作文档中的步骤配置阿里云的镜像加速器，如图 2-15 所示。

```
[root@master ~]# sudo mkdir -p /etc/docker
[root@master ~]# sudo tee /etc/docker/daemon.json <<-'EOF'
> {
>   "registry-mirrors": ["https://███████.mirror.aliyuncs.com"]
> }
> EOF
{
  "registry-mirrors": ["https://███████.mirror.aliyuncs.com"]
}
[root@master ~]# sudo systemctl daemon-reload
[root@master ~]# sudo systemctl restart docker
[root@master ~]#
```

图 2-15

（4）使用 "docker info" 命令确认加速器配置是否成功，如图 2-16 所示，从中可以看出，参数 Registry Mirrors 被设置成阿里云的镜像加速器地址。

```
[root@master ~]# docker info | tail
 WARNING: You're not using the default seccomp profile
Debug Mode (client): false
Debug Mode (server): false
Registry: https://index.docker.io/v1/
Experimental: false
Insecure Registries:
 127.0.0.0/8
Registry Mirrors:
 https://███████.mirror.aliyuncs.com
Live Restore Enabled: false
Registries: docker.io (secure)
[root@master ~]#
```

图 2-16

（5）重新拉取某个镜像，如 Tomcat，会发现速度快了很多。

```
docker pull tomcat
```

（7）使用拉取的 Tomcat 镜像创建一个容器。

```
docker run --name mytomcat -d -p 8080:8080 tomcat
```

（8）通过浏览器访问 "http://192.168.79.11:8080/" 会返回 404 错误，如图 2-17 所示。这是因为在 Tomcat 中没有部署任何应用。

图 2-17

（9）确定 Tomcat 容器 ID，进入容器内部署一个简单的 HTML 网页，如图 2-18 所示。

```
[root@master ~]# docker exec -it 2b093b3a529b /bin/bash
root@2b093b3a529b:/usr/local/tomcat# mkdir webapps/mydemo
root@2b093b3a529b:/usr/local/tomcat# echo "<h1>Hello Tomcat</h1>" > webapps/mydemo/index.html
root@2b093b3a529b:/usr/local/tomcat#
```

图 2-18

（10）在浏览器中访问"http://192.168.79.11:8080/mydemo/index.html"，可以成功看到 index.html，如图 2-19 所示。

图 2-19

2.3 使用命令行工具管理 Docker 的镜像

之前我们已经使用了若干条 Docker 命令来操作 Docker 的镜像。表 2-1 展示了与镜像相关的一些操作命令。

表 2-1

命　　令	说　　明
docker search [IMAGE NAME]	在拉取镜像之前，可以通过该命令搜索符合的镜像
docker images	列出本机上的所有镜像。该命令行还可以使用通配符，以找到符合条件的一系列镜像
docker inspect [IMAGE NAME]/[CONTAINER ID]	通过"docker images"命令只能查看镜像的基本信息，而通过"docker inspect"命令则可以查看镜像或者容器的详细信息
docker pull [IMAGE NAME]	将镜像拉到本地。镜像名必须包含命名空间和仓库名。如果在一个仓库中存在多个镜像，则必须指定 TAG，即指定版本的信息，否则使用默认的 TAG "latest"
docker push [IMAGE NAME]	将本地的镜像上传到镜像仓库中
docker rmi [IMAGE NAME]/[IMAGE ID]	将不需要的镜像删除。与移除容器的命令 rm 相比，删除镜像的命令多了一个 i，i 即 image 的意思。在删除镜像时，需要注意两点：①如果要删除多个镜像，则需要使用空格将它们隔开；②可以使用参数-f 强制删除镜像

下面演示这些命令的使用方法。

（1）在镜像仓库中搜索 Flink 的镜像，如图 2-20 所示。

```
docker search flink
```

```
[root@master ~]# docker search flink
INDEX       NAME                                      DESCRIPTION                                       STARS     OFFICIAL    AUTOMATED
docker.io   docker.io/flink                           Apache Flink® is a powerful open-source di...     298
docker.io   docker.io/bde2020/flink-master            Apache Flink master image                         5                     [OK]
docker.io   docker.io/melentye/flink                  Yet another Docker image for Apache Flink.        4                     [OK]
docker.io   docker.io/bde2020/flink-worker            Apache Flink worker image                         3                     [OK]
docker.io   docker.io/amd64/flink                     Apache Flink® is a powerful open-source di...     2
docker.io   docker.io/apache/flink                                                                      2
docker.io   docker.io/lyft/flinkk8soperator           Repository to store flinkk8soperator docke...     2
docker.io   docker.io/mesoshq/flink                   A Docker base image for creating Apache Fl...     2                     [OK]
docker.io   docker.io/arm32v7/flink                   Apache Flink® is a powerful open-source di...     1
docker.io   docker.io/arm64v8/flink                   Apache Flink® is a powerful open-source di...     1
docker.io   docker.io/fogguru/flink-job               Dummy Flink job for the FogGuru Lora platform     1
docker.io   docker.io/plucas/flink                    Docker images for Apache Flink                    1
docker.io   docker.io/trollin/flink                                                                     1
docker.io   docker.io/arm32v5/flink                   Apache Flink® is a powerful open-source di...     0
docker.io   docker.io/arm32v6/flink                   Apache Flink® is a powerful open-source di...     0
docker.io   docker.io/chethanuk/flink                 flink                                             0
docker.io   docker.io/grosinosky/flink                                                                  0
docker.io   docker.io/i386/flink                      Apache Flink® is a powerful open-source di...     0
docker.io   docker.io/iunera/flinkk8soperator         Based on https://github.com/lyft/flinkk8so...     0
docker.io   docker.io/morgel/flink                    custom flink image                                0
docker.io   docker.io/ppc64le/flink                                                                     0
docker.io   docker.io/rmetzger/flink-ci                                                                 0
docker.io   docker.io/s390x/flink                     Apache Flink® is a powerful open-source di...     0
docker.io   docker.io/tma75/flink-nginx-proxy                                                           0
docker.io   docker.io/wzorgdrager/flink-prometheus    Flink Docker image with Prometheus.               0
[root@master ~]#
```

图 2-20

（2）将 Flink 镜像拉取到本地。

```
docker pull flink
```

（3）查看本地的镜像信息，并使用通配符查看所有以"f"开头的镜像，如图 2-21 所示。

```
[root@master ~]# docker images f*
REPOSITORY          TAG          IMAGE ID          CREATED          SIZE
docker.io/flink     latest       28308bbc7b60      10 days ago      658 MB
[root@master ~]#
```

图 2-21

（4）使用"docker inspect"命令查看 Flink 镜像的详细信息（其中，28308bbc7b60 是图 2-21 中 Flink 镜像的 ID），如图 2-22 所示。

（5）删除 Flink 镜像，如图 2-23 所示。

```
[root@master ~]# docker inspect 28308bbc7b60
[
    {
        "Id": "sha256:28308bbc7b601d718e9adf53c5a5451a1cbeda4de7ff41e4e1eb5fc919330fbb",
        "RepoTags": [
            "docker.io/flink:latest"
        ],
        "RepoDigests": [
            "docker.io/flink@sha256:a15bad4083276dc295dd2a337265f1747b7e49ed490d3b1e4e10203bcfeadfea"
        ],
        "Parent": "",
        "Comment": "",
        "Created": "2021-12-22T13:12:50.744297261Z",
        "Container": "6b5ca7ce6b961689a59acc9baa8e6d42826ca8a39bbf226389ce49705150fe7b",
        "ContainerConfig": {
            "Hostname": "6b5ca7ce6b96",
            "Domainname": "",
            "User": "",
            "AttachStdin": false,
            "AttachStdout": false,
            "AttachStderr": false,
            "ExposedPorts": {
                "6123/tcp": {},
                "8081/tcp": {}
```

图 2-22

```
[root@master ~]# docker rmi -f 28308bbc7b60
Untagged: docker.io/flink:latest
Untagged: docker.io/flink@sha256:a15bad4083276dc295dd2a337265f1747b7e49ed490d3b1e4e10203bcfeadfea
Deleted: sha256:28308bbc7b601d718e9adf53c5a5451a1cbeda4de7ff41e4e1eb5fc919330fbb
Deleted: sha256:77cffbfd8693c3964faf8d14e3a5c82be3681ba6dcab6385f2a503a17415f1d5
Deleted: sha256:362243e967c2f7b3c97946558aacbc6e58b5455b3ea03979245b3a2e15adf1ec
Deleted: sha256:2330f0c4ba43e5261fa888f8298b5c91c2ade9d5f4252d565f0bd5176ed7c1a4
Deleted: sha256:abb2e9e0a288bc265d4c70c19aa415f96fa1cc721e7abcca9cf7f2b1e4278e51
Deleted: sha256:36238d6e2c5baafb5a008e3dba06353f6ab1468ea4ad07045c643e586ea33afe
Deleted: sha256:08ea2186cdadefb422e9c639ec04df10bf9f217eb3e12611e3cd4e5e6db7381c
Deleted: sha256:baf36e0a6449ebb2cef7c5dbd4421a9f8e19a226e72cc8b0add2f0f7d634fda6
Deleted: sha256:49f9490a86812df115acf62a5970ede49a03e2cc8a09d340008193b43ad42e38
Deleted: sha256:d8fdccae7a13f6ece167898bab3a601443abb2600bc703dfdb64165259137877
[root@master ~]#
```

图 2-23

2.4 构建自己的镜像

在前面的实战中，使用的镜像都是镜像仓库中已有的镜像，或者说是别人已经开发好的镜像。那么，是否可以构建自己的镜像，并把镜像上传到镜像仓库中呢？答案当然是可以的。

要构建自己的镜像有两种方式：①使用"docker commit"命令；②使用 Dockerfile 文件。

2.4.1 【实战】使用"docker commit"命令构建镜像

"docker commit"命令基于一个容器来创建镜像，图 2-24 展示了该命令的帮助信息。

```
[root@master ~]# docker help commit
Usage:  docker commit [OPTIONS] CONTAINER [REPOSITORY[:TAG]]

Create a new image from a container's changes

Options:
  -a, --author string    Author (e.g., "John Hannibal Smith <hannibal@a-team.com>")
  -c, --change list      Apply Dockerfile instruction to the created image (default [])
      --help             Print usage
  -m, --message string   Commit message
  -p, --pause            Pause container during commit (default true)
[root@master ~]#
```

图 2-24

在 2.2.2 节的最后已经成功基于 Tomcat 的镜像创建了一个容器，并在容器中部署了一个 Web 应用。接下来，使用"docker commit"命令基于该镜像生成一个新的镜像文件。

（1）执行"myps"命令确定该容器的 ID。可以确定该容器的 ID 是 33a3e34fcf0a，如图 2-25 所示。提示，这里的"myps"命令是在图 1-52 下面的提示框中配置的"docker ps"命令的别名。

```
[root@master ~]# myps
CONTAINER ID    IMAGE      NAMES       PORTS
33a3e34fcf0a    tomcat     mytomcat    0.0.0.0:8080->8080/tcp
[root@master ~]#
```

图 2-25

（2）使用"docker commit"命令基于该容器生成镜像，并将镜像保存到本地。

```
docker commit 33a3e34fcf0a tomcat_with_application
```

（3）查看本地的镜像文件，如图 2-26 所示。提示，这里的 myimages 是在图 1-52 下面的提示框中配置的"docker images"命令的别名。

```
[root@master ~]# myimages
IMAGE ID        REPOSITORY                  TAG
282f3f8b4d79    tomcat_with_application     latest
605c77e624dd    docker.io/nginx             latest
fb5657adc892    docker.io/tomcat            latest
3e3878acd190    docker.io/mysql             5.7.19
[root@master ~]#
```

图 2-26

（4）如果想将生成镜像上传到镜像仓库中，则需要在镜像名称前加上仓库的路径信息，如图 2-27 所示。

（5）登录 Docker Hub，输入注册的用户名和密码，如图 2-28 所示。

```
[root@master ~]# docker commit 33a3e34fcf0a collenzhao/tomcat_with_application
sha256:1a6b9823ce3f6fb86319188befb469ebb3dc7aad43eb92c6eee1364688e0043a
[root@master ~]# myimages
IMAGE ID            REPOSITORY                              TAG
1a6b9823ce3f        collenzhao/tomcat_with_application      latest
282f3f8b4d79        tomcat_with_application                 latest
605c77e624dd        docker.io/nginx                         latest
fb5657adc892        docker.io/tomcat                        latest
3e3878acd190        docker.io/mysql                         5.7.19
[root@master ~]#
```

图 2-27

```
[root@master ~]# docker login
Login with your Docker ID to push and pull images from Docker Hub. If you don't have a
 Docker ID, head over to https://hub.docker.com to create one.
Username: collenzhao
Password:
Login Succeeded
[root@master ~]#
```

图 2-28

（6）将镜像上传保存到镜像仓库中，默认保存到 Docker Hub 中，如图 2-29 所示。

```
[root@master ~]# docker push collenzhao/tomcat_with_application
The push refers to a repository [docker.io/collenzhao/tomcat_with_ap
plication]
86e200a3abdd: Pushed
3e2ed6847c7a: Pushed
bd2befca2f7e: Pushing   4.95 MB/20.15 MB
59c516e5b6fa: Pushed
3bb5258f46d2: Pushing 5.459 MB/342.7 MB
832e177bb500: Pushed
f9e18e59a565: Pushing 5.944 MB/11.31 MB
26a504e63be4: Pushing 5.489 MB/151.9 MB
8bf42db0de72: Pushing 210.4 kB/18.95 MB
31892cc314cb: Waiting
11936051f93b: Waiting
```

图 2-29

（7）上传完成后，登录 Docker Hub 检查上传的镜像，如图 2-30 所示。

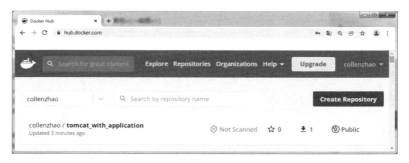

图 2-30

> 使用"docker commit"命令构建镜像比较简单,也比较直观。但是在实际环境中并不会这样使用,因为:
> - 只有生成镜像的人才知道执行过什么命令、怎么生成的镜像,而别人根本无从得知。这就是一种"黑箱"操作,不利于镜像的维护。
> - 在容器中的任何修改都是在当前层进行的,不会改动上一层。如果使用"docker commit"命令制作镜像,则会使镜像变得"臃肿",且无法使用"docker commit"命令实现复用。
>
> 既然"docker commit"命令创建镜像存在一些问题,那有没有更好的方式呢?答案是有的——可以使用 Dockerfile 文件来构建镜像。

2.4.2 【实战】使用 Dockerfile 文件构建镜像

Docker 镜像是一个特殊的分层文件系统,包含应用和必要的依赖环境,但并不包含任何的动态信息。构建一个镜像,实际上就是为镜像中的每一层创建相应的配置。因此,可以把构建的命令语句、参数配置等信息都写入一个脚本中。这样,"docker commit"命令的无法重复的问题、镜像"臃肿"的问题就都被解决了。这个脚本就是 Dockerfile 文件。

1. 什么是 Dockerfile

Dockerfile 是一个文本文件,其包含了一条条的指令,每一条指令都用于构建镜像中的一层。

Dockerfile 文件可以使用"docker build"命令进行编译。在编译过程中,每一条指令的内容描述了该层应如何进行构建。当我们需要定制自己额外的需求时,只需要在 Dockerfile 文件的基础上添加或者修改指令,重新生成新的镜像即可。

2. 示例

下面通过一个简单的示例来演示如何使用 Dockerfile 文件。在这个示例中,将基于 Nginx 的镜像来构建一个新的镜像,并在该镜像中部署一个简单的 Web 网页。

(1)创建一个 Dockerfile 文件。在该文件中输入以下命令。

```
FROM nginx
RUN echo '<h1>This is a Demo HTML</h1>' > /usr/share/nginx/html/index.html
```

(2)在 Dockerfile 文件所在的目录下执行"docker build"命令构建镜像。构建的过程如图 2-31 所示。

```
[root@master ~]# docker build -t mynginx .
Sending build context to Docker daemon 20.99 kB
Step 1/2 : FROM nginx
 ---> 605c77e624dd
Step 2/2 : RUN echo '<h1>This is a Demo HTML</h1>' > /usr/share/nginx/html/index.html
 ---> Using cache
 ---> a4a5d2ffc17a
Successfully built a4a5d2ffc17a
[root@master ~]#
```

图 2-31

 "docker build"命令会在当前目录下寻找名为"Dockerfile"的文件，然后对该文件进行编译生成镜像。如果文件名不是"Dockerfile"，则可以在使用"docker build"命令加上"-f"参数指定具体的文件名称。

（3）查看新生成的镜像，如图 2-32 所示。

```
[root@master ~]# myimages
IMAGE ID              REPOSITORY                         TAG
a4a5d2ffc17a          mynginx                            latest
1a6b9823ce3f          collenzhao/tomcat_with_application latest
282f3f8b4d79          tomcat_with_application            latest
605c77e624dd          docker.io/nginx                    latest
fb5657adc892          docker.io/tomcat                   latest
3e3878acd190          docker.io/mysql                    5.7.19
[root@master ~]#
```

图 2-32

（4）使用新生成的镜像创建容器。

```
docker run -d -p 7788:80 mynginx
```

（5）使用浏览器访问宿主机的 7788 端口，可以看到如图 2-33 所示界面。

图 2-33

2.4.3 Dockerfile 文件详解

图 2-34 展示了 Docker 镜像、Docker 容器和 Dockerfile 文件三者的关系。可以看出：使用 Dockerfile 文件定义镜像，然后运行镜像启动容器。

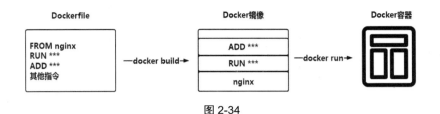

图 2-34

表 2-2 列出了一个完整的 Dockerfile 文件的组成部分。

表 2-2

组成部分	说 明
基础镜像信息	使用 FROM 关键字指定基础镜像信息，该命令必须是 Dockerfile 文件的第 1 条指令
维护者信息	使用 MAINTAINER 关键字指定，通常可以使用 Dockerfile 文件创建者的姓名或者电子邮件作为维护者信息
镜像操作指令	每执行一条镜像操作指令，将在镜像中添加新的一层。可以根据需要使用一条或者多条镜像操作指令。常见的镜像操作指令有：RUN、COPY、ADD、EXPOSE、WORKDIR、ONBUILD、USER、VOLUME 等
容器启动执行指令	用于指定在启动容器时需要执行的命令。通过关键字 CMD、ENTRYPOINT 指定

在完成 Dockerfile 文件的编写后执行"docker build"命令，则会根据 Dockerfile 文件中上下文的内容构建新 Docker 镜像。整个构建过程会被递归处理。因此，如果在 Dockerfile 文件中含有子路径或 URL 等信息，则它们都将被递归构建。

> 在使用"docker build"命令构建镜像时，还可以通过-t 参数指定生成镜像的仓库地址和标签等信息。

利用 Dockerfile 文件构建 Dockerfile 镜像的过程请参考图 2-35。

```
[root@master ~]# docker build -t mynginx .
Sending build context to Docker daemon 20.99 kB
Step 1/2 : FROM nginx
 ---> 605c77e624dd
Step 2/2 : RUN echo '<h1>This is a Demo HTML</h1>' > /usr/share/nginx/html/index.html
 ---> Using cache
 ---> a4a5d2ffc17a
Successfully built a4a5d2ffc17a
[root@master ~]#
```

图 2-35

"docker build"命令在使用 Dockerfile 文件生成镜像时，会通过 Docker 守护进程执行 Dockerfile 文件中的每一条指令，并在每一步执行完成后生成一个新镜像。当所有的指令执行完成后，会输出最终镜像的 ID。

当镜像最终生成后，Docker 守护进程会自动清理 Docker 的上下文环境，并自动重用已生成的中间镜像，以加速构建的速度。图 2-35 中的方框部分表明，在构建过程中使用到了 Dockerfile 文件的缓存机制。

2.4.4 【实战】使用 Dockerfile 文件的综合案例

下面将从一个 CentOS 的基础镜像开始，安装 JDK 和 Tomcat 环境，并完成一个 Web 应用的部署。整个过程通过一个 Dockerfile 文件来描述。通过该 Dockerfile 文件来构建一个镜像，并创建一个容器来运行 Web 应用。

表 2-3 列举了本实战需要用到的实验介质。

表 2-3

介 质	说 明
操作系统 CentOS	在编辑 Dockerfile 文件时，使用 FROM 关键字指定 CentOS 为基础镜像
jdk-8u144-linux-x64.tar.gz	由于该案例需要用 Tomcat 作为应用服务器，因此需要在 CentOS 中安装 JDK 环境，可以从 Oracle 的官方网站下载
Apache Tomcat	Java 的 Web 服务器，可以从 Apache 网站下载
MyDemoWeb.war	Java Web 应用

以下是具体的操作步骤。

（1）创建 Dockerfile 文件，输入以下内容。

```
01   FROM centos
02   MAINTAINER zhaoyuqiang collen7788@126.com
03   RUN mkdir /root/training
04   RUN mkdir /root/tools
05   COPY jdk-8u144-linux-x64.tar.gz /root/tools
06   RUN tar zxvf /root/tools/jdk-8u144-linux-x64.tar.gz -C /root/training/
07   ENV JAVA_HOME /root/training/jdk1.8.0_144
08   ENV PATH $JAVA_HOME/bin:$PATH
09   ADD https://dlcdn.apache.org/tomcat/tomcat-8/v8.5.73/bin/apache-tomcat-8.5.73.tar.gz  /root/tools
10   RUN tar zxvf /root/tools/apache-tomcat-8.5.73.tar.gz -C /root/training/
11   COPY MyDemoWeb.war /root/training/apache-tomcat-8.5.73/webapps
12   ENTRYPOINT ["/root/training/apache-tomcat-8.5.73/bin/catalina.sh","run"]
```

其中，

- 第 01 行：指定基础镜像为 CentOS。
- 第 02 行：执行维护者的信息。
- 第 03 行：在 CentOS 中创建一个"/root/training"目录。

- 第 04 行：在 CentOS 中创建一个"/root/tools"目录。
- 第 05 行：将宿主机上的 JDK 安装文件复制到 CentOS 的"/root/tools"目录下。
- 第 06 行：将 JDK 安装文件解压缩到 CentOS 的"/root/training"目录下。
- 第 07 行：在 CentOS 中设置 JAVA_HOME 的环境变量。
- 第 08 行：在 CentOS 中的 PATH 路径中添加 Java 的路径。
- 第 09 行：从 Apache 网站下载 Tomcat 的安装文件。
- 第 10 行：将 Tomcat 安装文件解压缩到 CentOS 的"/root/training"目录下。
- 第 11 行：将宿主机上的 MyDemoWeb.war 复制到 CentOS 的 Tomcat 的 webapps 目录下。
- 第 12 行：启动容器时运行 Tomcat。

（2）将 MyDemoWeb.war 应用下载到 Dockerfile 文件的同级目录下。

```
wget https://raw.githubusercontent.com/collenzhao/-docker-k8s-resources-chapter2/main/MyDemoWeb.war
```

（3）将 jdk-8u144-linux-x64.tar.gz 上传至 Dockerfile 文件的同级目录下。图 2-36 展示了该应用的目录结构。

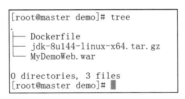

图 2-36

（4）使用"docker build"命令编译 Dockerfile 文件，并通过参数"-t"指定镜像的标签名称。提示，在这条"docker build"命令的最后有一个点（.）。

```
docker build -t mywebapp .
```

（5）执行"myimages"命令确定新构建的镜像，如图 2-37 所示。

图 2-37

（6）使用"docker run"命令基于 mywebapp 的镜像创建容器。

```
docker run -d -p 8080:8080 mywebapp
```

（7）使用浏览器访问"http://192.168.79.11:8080/MyDemoWeb/"就可以正常访问应用了，如图 2-38 所示。

图 2-38

2.5 搭建私有镜像仓库 Harbor

通过前面的学习，我们已经了解到 Docker 镜像与容器的开发和运行都离不开镜像管理。而要进行镜像管理就必须有镜像仓库。

前面的示例使用的都是 Docker 官方提供的镜像仓库 Docker Hub，但是从安全和效率等方面考虑，这样的公有镜像仓库无法在企业的私有环境中使用。因此，搭建和部署企业私有环境中的镜像仓库就非常有必要了。

 Harbor 是由 VMware 公司开发并开源的企业级的 Docker 镜像仓库的管理项目，它包括镜像的权限管理（RBAC）、目录访问（LDAP）、日志审核、管理界面、自我注册、镜像复制和中文支持等功能。

为了更好地演示 Harbor 的安装和部署，读者可以根据 1.4.1 节的内容单独准备一台虚拟机作为运行 Harbor 的主机。图 2-39 展示了 Harbor 主机的网络配置和主机名信息。

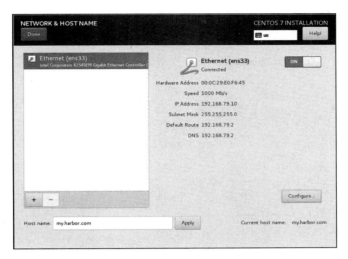

图 2-39

2.5.1　安装 Docker 和 Docker Compose

 Docker Compose 是 Docker 的服务编排工具，关于它的使用方法会在后续章节中介绍。通过 GitHub 网站可以查看 Docker Compose 的最新版本信息。

（1）在 Harbor 的主机上关闭 Linux 防火墙和 SELinux。

```
systemctl stop firewalld.service
systemctl disable firewalld.service
setenforce 0
```

（2）配置 Harbor 的主机的主机名和 IP 地址的映射，编辑文件 "/etc/hosts" 加入以下内容。

```
192.168.79.10 my.harbor.com
```

（3）使用 YUM 安装 Docker。

```
yum -y install docker
```

（4）启动 Docker 的服务。

```
systemctl start docker
systemctl enable docker
```

（5）安装 Docker Compose。以下指令将从 GitHub 下载服务编排工具 Docker Compose，并保存到本地的 "/usr/local/bin/docker-compose" 文件中。

```
curl -L https://github.com/docker/compose/releases/download/1.23.1/
docker-compose-`uname -s`-`uname -m` -o /usr/local/bin/docker-compose
```

（6）给 Docker Compose 加上可执行的权限。

```
chmod +x /usr/local/bin/docker-compose
```

（7）确定 Docker Compose 的版本信息，如图 2-40 所示。

```
[root@my ~]# docker-compose version
docker-compose version 1.23.1, build b02f1306
docker-py version: 3.5.0
CPython version: 3.6.7
OpenSSL version: OpenSSL 1.1.0f  25 May 2017
[root@my ~]#
```

图 2-40

2.5.2　安装与配置 Harbor

（1）从 GitHub 下载 Harbor 的安装文件，这里使用的是 Harbor 1.6.2 版本。

```
wget
https://storage.googleapis.com/harbor-releases/release-1.6.0/harbor-offline-
installer-v1.6.2.tgz
```

（2）将 Harbor 安装文件解压缩到 "/root/training/" 目录下。

```
mkdir /root/training/
tar -zxvf harbor-offline-installer-v1.6.2.tgz -C /root/training/
```

（3）进入 "/root/training/harbor" 目录下。

```
cd /root/training/harbor
```

（4）编辑 Harbor 的核心配置文件 harbor.cfg。下面展示了需要修改的参数值：将 Harbor 监听的地址设置为本机的主机名，并设置了 Harbor 管理员的登录密码。

```
# 监听的地址
hostname = my.harbor.com
# 管理员的登录密码
harbor_admin_password = Welcome_1
```

（5）执行 install 命令安装 Harbor。图 2-41 展示了安装完成后的界面。

```
./install.sh
```

（6）通过浏览器访问 Harbor 主页 "http://192.168.79.10/"，如图 2-42 所示。

```
Creating network "harbor_harbor" with the default driver
Creating harbor-log ... done
Creating harbor-adminserver ... done
Creating harbor-db          ... done
Creating registry           ... done
Creating redis              ... done
Creating harbor-ui          ... done
Creating nginx              ... done
Creating harbor-jobservice  ... done

✔----Harbor has been installed and started successfully.----

Now you should be able to visit the admin portal at http://my.harbor.com .
For more details, please visit https://github.com/goharbor/harbor .

[root@my harbor]#
```

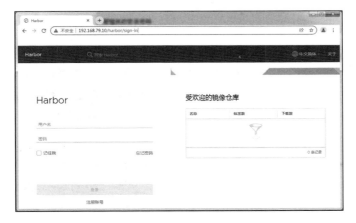

图 2-41

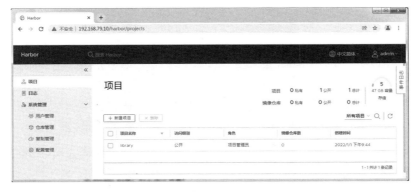

图 2-42

（7）使用 admin 账号进行登录，密码为第（4）步中设置的登录密码 Welcome_1。图 2-43 展示了 Harbor 主页。

图 2-43

2.5.3 【实战】在 Docker 中使用 Harbor

在配置好私有镜像仓库 Harbor 后,需要在 Docker 的客户机上进行设置才能使用它。这里以 master 主机为例进行设置。

(1)将 Harbor 主机的主机名和 IP 地址写入 Docker 客户机的 "/etc/hosts" 文件中。

```
192.168.79.10 my.harbor.com
```

(2)编辑文件 "/usr/lib/systemd/system/docker.service",输入以下内容。其中,my.harbor.com 是 Harbor 主机的主机名。

```
ExecStart=/usr/bin/dockerd --insecure-registry my.harbor.com
```

(3)执行以下语句创建 "/etc/docker/daemon.json" 文件,在该文件中指定私有镜像仓库的地址。

```
cat > /etc/docker/daemon.json << EOF
{ "insecure-registries":["my.harbor.com"] }
EOF
```

(4)重启 Docker 服务。

```
systemctl daemon-reload
systemctl restart docker
```

(5)使用 "docker login" 命令登录私有镜像仓库,如图 2-44 所示。

```
[root@master ~]# docker login my.harbor.com
Username: admin
Password:
WARNING! Your password will be stored unencrypted in /root/.docker/config.json.
Configure a credential helper to remove this warning. See
https://docs.docker.com/engine/reference/commandline/login/#credentials-store

Login Succeeded
[root@master ~]#
```

图 2-44

(6)创建一个简单的 Dockerfile 文件进行测试。

```
FROM centos
ENV TZ "Asia/Shanghai"
```

(7)使用 "docker build" 命令编译 Dockerfile 文件生成镜像,如图 2-45 所示。

(8)使用 "docker images" 命令查看生成的镜像,如图 2-46 所示。也可以使用 myimages 的别名来查看生成的镜像。

```
[root@master ~]# docker build -t my.harbor.com/library/mycentos .
Sending build context to Docker daemon  269.2MB
Step 1/2 : FROM centos
 ---> 5d0da3dc9764
Step 2/2 : ENV TZ "Asia/Shanghai"
 ---> Running in 42e98585f9cb
Removing intermediate container 42e98585f9cb
 ---> fdcdce75bd2d
Successfully built fdcdce75bd2d
Successfully tagged my.harbor.com/library/mycentos:latest
[root@master ~]#
```

图 2-45

```
[root@master ~]# myimages
IMAGE ID        REPOSITORY                        TAG
fdcdce75bd2d    my.harbor.com/library/mycentos    latest
605c77e624dd    nginx                             latest
5d0da3dc9764    centos                            latest
[root@master ~]#
```

图 2-46

（9）执行以下命令将镜像上传到私有镜像仓库 harbor 中。上传完成后的效果如图 2-47 所示。

```
docker push my.harbor.com/library/mycentos
```

```
[root@master ~]# docker push my.harbor.com/library/mycentos
Using default tag: latest
The push refers to repository [my.harbor.com/library/mycentos]
74ddd0ec08fa: Pushed
latest: digest: sha256:e99d34f640a0f0f55ac0ed11c159345a8bdbc259c2269120c880c9abb08a512
a size: 529
[root@master ~]#
```

图 2-47

（10）在 Harbor 的 Web 主页查看上传的镜像信息。可以看到成功将生成的 mycentos 镜像保存到了私有镜像仓库中了，如图 2-48 所示。

图 2-48

第 3 章
Docker 的容器

有了 Docker 的镜像，就可以创建 Docker 容器了。Docker 容器是一种轻量级的虚拟化方式，可以在"秒"级时间内快速启动和停止。另外，Docker 容器对系统资源（如 CPU、内存、I/O 带宽）的要求很低。除运行在容器中的应用外，Docker 容器基本不会消耗其他的系统资源。

 如果一个 Docker 容器被删除了，则该容器本身的状态是不会被保留的。因此，要实现数据的持久化保存，则需要借助数据卷的方式来实现。

3.1 Docker 容器的基本概念与操作

Docker 的容器类似操作系统的文件夹，其中包含应用和应用运行时所需要的依赖环境。每一个 Docker 容器都是从 Docker 镜像创建的。

Docker 容器的操作包括：创建容器、停止容器、进入容器、删除容器、导入和导出容器、查看容器等。

下面演示如何进行这些操作。

（1）使用"docker create"创建容器，如以下命令所示。这里基于 Nginx 的镜像创建了一个容器，名字为"mycontainer1"。但通过"docker ps"命令（或使用第 1 章中创建的"myps"命令）无法看到创建的容器，因为此时容器还没有启动。

```
docker create --name mycontainer1 nginx
```

（2）在"myps"命令后使用-a 参数可以查看所有的容器，如图 3-1 所示。注意，这时容器不一定是运行状态。

```
[root@master ~]# myps -a
CONTAINER ID        IMAGE          NAMES              PORTS
605ef253eedf        nginx          mycontainer1
[root@master ~]#
```

图 3-1

（3）使用"docker start"命令可以启动容器。

```
docker start 605ef253eedf
```

（4）使用"docker run"命令可以直接创建容器并启动容器。

```
docker run -d --name mycontainer2 nginx
```

（5）使用"docker exec"命令可以进入容器内，例如：

```
docker exec -it mycontainer1 /bin/bash
```

其中的参数说明如下。

- -t：为 Docker 分配一个伪终端。
- -i：为 Docker 打开标准输入。

也可以使用容器的 ID 进入容器，如图 3-2 所示。

```
[root@master ~]# docker exec -it 605ef253eedf /bin/bash
root@605ef253eedf:/#
```

图 3-2

（6）在停止并删除容器时需要注意：当容器正在运行时，是不能直接将其删除的。但可以使用 -f 参数进行强制删除，如图 3-3 所示。

```
docker stop 4bed9e655cb3
docker rm mycontainer2
```

```
[root@master ~]# docker rm -f mycontainer2
mycontainer2
[root@master ~]#
```

图 3-3

如果要一次性删除所有的容器，则执行以下命令。

docker rm -f $(docker ps -aq)

（7）可以导出一个已经创建的容器到一个文件中。不管容器处于什么状态，都可以使用"docker export"命令进行导出，如图 3-4 所示。通过这种方式可以实现容器的迁移。

```
docker export -o myexportcontainer1.tar 0ccc364cdfc7
```

其中的参数说明如下。

- myexportcontainer1.tar：导出的容器文件。
- 0ccc364cdfc7：容器的 ID。

```
[root@master ~]# ls
anaconda-ks.cfg  a.txt  demo  Dockerfile  myexportcontainer1.tar
[root@master ~]#
```

图 3-4

（8）从容器导出生成的文件，可以使用"docker import"命令重新导入容器。

```
docker import myexportcontainer1.tar mynginx_imported
```

这里的 mynginx_imported 是一个镜像。

（9）使用 mynginx_imported 镜像来创建一个容器。这里需要注意的是：如果要使用导入的镜像直接启动容器，则需要在启动时跟上具体的命令，否则会出现的错误，如图 3-5 所示。

```
[root@master ~]# docker run -d mynginx_imported
/usr/bin/docker-current: Error response from daemon: No command specified.
See '/usr/bin/docker-current run --help'.
[root@master ~]#
```

图 3-5

下面给出了正确的做法。命令中的"/bin/bash"表示在启动容器时需要在容器中执行的指令。

```
docker run -d mynginx_imported /bin/bash
```

3.2 Docker 的日志

在 Docker 的整个生命周期中，超过 70%的时间我们都是在做 Docker 的运维工作。查看 Docker 日志，是运维工作中非常重要的一个环节。

Docker 的日志分为两种类型：Docker 引擎的日志、应用的日志。下面分别对这两种日志进行介绍。

3.2.1 【实战】访问 Docker 引擎的日志

Docker 引擎的日志是指，在 Docker 守护进程执行过程中产生的日志信息。如果 Docker 引擎出现了问题，则需要在不同操作系统的宿主机上使用不同的方式进行查看。

表 3-1 中列举了不同操作系统上 Docker 引擎日志的位置。

表 3-1

操作系统	日志的位置
CentOS 7	journalctl –u docker.service
RHEL 7	journalctl –u docker.service
Ubuntu(16.04)	journalctl –u docker.service
Ubuntu(14.04)	/var/log/upstart/docker.log
OpenSuSE	journalctl –u docker.service
Boot2Docker	/var/log/docker.log

以 master 宿主机为例，操作系统是 CentOS 7（64 位），因此，可以使用以下命令查看 Docker 引擎的日志。为了方便查看，可以使用 Linux 的管道进行分页显示，如图 3-6 所示。

```
journalctl -u docker.service | more
```

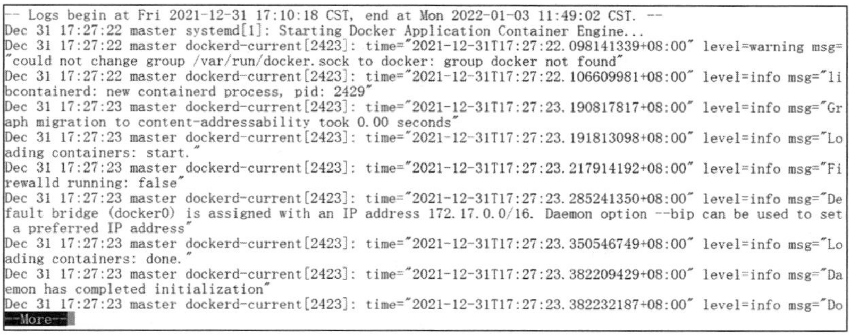

```
-- Logs begin at Fri 2021-12-31 17:10:18 CST, end at Mon 2022-01-03 11:49:02 CST. --
Dec 31 17:27:22 master systemd[1]: Starting Docker Application Container Engine...
Dec 31 17:27:22 master dockerd-current[2423]: time="2021-12-31T17:27:22.098141339+08:00" level=warning msg=
"could not change group /var/run/docker.sock to docker: group docker not found"
Dec 31 17:27:22 master dockerd-current[2423]: time="2021-12-31T17:27:22.106609981+08:00" level=info msg="li
bcontainerd: new containerd process, pid: 2429"
Dec 31 17:27:23 master dockerd-current[2423]: time="2021-12-31T17:27:23.190817817+08:00" level=info msg="Gr
aph migration to content-addressability took 0.00 seconds"
Dec 31 17:27:23 master dockerd-current[2423]: time="2021-12-31T17:27:23.191813098+08:00" level=info msg="Lo
ading containers: start."
Dec 31 17:27:23 master dockerd-current[2423]: time="2021-12-31T17:27:23.217914192+08:00" level=info msg="Fi
rewalld running: false"
Dec 31 17:27:23 master dockerd-current[2423]: time="2021-12-31T17:27:23.285241350+08:00" level=info msg="De
fault bridge (docker0) is assigned with an IP address 172.17.0.0/16. Daemon option --bip can be used to set
 a preferred IP address"
Dec 31 17:27:23 master dockerd-current[2423]: time="2021-12-31T17:27:23.350546749+08:00" level=info msg="Lo
ading containers: done."
Dec 31 17:27:23 master dockerd-current[2423]: time="2021-12-31T17:27:23.382209429+08:00" level=info msg="Da
emon has completed initialization"
Dec 31 17:27:23 master dockerd-current[2423]: time="2021-12-31T17:27:23.382232187+08:00" level=info msg="Do
--More--
```

图 3-6

3.2.2 【实战】访问 Docker 应用的日志

Docker 将应用运行在容器中，应用输出日志也就输出到容器中了，访问 Docker 应用的日志也就变成了访问 Docker 容器的日志。要访问容器的日志，首先需要了解 Docker 的日志引擎。表 3-2 列出了 Docker 支持的日志引擎。

表 3-2

日志引擎	说 明
journald	Docker 默认的日志引擎。这种引擎把所有容器的回显日志输出到系统的 journald 服务中
json-file	把每个容器的回显日志以 JSON 文件格式输出到每个容器的内部。如果在实际应用中某些应用产生了大量的日志信息，则可能导致容器的 JSON 日志文件过大而占满宿主机的磁盘
syslog	把所有容器的回显日志输出到系统的 syslog 服务中
fluentd	把所有容器的回显日志输出到系统的 fluentd 服务中
gelf	把所有容器的日志输出到支持 GELF（Graylog Extended Log Format）格式的服务中，如 Logstash
none	关闭 Docker 容器的日志。使用这种方式，则意味着无法通过"docker logs"命令查看任何容器输出的日志

利用以下命令可以查看 Docker 默认的日志引擎，如图 3-7 所示。

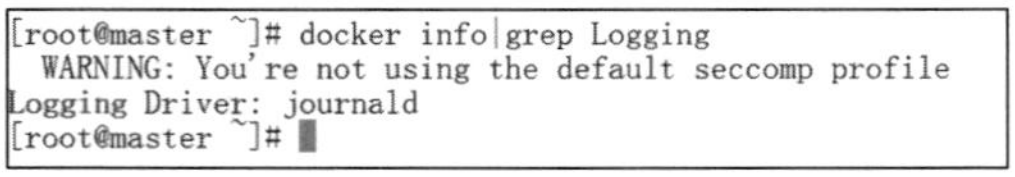

```
[root@master ~]# docker info|grep Logging
 WARNING: You're not using the default seccomp profile
 Logging Driver: journald
[root@master ~]#
```

图 3-7

下面通过具体的示例来说明。

示例 1 修改 Docker 默认的日志引擎

（1）修改文件"/etc/docker/daemon.json"的内容，输入以下配置。

```
{
    "log-driver": "json-file"
}
```

（2）重新加载 Docker 的服务，并重启 Docker。

```
systemctl daemon-reload
systemctl restart docker
```

（3）查看 Docker 的日志引擎。此时 Docker 的日志引擎被修改为 json-file，如图 3-8 所示。

```
docker info|grep Logging
```

```
[root@master ~]# docker info|grep Logging
 Logging Driver: json-file
[root@master ~]#
```

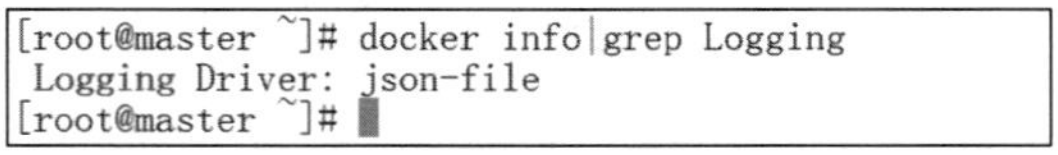

图 3-8

（4）启动一个容器进行简单的测试，可以看到这时容器的日志默认以 JSON 的格式存储于"/var/lib/docker/containers/<容器 id>/"目录下。

```
docker run -d -p 1234:80 nginx
```

（5）进入目录"/var/lib/docker/containers/<容器 id>"下查看容器的日志，如图 3-9 所示。

```
[root@master ~]# myps
CONTAINER ID   IMAGE     NAMES            PORTS
1505f5a02f60   nginx     busy_goldwasser  0.0.0.0:1234->80/tcp, :::1234->80/tcp
[root@master ~]# cd /var/lib/docker/containers/1505f5a02f60486c23cc8a189558e3c044004deaf3d
8aa7065642f7369c7f0b1/
[root@master 1505f5a02f60486c23cc8a189558e3c044004deaf3d8aa7065642f7369c7f0b1]# pwd
/var/lib/docker/containers/1505f5a02f60486c23cc8a189558e3c044004deaf3d8aa7065642f7369c7f0b
1
[root@master 1505f5a02f60486c23cc8a189558e3c044004deaf3d8aa7065642f7369c7f0b1]# ls *.log
1505f5a02f60486c23cc8a189558e3c044004deaf3d8aa7065642f7369c7f0b1-json.log
[root@master 1505f5a02f60486c23cc8a189558e3c044004deaf3d8aa7065642f7369c7f0b1]#
```

图 3-9

并不推荐直接读取 Docker 日志的内容，原因是 Docker 提供了"docker logs"命令来帮助我们读取日志的信息。图 3-10 中显示的是"docker logs"命令的帮助信息。

```
[root@master ~]# docker help logs

Usage:  docker logs [OPTIONS] CONTAINER

Fetch the logs of a container

Options:
      --details        Show extra details provided to logs
  -f, --follow         Follow log output
      --since string   Show logs since timestamp (e.g. 2013-01-02T13:23:37Z) or
                       relative (e.g. 42m for 42 minutes)
  -n, --tail string    Number of lines to show from the end of the logs (default "all")
  -t, --timestamps     Show timestamps
      --until string   Show logs before a timestamp (e.g. 2013-01-02T13:23:37Z) or
                       relative (e.g. 42m for 42 minutes)
[root@master ~]#
```

图 3-10

示例 2　使用"docker logs"命令查看容器的日志

（1）查看指定时间后的日志，只显示最后 5 行，如图 3-11 所示。

```
docker logs -f -t --since="2021-01-03" --tail=5 1505f5a02f60
```

其中，1505f5a02f60 是容器的 ID。

```
[root@master ~]# docker logs -f -t --since="2021-01-03" --tail=5 1505f5a02f60
2022-01-03T06:29:22.426899538Z 2022/01/03 06:29:22 [notice] 1#1: built by gcc 10.2.1 20210
110 (Debian 10.2.1-6)
2022-01-03T06:29:22.426901330Z 2022/01/03 06:29:22 [notice] 1#1: OS: Linux 3.10.0-693.el7.
x86_64
2022-01-03T06:29:22.426903558Z 2022/01/03 06:29:22 [notice] 1#1: getrlimit(RLIMIT_NOFILE):
 65536:65536
2022-01-03T06:29:22.426905218Z 2022/01/03 06:29:22 [notice] 1#1: start worker processes
2022-01-03T06:29:22.426906818Z 2022/01/03 06:29:22 [notice] 1#1: start worker process 31
```

图 3-11

（2）查看容器最近 30 分钟的日志。

```
docker logs --since 30m 1505f5a02f60
```

（3）查看某个时间之后的所有日志，如图 3-12 所示。

```
docker logs -t --since="2021-01-03" 1505f5a02f60
```

```
[root@master ~]# docker logs -t --since="2021-01-03" 1505f5a02f60
2022-01-03T06:29:22.403231651Z /docker-entrypoint.sh: /docker-entrypoint.d/ is not empty,
will attempt to perform configuration
2022-01-03T06:29:22.403269692Z /docker-entrypoint.sh: Looking for shell scripts in /docker
-entrypoint.d/
2022-01-03T06:29:22.403272994Z /docker-entrypoint.sh: Launching /docker-entrypoint.d/10-li
sten-on-ipv6-by-default.sh
2022-01-03T06:29:22.411006008Z 10-listen-on-ipv6-by-default.sh: info: Getting the checksum
 of /etc/nginx/conf.d/default.conf
2022-01-03T06:29:22.416918185Z 10-listen-on-ipv6-by-default.sh: info: Enabled listen on IP
v6 in /etc/nginx/conf.d/default.conf
2022-01-03T06:29:22.417961999Z /docker-entrypoint.sh: Launching /docker-entrypoint.d/20-en
vsubst-on-templates.sh
2022-01-03T06:29:22.420284140Z /docker-entrypoint.sh: Launching /docker-entrypoint.d/30-tu
ne-worker-processes.sh
2022-01-03T06:29:22.426771782Z /docker-entrypoint.sh: Configuration complete; ready for st
art up
2022-01-03T06:29:22.426894700Z 2022/01/03 06:29:22 [notice] 1#1: using the "epoll" event m
ethod
2022-01-03T06:29:22.426897634Z 2022/01/03 06:29:22 [notice] 1#1: nginx/1.21.5
```

图 3-12

3.3 管理容器的资源

在一台 Docker 宿主机上可以同时启动多个容器。在默认情况下，Docker 没有限制其中运行的容器使用硬件资源。而在实际环境中，容器的负载过高会占用宿主机的大量资源。这里的资源主要是指宿主机的 CPU、内存和 I/O 带宽这三个方面。

本节将介绍如何使用 Docker 的资源管理给容器的资源使用设置一个阈值，以控制容器对宿主机 CPU、内存和 I/O 带宽的使用。

3.3.1 什么是 Linux CGroup

由于 Docker 构建在 Linux 的基础之上，因此从 Linux 底层来看，Docker 是利用 Linux Control

Group（简称 Linux CGroup）来实现对资源使用的控制。因此，要掌握 Docker 容器的资源管理，有必要先了解一下什么是 Linux CGroup。

> Linux CGroup 是 Linux 中的一些进程，通过这些进程可以限制应用对资源的使用。并且，通过 Linux CGroup 可以对系统资源做精细化控制。例如，可以实现对每个容器使用的 CPU 比率进行限制。

Linux CGroup 主要提供了以下功能。

- Resource limitation：限制资源的使用，例如，使用 CPU 及内存的上限。
- Prioritization：应用的优先级控制，例如，控制任务的调度。
- Accounting：应用的审计和统计，例如，实现应用的计费。
- Control：实现对应用的控制，例如，应用的挂起、恢复和执行等。

要使用 Linux CGroup，则需要先通过执行以下步骤确定 Linux 的内核是否启用了 Linux CGroup。

（1）确定操作系统的发行版本，如图 3-13 所示。

```
[root@master ~]# uname -r
3.10.0-693.el7.x86_64
[root@master ~]#
```

图 3-13

（2）根据操作系统的发行版本，可以确定是否启用了 Linux CGroup。图 3-14 中的 CGROUP 参数的值是"y"，表示已经启动 Linux CGroup。

```
cat /boot/config-3.10.0-693.el7.x86_64 | grep CGROUP
```

```
[root@master ~]# cat /boot/config-3.10.0-693.el7.x86_64 | grep CGROUP
CONFIG_CGROUPS=y
# CONFIG_CGROUP_DEBUG is not set
CONFIG_CGROUP_FREEZER=y
CONFIG_CGROUP_PIDS=y
CONFIG_CGROUP_DEVICE=y
CONFIG_CGROUP_CPUACCT=y
CONFIG_CGROUP_HUGETLB=y
CONFIG_CGROUP_PERF=y
CONFIG_CGROUP_SCHED=y
CONFIG_BLK_CGROUP=y
# CONFIG_DEBUG_BLK_CGROUP is not set
CONFIG_NETFILTER_XT_MATCH_CGROUP=m
CONFIG_NET_CLS_CGROUP=y
CONFIG_NETPRIO_CGROUP=y
[root@master ~]#
```

图 3-14

下面通过 3 个示例来演示如何使用 Linux Group 实现对系统资源的控制。

示例 1 通过 Linux CGroup 限制应用的 CPU 使用率

在本示例中，利用 C 语言开发一段执行死循环的代码。由于是死循环，所以代码的 CPU 使用率将很高。然后通过使用 Linux CGroup，将代码的 CPU 使用率限制在一定范围内，如 20%。下面是具体的操作步骤。

（1）开发一段 C 语言程序代码产生一个死循环，并将代码保存为 hello.c。

```c
//hello.c
int main(void)
{
    int i = 0;
    for(;;) i++;
    return 0;
}
```

（2）将程序代码进行编译。

```
gcc -o hello hello.c
```

（3）执行程序代码，这时程序将产生死循环无法退出，如图 3-15 所示。

```
./hello
```

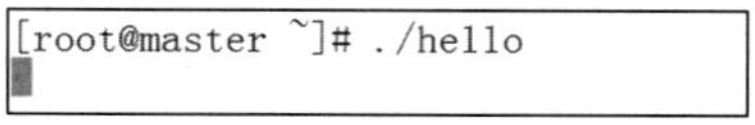

```
[root@master ~]# ./hello
```

图 3-15

（4）在一个新的命令行窗口中，使用"top"命令监控应用 hello 的 CPU 使用率，可以看到已经达到了 99.7%，如图 3-16 所示。

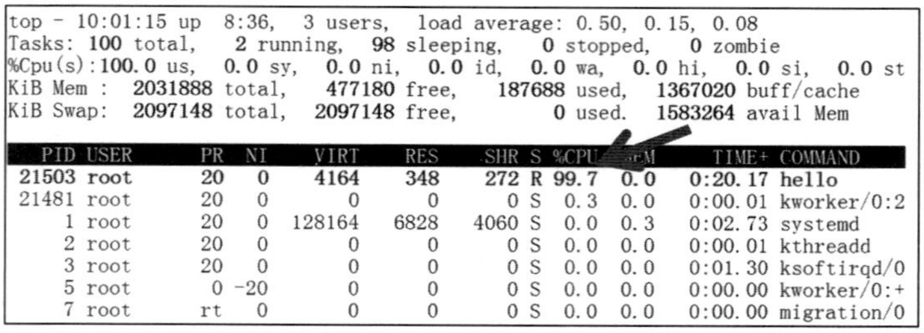

```
top - 10:01:15 up  8:36,  3 users,  load average: 0.50, 0.15, 0.08
Tasks: 100 total,   2 running,  98 sleeping,   0 stopped,   0 zombie
%Cpu(s):100.0 us,  0.0 sy,  0.0 ni,  0.0 id,  0.0 wa,  0.0 hi,  0.0 si,  0.0 st
KiB Mem :  2031888 total,   477180 free,   187688 used,  1367020 buff/cache
KiB Swap:  2097148 total,  2097148 free,        0 used.  1583264 avail Mem

   PID USER      PR  NI    VIRT    RES    SHR S %CPU %MEM     TIME+ COMMAND
 21503 root      20   0    4164    348    272 R 99.7  0.0   0:20.17 hello
 21481 root      20   0       0      0      0 S  0.3  0.0   0:00.01 kworker/0:2
     1 root      20   0  128164   6828   4060 S  0.0  0.3   0:02.73 systemd
     2 root      20   0       0      0      0 S  0.0  0.0   0:00.01 kthreadd
     3 root      20   0       0      0      0 S  0.0  0.0   0:01.30 ksoftirqd/0
     5 root       0 -20       0      0      0 S  0.0  0.0   0:00.00 kworker/0:+
     7 root      rt   0       0      0      0 S  0.0  0.0   0:00.00 migration/0
```

图 3-16

（5）进入"/sys/fs/cgroup/cpu/"目录下，创建一个新的子目录 hello。该目录用于设置 CPU 使用率的阈值，如图 3-17 所示。

```
cd /sys/fs/cgroup/cpu/
mkdir hello
```

```
cd hello/
ls
```

图 3-17

（6）查看文件 cpu.cfs_quota_us 的内容为"-1"，表示没有对其 CPU 使用率进行限制。

（7）执行以下语句将 CPU 使用率的阈值设置为 20%。

```
echo 20000 > cpu.cfs_quota_us
```

（8）将应用 hello 的进程 ID 写入 tasks 文件，如图 3-18 所示。

```
echo 21503 > tasks
```

图 3-18

（9）再次观察"top"命令的输出信息，发现应用 hello 的 CPU 使用率降到了 20%，如图 3-19 所示。

图 3-19

（10）重新启动一个 hello 应用，并按照上面的步骤将进程 ID 写入 tasks 文件。这时观察"top"命令的输出会发现，两个 hello 应用各自占用 10%的 CPU 使用率，如图 3-20 所示。

```
top - 10:03:13 up   8:38,   4 users,   load average: 1.10, 0.52, 0.23
Tasks: 103 total,    3 running, 100 sleeping,    0 stopped,   0 zombie
%Cpu(s): 20.3 us,   0.0 sy,   0.0 ni, 79.7 id,   0.0 wa,   0.0 hi,   0.0 si,   0.0 st
KiB Mem :  2031888 total,   474900 free,   189716 used,  1367272 buff/cache
KiB Swap:  2097148 total,  2097148 free,        0 used.  1580988 avail Mem

    PID USER      PR  NI    VIRT    RES    SHR S %CPU %MEM     TIME+ COMMAND
  21503 root      20   0    4164    348    272 R 10.0  0.0   1:01.57 hello
  21538 root      20   0    4164    344    272 R 10.0  0.0   0:12.53 hello
  18726 root      20   0 1254384  27912  11160 S  0.3  1.4   0:24.19 containerd
      1 root      20   0  128164   6828   4060 S  0.0  0.3   0:02.74 systemd
      2 root      20   0       0      0      0 S  0.0  0.0   0:00.01 kthreadd
      3 root      20   0       0      0      0 S  0.0  0.0   0:01.30 ksoftirqd/0
      5 root       0 -20       0      0      0 S  0.0  0.0   0:00.00 kworker/0:+
```

图 3-20

示例 2　通过 Linux CGroup 限制应用使用系统内存

（1）进入"cd /sys/fs/cgroup/memory"目录下，创建子目录 hello。

```
cd /sys/fs/cgroup/memory
mkdir hello
cd hello
```

（2）查看文件 memory.limit_in_bytes 的内容。

```
more memory.limit_in_bytes
```

输出的文件内容如下：

```
9223372036854771712
```

这里设定的值是 9223372036854771712，表示没有对内存进行任何限制。

（3）以下语句会将内存的阈值设置为 64KB。如果应用使用的内存超过了该值，则该应用会被操作系统自动"杀掉"。

```
echo 64k > memory.limit_in_bytes
```

（4）生效配置，将应用 hello 的进程 ID 写入 tasks 文件。

```
echo 21503 > tasks
```

示例 3　通过 Linux CGroup 限制应用使用 I/O 带宽

为了更好地观察结果，首先安装 iotop 工具，如图 3-21 所示。

```
yum -y install iotop
```

```
[root@master ~]# yum -y install iotop
Loaded plugins: fastestmirror
base                                                    | 3.6 kB     00:00
extras                                                  | 2.9 kB     00:00
updates                                                 | 2.9 kB     00:00
Loading mirror speeds from cached hostfile
 * base: mirrors.bupt.edu.cn
 * extras: mirrors.bfsu.edu.cn
 * updates: mirrors.bupt.edu.cn
Resolving Dependencies
--> Running transaction check
---> Package iotop.noarch 0:0.6-4.el7 will be installed
--> Finished Dependency Resolution

Dependencies Resolved
```

图 3-21

> iotop 工具用来监视磁盘 I/O 使用状况，包括 pid、user、I/O、进程等相关信息。

以下演示如何使用 Linux CGroup 限制应用使用 I/O 带宽。

（1）使用 Linux 的“dd”命令从磁盘持续读写数据。

```
dd if=/dev/sda of=/dev/null
```

（2）通过 iotop 工具查看 I/O 读取的速度为 569.04 MB/s，如图 3-22 所示。

```
Total DISK READ :    569.04 M/s | Total DISK WRITE :      0.00 B/s
Actual DISK READ:    569.04 M/s | Actual DISK WRITE:      0.00 B/s
  TID  PRIO  USER     DISK READ  DISK WRITE  SWAPIN     IO>    COMMAND
21681 be/4 root     569.04 M/s    0.00 B/s  0.00 % 21.13 % dd if=/dev/sda of=/dev/null
  512 be/0 root       0.00 B/s    0.00 B/s  0.00 %  0.00 % [nfit]
    1 be/4 root       0.00 B/s    0.00 B/s  0.00 %  0.00 % systemd --swit~deserialize 21
    2 be/4 root       0.00 B/s    0.00 B/s  0.00 %  0.00 % [kthreadd]
    3 be/4 root       0.00 B/s    0.00 B/s  0.00 %  0.00 % [ksoftirqd/0]
 1028 be/4 root       0.00 B/s    0.00 B/s  0.00 %  0.00 % python -Es /us~in/tuned -l -P
    5 be/0 root       0.00 B/s    0.00 B/s  0.00 %  0.00 % [kworker/0:0H]
    7 rt/4 root       0.00 B/s    0.00 B/s  0.00 %  0.00 % [migration/0]
    8 be/4 root       0.00 B/s    0.00 B/s  0.00 %  0.00 % [rcu_bh]
    9 be/4 root       0.00 B/s    0.00 B/s  0.00 %  0.00 % [rcu_sched]
   10 rt/4 root       0.00 B/s    0.00 B/s  0.00 %  0.00 % [watchdog/0]
```

图 3-22

（3）查看设备“/dev/sda”的信息。从图 3-23 中可以看到，设备“/dev/sda”的设备号是“disk 8，0”。

```
ls -l /dev/sda
```

```
[root@master ~]# ls -l /dev/sda
brw-rw----. 1 root disk 8, 0 Dec 31 18:54 /dev/sda
[root@master ~]#
```

图 3-23

（4）使用 Linux CGroup 限制 I/O 对设备 "/dev/sda" 的读取速率。

```
mkdir /sys/fs/cgroup/blkio/io
cd /sys/fs/cgroup/blkio/io
echo '8:0 1048576' > blkio.throttle.read_bps_device
```

通过这样，I/O 对该设备的读取速率被限制在 1MB/s 之内了。

（5）将 "dd" 命令的进程 ID（21681）写入 tasks 文件。

```
echo 21681 > tasks
```

（6）再次观察 iotop 工具的监控输出信息会发现，这时 "dd" 命令对该设备的读取速率已经被设置为 1.00 MB/s 了，如图 3-24 所示。

图 3-24

了解了 Linux CGroup 后，再来讨论 "Docker 是如何对容器使用的资源进行设定的" 就变得非常简单了。Docker 只是对 Linux CGroup 进行了封装，从而简化了调用操作的方式。

3.3.2 【实战】Docker 对 CPU 的使用

Docker 的容器可以被看成是一个虚拟机，或者一个运行在 Linux 之上的进程。Docker 的引擎可以通过参数 -c 或者 --cpu-shares，为每一个容器分配一个 "CPU 使用率的相对权重"。该权重与实际的处理速度无关。每个容器默认有 1024 个 CPU 配额的权重。

> 如果启动了两个容器，并且两个都使用默认的权重值 1024，则 Docker 引擎将把 CPU 使用率平均分配给这两个容器，即这两个容器各自占用 50% 的 CPU 使用率。
> 但是，如果给容器 A 分配的是 1024 权重，而给容器 B 分配的是 512 权重，则会发现容器 A 使用 CPU 的比例比容器 B 使用 CPU 的比例多一倍。

下面通过具体步骤来演示。这里基于 CentOS 的基础镜像安装压力测试工具 stress 来进行测试。

（1）创建一个 Dockerfile 文件，输入以下内容。

```
FROM centos:7
RUN yum install -y epel-release && yum install -y stress
ENTRYPOINT ["stress"]
```

（2）执行"docker build"命令将其编译成镜像，镜像的名称是 mycentos。

```
docker build -t mycentos .
```

（3）使用 mycentos 镜像创建两个容器。注意：第 1 个容器占用 CPU 的权重是默认的 1024；而第 2 个容器占用 CPU 的权重是 512。

```
docker run -it mycentos --cpu 4
docker run -it -c 512 mycentos --cpu 4
```

其中，参数--cpu 4 用于指定压力测试工具 stress 模拟创建 4 个 CPU。

（4）通过"docker stats"命令观察这两个容器占用 CPU 的情况，可以看到容器 4ca2db2be120 占用的 CPU 比容器 a395e52f60cc 占用的 CPU 多近一倍，如图 3-25 所示。

```
docker stats
```

```
CONTAINER       CPU %      MEM USAGE / LIMIT     MEM %     NET I/O          BLOCK I/O       PIDS
a395e52f60cc    84.69%     192 KiB / 1.938 GiB   0.01%     648 B / 648 B    0 B / 0 B       5
4ca2db2be120    164.04%    192 KiB / 1.938 GiB   0.01%     1.34 kB / 690 B  24.6 kB / 0 B   5
```

图 3-25

（5）如果要变更一个正在运行的容器的配额（例如修改容器 a395e52f60cc 的 CPU 权重），则需要先使用"docker inspect CONTAINER"命令获取容器 ID。获取的这个 ID 会很长，如图 3-26 所示。

```
docker inspect a395e52f60cc | grep Id
```

```
[root@master ~]# docker inspect a395e52f60cc | grep Id
        "Id": "a395e52f60ccd02e4ef7f6e4abdaca4a72a65f8f
ed10372e5559502e4f6ae13e",
[root@master ~]#
```

图 3-26

（6）使用"systemctl set-property"命令将容器 a395e52f60cc 占用 CPU 的权重设置为 1024，如图 3-27 所示。

```
systemctl set-property \
  docker-a395e52f60ccd02e4ef7f6e4abdaca4a72a65f8fed10372e5559502e4f6ae13e.scope \
  CPUShares=1024
```

```
[root@master ~]# systemctl set-property \
> docker-a395e52f60ccd02e4ef7f6e4abdaca4a72a65f8fed10372e5559502e4f6ae13e.scope \
> CPUShares=1024
[root@master ~]#
```

图 3-27

（7）通过"docker stats"命令观察这两个容器的 CPU 占用率，会发现两个容器基本平均地占用了 CPU，如图 3-28 所示。

```
CONTAINER      CPU %      MEM USAGE / LIMIT    MEM %    NET I/O         BLOCK I/O      PIDS
a395e52f60cc   224.65%    192 KiB / 1.938 GiB  0.01%    648 B / 648 B   0 B / 0 B      5
4ca2db2be120   221.98%    192 KiB / 1.938 GiB  0.01%    1.34 kB / 690 B 24.6 kB / 0 B  5
```

图 3-28

3.3.3 【实战】Docker 对内存的使用

利用底层的 Linux CGroup，能够很方便地通过参数-m 来设定容器所使用的内存。

默认情况下，一个容器可以使用主机上的所有内存。

在使用-m 参数时，可以指定具体的参数值后缀，如 K、M 或 G。

以下语句使用了 mycentos 来进行测试。

```
docker run -it --rm -m 256m mycentos --vm 1 --vm-bytes 128M --vm-hang 0
```

其中的参数说明如下。

- -m 256m：限制容器使用的内存大小。
- --vm 1：给压力测试工具 stress 产生一个内存分配的进程。
- --vm-bytes 128M：压力测试工具 stress 每次分配大小为 128MB 的内存空间。
- --vm-hang 0：压力测试工具 stress 分配完内存后立即释放分配的内存；如果该值为非零值，如 100，则表示分配内存后，等待 100 秒后再释放分配的内存。

使用"docker stats"命令观察容器占用的内存，会发现容器占用的内存被限定在 256MB，如图 3-29 所示。

```
docker stats --format "table {{.Container}}\t{{.MemUsage}}"
```

```
CONTAINER              MEM USAGE / LIMIT
36d1dee3502e           128.1 MiB / 256 MiB
```

图 3-29

3.3.4 【实战】Docker 对 I/O 带宽的使用

要在 Docker 中对容器使用的 I/O 带宽进行限定，则需要先了解 Docker I/O 管理相关的参数。执行以下命令可以获取这些参数及其含义，如图 3-30 所示。

```
[root@master ~]# docker help run | grep -E 'bps|IO'
Usage:  docker run [OPTIONS] IMAGE [COMMAND] [ARG...]
      --blkio-weight uint16            Block IO (relative weight), between 10 and 1000, or 0 to disable (default 0)
      --blkio-weight-device weighted-device   Block IO weight (relative device weight) (default [])
      --device-read-bps throttled-device      Limit read rate (bytes per second) from a device (default [])
      --device-read-iops throttled-device     Limit read rate (IO per second) from a device (default [])
      --device-write-bps throttled-device     Limit write rate (bytes per second) to a device (default [])
      --device-write-iops throttled-device    Limit write rate (IO per second) to a device (default [])
      --io-maxbandwidth string                Maximum IO bandwidth limit for the system drive (Windows only)
      --io-maxiops uint                       Maximum IOps limit for the system drive (Windows only)
[root@master ~]#
```

图 3-30

表 3-3 中列出了 Docker I/O 管理相关的参数。

表 3-3

参数名称	说　明
--blkio-weight	可以通过--blkio-weight 修改容器 blkio 的权重，权重值为 10～1000
--blkio-weight-device weighted-device	指定某个设备的权重
--device-read-bps throttled-device	按每秒读取块设备的数据量设定上限
--device-read-iops throttled-device	按照每秒读操作的次数设定上限
--device-write-bps throttled-device	按每秒写入块设备的数据量设定上限
--device-write-iops throttled-device	按照每秒写操作的次数设定上限

下面通过两个示例来演示如何在启动了的 Docker 容器中使用上面的参数来对容器使用的 I/O 带宽进行限定。

示例 1　使用--device-read-bps 和--device-write-bps 参数限定容器的读写速率

（1）使用以下语句创建并启动容器。这里将限制容器的写入速度设为 1MB/s。

```
docker run -it --rm --device-write-bps /dev/sda:1mb centos /bin/bash
```

（2）在容器中执行以下命令。其中，oflag=direct 表示读写数据采用直接 I/O 方式。

```
dd if=/dev/zero of=test.out bs=1M count=200 oflag=direct
```

（3）在宿主机上使用 iotop 命令观察容器的写入速度，会发现容器的写入速度被限定在 1011.77 KB/s（即 1MB/s 左右）如图 3-31 所示。

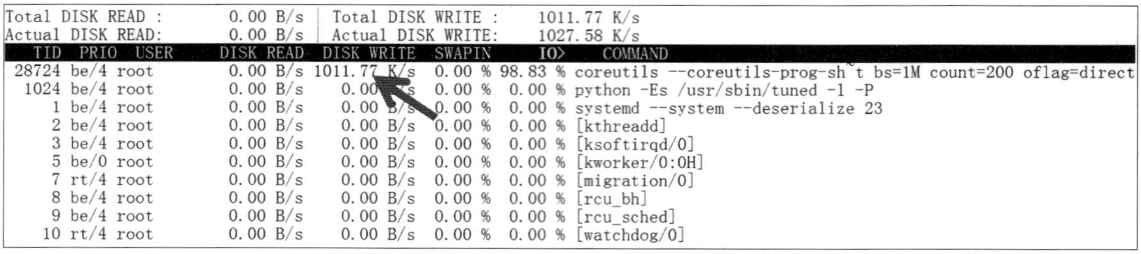

```
Total DISK READ :      0.00 B/s | Total DISK WRITE :     1011.77 K/s
Actual DISK READ:      0.00 B/s | Actual DISK WRITE:     1027.58 K/s
   TID  PRIO  USER     DISK READ  DISK WRITE  SWAPIN     IO>    COMMAND
 28724 be/4 root       0.00 B/s  1011.77 K/s  0.00 %  98.83 %  coreutils --coreutils-prog-sh t bs=1M count=200 oflag=direct
  1024 be/4 root       0.00 B/s     0.00 B/s  0.00 %   0.00 %  python -Es /usr/sbin/tuned -l -P
     1 be/4 root       0.00 B/s     0.00 B/s  0.00 %   0.00 %  systemd --system --deserialize 23
     2 be/4 root       0.00 B/s     0.00 B/s  0.00 %   0.00 %  [kthreadd]
     3 be/4 root       0.00 B/s     0.00 B/s  0.00 %   0.00 %  [ksoftirqd/0]
     5 be/0 root       0.00 B/s     0.00 B/s  0.00 %   0.00 %  [kworker/0:0H]
     7 rt/4 root       0.00 B/s     0.00 B/s  0.00 %   0.00 %  [migration/0]
     8 be/4 root       0.00 B/s     0.00 B/s  0.00 %   0.00 %  [rcu_bh]
     9 be/4 root       0.00 B/s     0.00 B/s  0.00 %   0.00 %  [rcu_sched]
    10 rt/4 root       0.00 B/s     0.00 B/s  0.00 %   0.00 %  [watchdog/0]
```

图 3-31

（4）以下语句将限制容器的读取速度为 1MB/s。

```
docker run -it --rm --device-read-bps /dev/sda:1mb centos /bin/bash
```

示例 2　使用--device-read-iops 和--device-write-iops 参数限定容器的读写次数

参数--device-read-iops 和--device-write-iops 的格式如下，其中，limit 必须是正整数。

```
--device-read-iops <device-path>:<limit>
--device-write-iops <device-path>:<limit>
```

以下是操作步骤。

（1）为了观察结果，先在宿主机上安装 sysstat 软件包。

```
yum install sysstat -y
```

（2）创建并启动容器，通过使用"--device-write-iops"参数将容器写数据的速率设定为每秒 5 次。

```
docker run -it --rm --device-write-iops /dev/sda:5 centos /bin/bash
```

（3）在容器中执行以下命令，其中，oflag=direct 表示读写数据采用直接 I/O 方式。

```
dd if=/dev/zero of=test.out bs=1M count=200 oflag=direct
```

（4）在宿主机上使用以下命令观察输出的结果，这个命令将每隔 1 秒刷新 1 次。从图 3-32 中可以看到容器的 TPS 被设定为 5。

```
iostat 1
```

```
dm-1              0.00        0.00        0.00           0           0
avg-cpu:  %user   %nice %system %iowait  %steal   %idle
           0.00    0.00    1.00   99.00    0.00    0.00

Device:            tps    kB_read/s    kB_wrtn/s    kB_read    kB_wrtn
sda               5.00        0.00      2560.00          0       2560
dm-0              6.00        0.00      3072.00          0       3072
dm-1              0.00        0.00        0.00           0          0
```

图 3-32

（5）以下语句会将容器读数据的速率设定为每秒 5 次。

```
docker run -it --rm --device-read-iops /dev/sda:5 centos /bin/bash
```

3.4 管理 Docker 容器中的数据

在生产环境中使用 Docker，一方面，需要对数据进行保存或者在多个容器之间进行数据共享；另一方面，在 Docker 的容器被删除后，并不会保留容器的状态信息。

那么如何实现信息的持久化呢？这必然涉及容器的数据管理。

3.4.1 在 Docker 容器中实现数据管理的两种方式

在 Docker 容器中实现数据管理（或者说实现数据的持久化）主要有以下两种方式。

1. 数据卷（Data Volumes）

数据卷本质上是一个挂载目录，类似使用 Linux 的 mount 命令挂载的目录。数据卷可以供容器使用，并且可以在不同的容器之间共享和重用数据卷。对数据卷的修改会立即生效。数据卷与容器彼此独立，对数据卷的更新不会影响镜像。

即使容器被删除，数据卷默认也会一直存在，直到数据卷被删除为止。

在 Docker 中可以使用 -mount 和 -v 两种方式给容器挂载数据卷。图 3-33 展示了数据卷与容器的关系。

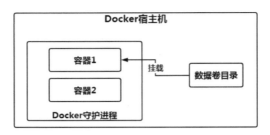

图 3-33

2. 数据卷容器（Data Volume Containers）

数据卷容器是一种特殊的容器，用来维护数据卷。它可以在多个容器之间共享数据信息。利用数据卷容器可以很方便地完成数据迁移。

图 3-34 展示了数据卷、数据卷容器和容器之间的关系。

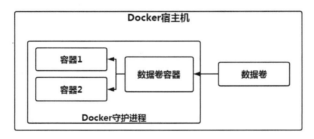

图 3-34

3.4.2　【实战】使用数据卷管理 Docker 容器中的数据

在 Docker 中，可以使用-mount 和-v 两种方式给容器挂载数据卷。下面通过具体的实战来演示。

（1）创建一个名为"myvolume"的数据卷。

```
docker volume create myvolume
```

（2）查看所有的数据卷，如图 3-35 所示。

```
docker volume ls
```

```
[root@master ~]# docker volume ls
DRIVER    VOLUME NAME
local     myvolume
[root@master ~]#
```

图 3-35

(3)使用"docker inspect volume"命令查看数据卷的详细信息,如图3-36所示。

```
docker inspect myvolume
```

```
[root@master ~]# docker inspect myvolume
[
    {
        "CreatedAt": "2022-01-03T19:10:40+08:00",
        "Driver": "local",
        "Labels": {},
        "Mountpoint": "/var/lib/docker/volumes/myvolume/_data",
        "Name": "myvolume",
        "Options": {},
        "Scope": "local"
    }
]
[root@master ~]#
```

图 3-36

把复杂度很高的产品拆分成一些较小的模块(每一个模块用 5~9 个小团队来维护),并遵循康威定律,这样可以减少沟通成本,提高协作效率,更好地实现产品快速迭代和弹性扩展。

(4)启动一个容器,并使用 myvolume 数据卷。这里使用 Nginx 镜像创建了一个容器,并将容器的 80 端口映射到了宿主机的 1234 端口,容器的名称为"mynginx"。

```
docker run -d -p 1234:80 --name mynginx \
--mount type=volume,source=myvolume,target=/usr/share/nginx/html/ nginx
```

其中的参数说明如下。

- --mount:指定在容器启动时挂载数据卷。
- type:指定数据卷挂载的方式,它有 3 个值,见表 3-4。

表 3-4

type 的取值	说　　明
volume	普通数据卷,这是默认的 type 类型。其函数映射到主机"/var/lib/docker/volumes"目录下
bind	绑定数据卷。使用这种类型可以在挂载数据卷时将其映射到主机的指定目录下
tmpfs	临时数据卷,只将容器的目录挂载到宿主机的内存中。一般在实际环境中不会使用这种方式

- source:指定宿主机上的目录或者数据卷。这里使用的是第(1)步所创建的数据卷 myvolume。
- target:将容器中的"/usr/share/nginx/html/"目录挂载到宿主机。

> 通过宿主机的"/var/lib/docker/volumes/myvolume/_data"目录可以访问容器内部的"/usr/share/nginx/html/"目录。

(5)切换到"/var/lib/docker/volumes/myvolume/_data"目录下,并修改 Nginx 的首页 index.html 文件的内容,如图 3-37 所示。

```
cd /var/lib/docker/volumes/myvolume/_data
echo "<h1>New Nginx Home Page</h1>" > index.html
```

这里将 Nginx 的首页修改为自定义的内容。对数据卷的修改会立即反映到容器的内部,并且会立即生效。

```
[root@master ~]# cd /var/lib/docker/volumes/myvolume/_data
[root@master _data]# pwd
/var/lib/docker/volumes/myvolume/_data
[root@master _data]# ls
50x.html   index.html
[root@master _data]# echo "<h1>New Nginx Home Page</h1>" > index.html
[root@master _data]#
```

图 3-37

(6)通过浏览器访问宿主机的 1234 端口,打开 Nginx 的主页会看到显示的就是新修改的 Nginx 主页,如图 3-38 所示。

图 3-38

(7)在挂载数据卷时,也可以使用 -v 参数。以下命令是将宿主机的"/root"目录挂载到了容器内的"/root/container/mydatavolume"目录下,从图 3-39 中可以看到挂载成功了。

```
docker run -it -v /root:/root/container/mydatavolume centos bash
```

图 3-39

其中，-v 参数用于指定在创建容器时挂载的数据卷，格式为：

-v 宿主机目录:容器内部目录

需要说明的是，使用这样方式挂载的数据卷默认的权限是"读写"（rw），用户也可以通过 ro 将其指定为"只读"。例如：

```
docker run -it -v /root:/root/container/mydatavolume:ro centos bash
```

要使用 Docker 数据卷，则需要先关闭 SELinux，否则会出现 Permission denied 的问题。以下语句可以临时关闭 SELinux。要永久关闭 SELinux，则需要修改 Linux 配置文件。

setenforce 0

另外，也可以在创建容器时使用--privileged=true 参数来解决权限的问题。例如：

```
docker run -it -v /root:/root/container/mydatavolume --privileged=true centos bash
```

3.4.3 【实战】使用数据卷容器管理 Docker 容器中的数据

数据卷容器也是一个容器，专门用来提供数据卷供其他容器挂载。如果用户需要在多个容器之间共享一些持续更新的数据，则最简单的方式是使用数据卷容器。

1. 使用数据卷容器

下面演示如何使用数据卷容器。

（1）创建一个数据卷容器 dbdata，并在其中创建一个数据卷挂载到"/dbdata"下。

```
docker run -it -v /dbdata --name dbdata centos
```

（2）执行以下语句在数据卷容器 dbdata 中生成一些测试文件，如图 3-40 所示。

```
[root@a8901080c225 /]# cd /dbdata/
[root@a8901080c225 dbdata]# echo Hello World > a.txt
[root@a8901080c225 dbdata]# ls
a.txt
[root@a8901080c225 dbdata]# more a.txt
Hello World
[root@a8901080c225 dbdata]#
```

图 3-40

（3）创建一个容器 db1，并使用--volumes-from 挂载 dbdata 容器中的数据卷。

```
docker run -it --volumes-from dbdata --name db1 centos
```

（4）在容器 db1 中查看目录"/dbdata"就可以看到数据卷容器中的数据文件了，如图 3-41 所示。

```
[root@master ~]# docker run -it --volumes-from dbdata --name db1 centos
[root@1160e0d77199 /]# cd /dbdata/
[root@1160e0d77199 dbdata]# ls
a.txt
[root@1160e0d77199 dbdata]# more a.txt
Hello World
[root@1160e0d77199 dbdata]#
```

图 3-41

（5）在容器 db1 的"/dbdata"目录下生成了一个新的文件 b.txt。

```
echo "Hello Docker" > b.txt
```

（6）创建一个容器 db2，并使用参数--volumes-from 来挂载 dbdata 容器中的数据卷。

```
docker run -it --volumes-from dbdata --name db2 centos
```

（7）在容器 db2 中查看目录"/dbdata"，如图 3-42 所示。

```
[root@master ~]# docker run -it --volumes-from dbdata --name db2 centos
[root@d22ba6b4dd8c /]# cd /dbdata/
[root@d22ba6b4dd8c dbdata]# ls
a.txt  b.txt
[root@d22ba6b4dd8c dbdata]# more a.txt
Hello World
[root@d22ba6b4dd8c dbdata]# more b.txt
Hello Docker
[root@d22ba6b4dd8c dbdata]#
```

图 3-42

从上面的示例可以看出，容器 db1 和 db2 挂载了同一个数据卷，并且数据卷都在相同的"/dbdata"目录下。这些容器的任何一方在该目录下的写入，其他容器都可以看到。这样可以很方便地实现不同容器之间的数据共享，并且利用这种方式能很容易地实现容器数据迁移。

2. 利用数据卷容器实现数据迁移

下面演示如何利用数据卷容器实现数据迁移。

（1）使用以下命令备份数据卷容器 dbdata 中的数据，如图 3-43 所示。

```
docker run --volumes-from dbdata -v $(pwd):/backup --name worker centos \
tar cvf /backup/backup.tar /dbdata
```

```
[root@master ~]# docker run --volumes-from dbdata -v $(pwd):/backup --name worker centos \
> tar cvf /backup/backup.tar /dbdata
tar: Removing leading `/' from member names
/dbdata/
/dbdata/a.txt
/dbdata/b.txt
[root@master ~]# pwd
/root
[root@master ~]# ls backup.tar
backup.tar
[root@master ~]#
```

图 3-43

这条命令首先利用 centos 镜像创建了一个容器 worker。

- 使用--volumes-from dbdata 参数，让 worker 容器挂载 dbdata 容器的数据卷（即 dbdata 数据卷）。
- 使用-v $(pwd):/backup 参数，挂载本地的当前目录到 worker 容器的"/backup"目录下。

> 在 worker 容器启动后，可以使用"tar cvf /backup/backup.tar /dbdata"命令将 "/dbdata"下的内容备份为容器内的"/backup/backup.tar"文件（即宿主机当前目录下的 backup.tar）。

下面执行恢复。

（2）创建一个带有数据卷的容器 dbdata2。

```
docker run -v /dbdata --name dbdata2 centos /bin/bash
```

（3）创建另一个新的容器 newworker，挂载 dbdata2，并使用 tar 命令解压缩备份文件到所挂载的容器卷中，如图 3-44 所示。可以看出，数据文件 a.txt 和 b.txt 成功迁移到新容器 newworker 中了。

```
docker run -it --volumes-from dbdata2 -v $(pwd):/backup --name newworker centos bash
tar xvf /backup/backup.tar
```

```
[root@master ~]# docker run -v /dbdata --name dbdata2 centos /bin/bash
[root@master ~]# docker run -it --volumes-from dbdata2 -v $(pwd):/backup --name newworker centos bash
[root@5bb6d6d54421 /]# tar xvf /backup/backup.tar
dbdata/
dbdata/a.txt
dbdata/b.txt
[root@5bb6d6d54421 /]# cd /dbdata/
[root@5bb6d6d54421 dbdata]# ls
a.txt  b.txt
[root@5bb6d6d54421 dbdata]#
```

图 3-44

第 4 章
Docker 的网络通信

Docker 的容器运行在宿主机的虚拟机上。这些虚拟机彼此独立，彼此之间没有任何接口，即容器彼此之间是逻辑隔离的。

那么，如何实现容器的相互通信呢？容器又如何访问外部的网络呢？外部的网络如何才能访问部署在容器内的应用呢？本章将带领读者详细了解这些问题。

4.1 Docker 容器网络通信的基本原理

Docker 容器中的网络接口默认都是虚拟接口。虚拟接口的最大优势是转发效率极高。这是因为，Linux 通过在内核中进行数据复制来实现虚拟接口之间的数据转发，即发送接口缓存中的数据包会被直接复制到接收接口的缓存中，而无须通过外部的物理网络设备进行交换。

> 虚拟接口和一个正常的以太网卡并无太大区别，只是它的速度比以太网卡快得多。

Docker 的网络很好地利用了 Linux 虚拟网络技术，在宿主机的物理网卡和容器内分别创建一个虚拟接口（veth），并让它们通过宿主机的 docker0 网桥进行连接，如图 4-1 所示。我们把这样的一对 veth 叫作 veth pair。

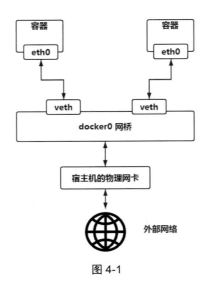

图 4-1

1. 查看 Docker 容器网络

（1）基于 Centos 的镜像创建一个容器，并进入该容器内。

```
docker run -it centos /bin/bash
```

（2）为了查看容器的网络信息，在容器内安装 "net-tools" 网络工具。

```
yum install -y net-tools
```

（3）在宿主机上打开一个命令行窗口，执行以下命令查看宿主机的 docker0 网桥的信息，如图 4-2 所示。

```
[root@master /]# ifconfig
docker0: flags=4163<UP,BROADCAST,RUNNING,MULTICAST>  mtu 1500
        inet 172.17.0.1  netmask 255.255.0.0  broadcast 0.0.0.0
        inet6 fe80::42:99ff:fe73:fd1c  prefixlen 64  scopeid 0x20<link>
        ether 02:42:99:73:fd:1c  txqueuelen 0  (Ethernet)
        RX packets 10515  bytes 457557 (446.8 KiB)
        RX errors 0  dropped 0  overruns 0  frame 0
        TX packets 11723  bytes 56988929 (54.3 MiB)
        TX errors 0  dropped 0 overruns 0  carrier 0  collisions 0

ens33: flags=4163<UP,BROADCAST,RUNNING,MULTICAST>  mtu 1500
        inet 192.168.79.11  netmask 255.255.255.0  broadcast 192.168.79.255
        inet6 fe80::9927:4703:e66d:3c67  prefixlen 64  scopeid 0x20<link>
        ether 00:0c:29:95:e8:e9  txqueuelen 1000  (Ethernet)
        RX packets 1581719  bytes 1738668379 (1.6 GiB)
        RX errors 0  dropped 0  overruns 0  frame 0
        TX packets 504974  bytes 401999657 (383.3 MiB)
        TX errors 0  dropped 0 overruns 0  carrier 0  collisions 0

vetha66e425: flags=4163<UP,BROADCAST,RUNNING,MULTICAST>  mtu 1500
        inet6 fe80::9c49:7fff:fe9e:2a11  prefixlen 64  scopeid 0x20<link>
        ether 9e:49:7f:9e:2a:11  txqueuelen 0  (Ethernet)
        RX packets 1968  bytes 111794 (109.1 KiB)
        RX errors 0  dropped 0  overruns 0  frame 0
        TX packets 2320  bytes 14138386 (13.4 MiB)
        TX errors 0  dropped 0 overruns 0  carrier 0  collisions 0
```

图 4-2

宿主机上的 docker0 的 flag 地址是 4163。

（4）在容器内执行以下命令查看容器网络信息，如图 4-3 所示。

```
ifconfig
```

```
[root@133aa3b5ee19 /]# ifconfig
eth0: flags=4163<UP,BROADCAST,RUNNING,MULTICAST>  mtu 1500
        inet 172.17.0.2  netmask 255.255.0.0  broadcast 0.0.0.0
        inet6 fe80::42:acff:fe11:2  prefixlen 64  scopeid 0x20<link>
        ether 02:42:ac:11:00:02  txqueuelen 0  (Ethernet)
        RX packets 2320  bytes 14138386 (13.4 MiB)
        RX errors 0  dropped 0  overruns 0  frame 0
        TX packets 1968  bytes 111794 (109.1 KiB)
        TX errors 0  dropped 0  overruns 0  carrier 0  collisions 0
```

图 4-3

对比图 4-2 与图 4-3 可以发现，容器内的网络地址与宿主机的网络地址具有相同的 flag，即：

flags=4163<UP,BROADCAST,RUNNING,MULTICAST> mtu 1500

这说明了在创建容器时，会成对地创建 veth pair，并且它们通过宿主机的 docker0 网桥进行通信。

2. 宿主机与 Docker 容器建立网络通信的过程

Docker 默认采用的是 bridge 网络通信模式。图 4-4 说明了在创建 Docker 容器时建立网络通信的过程。

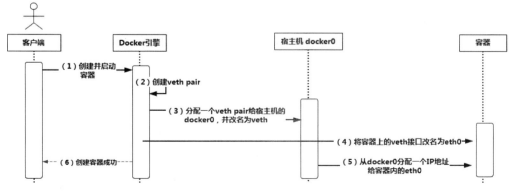

图 4-4

下面用文字描述了这过程。

（1）客户端执行"docker run"命令创建并启动容器。

（2）Docker 引擎创建一对虚拟接口 veth pair；并把它们分别放到宿主机和新容器的网络命名空间中。

（3）Docker 引擎将宿主机上的 veth 接口连接到宿主机的 docker0 网桥上，并且给它分配一个以"veth"开头的名字，如 vetha66e425。

（4）Docker 引擎将容器上的 veth 接口改名为"eth0"，并且该接口只有在容器内网络命名空间中是可见的。

（5）Docker 引擎从宿主机的 docker0 网桥上分配一个空闲的 IP 地址给容器内的 eth0，例如：172.17.0.2。并将容器内 eth0 的路由网关设置为 docker0 的内部 IP 地址，例如：172.17.0.1。

完成以上的这些步骤后，容器就可以使用其内部的虚拟接口"eth0"来连接到其他的容器和访问外部的网络了。

4.2 使用命令查看 Docker 的网络配置信息

"docker network"命令用于查看 Docker 的网络配置信息。图 4-5 展示了该命令的帮助信息。

```
[root@master /]# docker help network

Usage:  docker network COMMAND

Manage networks

Options:
      --help    Print usage

Commands:
  connect      Connect a container to a network
  create       Create a network
  disconnect   Disconnect a container from a network
  inspect      Display detailed information on one or more networks
  ls           List networks
  prune        Remove all unused networks
  rm           Remove one or more networks

Run 'docker network COMMAND --help' for more information on a command.
[root@master /]#
```

图 4-5

下面通过例子来说明这些命令的使用方式。

（1）利用以下命令查看 Docker 的网络通信模式，如图 4-6 所示。

```
docker network ls
```

```
[root@master /]# docker network ls
NETWORK ID          NAME                DRIVER              SCOPE
20ca9d84d5b4        bridge              bridge              local
27f7740cecd7        host                host                local
aa7170076be3        none                null                local
[root@master /]#
```

图 4-6

通过输出信息可以看到，Docker 有 3 种网络通信模式：bridge、host 和 none。在默认情况下，Docker 使用 bridge 模式。

除这里列出的 3 种网络通信模式外，Docker 还提供了 container 网络通信模式用于容器间的相互通信，将在 4.3 节中介绍。

（2）查看 bridge 模式的详细信息，如图 4-7 所示。

```
docker network inspect 20ca9d84d5b4
```

其中的参数 "20ca9d84d5b4" 指 Docker 网络的 ID。

```
[root@master /]# docker network inspect 20ca9d84d5b4
[
    {
        "Name": "bridge",
        "Id": "20ca9d84d5b42ae50ee8a77a7443a998328eaf9a9bce1e23509337ff38268dab",
        "Created": "2022-01-03T12:23:06.386329099+08:00",
        "Scope": "local",
        "Driver": "bridge",
        "EnableIPv6": false,
        "IPAM": {
            "Driver": "default",
            "Options": null,
            "Config": [
                {
                    "Subnet": "172.17.0.0/16",
                    "Gateway": "172.17.0.1"
                }
            ]
        }
```

图 4-7

从图 4-7 中可以看出，bridge 模式默认的子网地址是 "172.17.0.0/16"。对比图 4-3 可以发现，容器的 IP 地址是 172.17.0.2，属于 bridge 网络的子网地址。

4.3 Docker 的 4 种网络通信模式

由于 Docker 容器彼此之间是逻辑隔离的，所以，在安装 Docker 时会在容器中创建隔离的网络环境。在该隔离的网络模式环境中，运行在宿主机上的各个容器具有完全独立的网络栈，并且 Docker 容器的网络环境与宿主机相互隔离。通过使用 Docker 的不同网络模式，可以使 Docker 容器共享宿主机的网络命名空间，也可以实现 Docker 容器间的相互访问。

Docker 一共提供了 4 种网络通信模式：bridge、container、host 和 none。表 4-1 对比了这 4 种模式的特点。

表 4-1

网络通信模式	是否支持多主机	纵向通信机制	横向通信机制
bridge	否	绑定宿主机端口	通过 Linux 桥接进行通信
container	否	绑定宿主机端口	通过 Linux 连接进行通信
host	是	通过宿主机网络进行通信	通过宿主机网络进行通信
none	否	无法通信	只能通过 Linux 连接进行通信

下面分别介绍这四种模式。

4.3.1 bridge 模式

bridge 模式是 Docker 默认的网络通信模式，是开发者最常用的模式。

在 bridge 模式下，Docker 引擎会创建独立的网络命名空间。这样就可以保证运行在每一个命名空间中的容器具有独立的网卡等网络资源。

利用 bridge 模式，可以非常方便地实现容器与容器之间、容器与宿主机之间的网络隔离。通过使用宿主机上的 docker0 网桥，可以实现 Docker 容器与宿主机（乃至外部网络）的网络通信。下面通过示例演示如何使用 bridge 模式。

1. 使用 bridge 模式创建容器

（1）使用 busybox 的镜像创建容器。

```
docker run -it --network=bridge busybox /bin/sh
```

> 这里的--network=bridge 可以不写，默认就是 bridge 模式。
> busybox 是一个集成了一百多个最常用 Linux 命令和工具（如 cat、echo 等）的软件工具箱。

（2）在容器内执行"ifconfig"命令查看容器的网络信息，如图 4-8 所示。

```
[root@master ~]# docker run -it --network=bridge busybox /bin/sh
/ # ifconfig
eth0      Link encap:Ethernet  HWaddr 02:42:AC:11:00:02
          inet addr:172.17.0.2  Bcast:0.0.0.0  Mask:255.255.0.0
          inet6 addr: fe80::42:acff:fe11:2/64 Scope:Link
          UP BROADCAST RUNNING MULTICAST  MTU:1500  Metric:1
          RX packets:6 errors:0 dropped:0 overruns:0 frame:0
          TX packets:5 errors:0 dropped:0 overruns:0 carrier:0
          collisions:0 txqueuelen:0
          RX bytes:508 (508.0 B)  TX bytes:418 (418.0 B)

lo        Link encap:Local Loopback
          inet addr:127.0.0.1  Mask:255.0.0.0
          inet6 addr: ::1/128 Scope:Host
          UP LOOPBACK RUNNING  MTU:65536  Metric:1
          RX packets:0 errors:0 dropped:0 overruns:0 frame:0
          TX packets:0 errors:0 dropped:0 overruns:0 carrier:0
          collisions:0 txqueuelen:1
          RX bytes:0 (0.0 B)  TX bytes:0 (0.0 B)

/ #
```

图 4-8

2. 实例：用户自定义 bridge 网络

在默认情况下，Docker 引擎会自动创建一个 bridge 网络。Docker 引擎也为用户提供了自定义 bridge 网络的方式。利用该方式，用户可以自定义 bridge 网络的子网地址和网关等参数。

用户自定义 bridge 网络是在生产环境中最推荐的方式。

（1）执行以下命令自定义 bridge 网络。

```
docker network create -d bridge --ip-range=192.168.1.0/24 \
--gateway=192.168.1.1 --subnet=192.168.1.0/24 bridge2
```

其中的参数说明如下。

- -d：指定网络通信模式，默认是 bridge。
- --ip-range：指定子网 IP 地址的范围。
- --gateway：指定网关的 IP 地址。
- --subnet：指定子网的 IP 地址。
- bridge2：指定 bridge 网络的名称。

（2）查看 Docker 的网络，可以看到新创建的 bridge2，如图 4-9 所示。

```
docker network ls
```

```
[root@master ~]# docker network ls
NETWORK ID          NAME                DRIVER              SCOPE
20ca9d84d5b4        bridge              bridge              local
71724ff94f5a        bridge2             bridge              local
27f7740cecd7        host                host                local
aa7170076be3        none                null                local
[root@master ~]#
```

图 4-9

（3）使用 bridge2 创建一个容器，这里通过参数--ip 指定了容器的 IP 地址。

```
docker run -it --network=bridge2 --ip=192.168.1.3 busybox
```

（4）在容器内执行"ifconfig"命令查看网络信息，如图 4-10 所示。

```
/ # ifconfig
eth0      Link encap:Ethernet  HWaddr 02:42:C0:A8:01:03
          inet addr:192.168.1.3  Bcast:0.0.0.0  Mask:255.255.255.0
          inet6 addr: fe80::42:c0ff:fea8:103/64 Scope:Link
          UP BROADCAST RUNNING MULTICAST  MTU:1500  Metric:1
          RX packets:12 errors:0 dropped:0 overruns:0 frame:0
          TX packets:5 errors:0 dropped:0 overruns:0 carrier:0
          collisions:0 txqueuelen:0
```

图 4-10

4.3.2　host 模式

在使用 host 模式时，容器与宿主机共享同一个网络命名空间，容器的 IP 地址与宿主机的 IP 地址相同。如果宿主机具有公网的 IP 地址，则容器也拥有这个公网的 IP 地址。即这时容器可以直接使用宿主机的 IP 地址与外界进行通信，且容器内服务的端口也可以直接使用宿主机的端口，无须进行任何的转换。

由于在 host 模式下不再需要宿主机的转发，因此其性能得到了极大的提高。图 4-11 说明了 host 模式的工作机制。

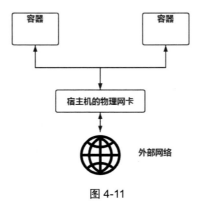

图 4-11

使用 host 模式来创建容器，如以下命令所示。

```
docker run -it --network=host busybox /bin/sh
```

对比一下容器内的网络信息和宿主机的网络信息，如图 4-12 所示，可以发现，容器与宿主机共享了同一个网络命名空间，即容器使用了宿主机的网络配置信息。

图 4-12

尽管使用 host 模式可以很方便地通过 localhost 或者 127.0.0.1 实现容器与宿主机的相互访问，并且性能也比较好。但是这种模式也存在以下两个问题：

- 由于容器使用了宿主机的网络环境，因此网络环境的隔离性功能被减弱，从而造成宿主机和容器争用网络资源。容器本身也不再拥有所有的网络资源，而是与宿主机共享网络资源。
- 宿主机和容器使用了相同的 IP 地址，这不利于网络的配置和管理。

4.3.3　container 模式

在 container 模式下，容器之间会共享网络环境。即一个容器会使用另一个容器的网络命名空间。因此，在这种模式下，容器之间可以通过 localhost 或者 127.0.0.1 进行相互间的访问，从而提高了传输的效率。

> container 模式节约了网络资源，但是运行在这种模式下的容器不存在网络隔离。Container 网络的隔离性处于 bridge 网络与 host 网络之间。

container 模式在一些特殊场景中非常有用。例如：在 Kubernetes 中创建 Pod 时，会首先创建 Pod 的基础容器；而 Pod 中的其他容器则采用 container 模式与基础容器进行通信。Pod 中的各个容器采用 localhost 或者 127.0.0.1 进行通信，从而将 Pod 中的所有容器形成一个逻辑整体。

container 模式的工作机制如图 4-13 所示。

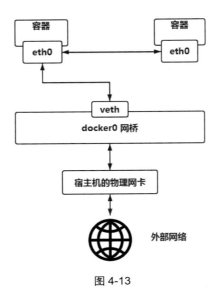

图 4-13

下面来演示如何使用 container 模式。

（1）使用 busybox 的镜像创建一个容器 A，并查看容器的网络信息，如图 4-14 所示。

```
docker run -it busybox /bin/sh
ifconfig
```

```
[root@master ~]# docker run -it busybox /bin/sh
/ # ifconfig
eth0      Link encap:Ethernet  HWaddr 02:42:AC:11:00:02
          inet addr:172.17.0.2  Bcast:0.0.0.0  Mask:255.255.0.0
          inet6 addr: fe80::42:acff:fe11:2/64 Scope:Link
          UP BROADCAST RUNNING MULTICAST  MTU:1500  Metric:1
          RX packets:6 errors:0 dropped:0 overruns:0 frame:0
          TX packets:6 errors:0 dropped:0 overruns:0 carrier:0
          collisions:0 txqueuelen:0
          RX bytes:508 (508.0 B)  TX bytes:508 (508.0 B)

lo        Link encap:Local Loopback
          inet addr:127.0.0.1  Mask:255.0.0.0
          inet6 addr: ::1/128 Scope:Host
          UP LOOPBACK RUNNING  MTU:65536  Metric:1
```

图 4-14

（2）开启一个新的命令行仓库查看容器 A 的 ID。从图 4-15 中可以看到容器 A 的 ID 是 d71751ade532。

```
[root@master ~]# myps
CONTAINER ID            IMAGE                   NAMES                   PORTS
d71751ade532            busybox                 brave_kirch
[root@master ~]#
```

图 4-15

（3）开启一个新的命令行窗口，并使用 container 模式创建一个新的容器 B。

```
docker run -it --network=container:d71751ade532 busybox /bin/sh
```

其中，参数 --network 用于指定新容器使用哪一个容器的网络信息，这里指定的是容器 A。

（4）查看容器 B 的网络信息，如图 4-16 所示。

```
[root@master ~]# docker run -it --network=container:d71751ade532 busybox /bin/sh
/ # ifconfig
eth0      Link encap:Ethernet  HWaddr 02:42:AC:11:00:02
          inet addr:172.17.0.2  Bcast:0.0.0.0  Mask:255.255.0.0
          inet6 addr: fe80::42:acff:fe11:2/64 Scope:Link
          UP BROADCAST RUNNING MULTICAST  MTU:1500  Metric:1
          RX packets:8 errors:0 dropped:0 overruns:0 frame:0
          TX packets:8 errors:0 dropped:0 overruns:0 carrier:0
          collisions:0 txqueuelen:0
          RX bytes:648 (648.0 B)  TX bytes:648 (648.0 B)

lo        Link encap:Local Loopback
          inet addr:127.0.0.1  Mask:255.0.0.0
          inet6 addr: ::1/128 Scope:Host
          UP LOOPBACK RUNNING  MTU:65536  Metric:1
          RX packets:0 errors:0 dropped:0 overruns:0 frame:0
          TX packets:0 errors:0 dropped:0 overruns:0 carrier:0
          collisions:0 txqueuelen:1
```

图 4-16

（5）对比图 4-14 和图 4-16 会发现，容器 A 和容器 B 使用了相同的网络命名空间。这是因为，在创建容器 B 时使用了 container 模式，使得容器 B 不再创建自己的网络命名空间，而直接使用容器 A 的网络命名空间。

4.3.4　none 模式

none 模式下的容器具有独立的网络命名空间，但不包含任何网络配置，只能通过 Local Loopback 网卡与容器进行通信，即只能使用 localhost 或者 127.0.0.1 访问容器。

在 none 模式下需要手动进行网络配置，例如使用 pipwork 工具指定容器的 IP 地址等。

下面使用 none 模式来创建一个容器。

```
docker run -it --network=none busybox /bin/sh
```

由于 none 模式不包含任何网络配置，所以在其网络配置信息中就只包含一个 127.0.0.1 的 IP 地址，如图 4-17 所示。

```
[root@master ~]# docker run -it --network=none busybox /bin/sh
/ # ifconfig
lo        Link encap:Local Loopback
          inet addr:127.0.0.1  Mask:255.0.0.0
          inet6 addr: ::1/128 Scope:Host
          UP LOOPBACK RUNNING  MTU:65536  Metric:1
          RX packets:0 errors:0 dropped:0 overruns:0 frame:0
          TX packets:0 errors:0 dropped:0 overruns:0 carrier:0
          collisions:0 txqueuelen:1
          RX bytes:0 (0.0 B)  TX bytes:0 (0.0 B)

/ #
```

图 4-17

4.4 容器间的通信

Docker 提供了不同的网络通信模式，运行在宿主机上的容器可以相互通信。

容器之间的相互通信方式主要分为：通过 IP 地址进行通信、通过 Docker DNS Server 进行通信和通过 Joined 方式进行通信。

如果容器运行在不同的宿主机上，那该如何实现容器间的相互通信呢？本节将详细容器间的通信方式。

4.4.1 【实战】通过 IP 地址进行通信

在宿主机上创建一个容器时，Docker 守护进程会为每一个新创建的容器自动分配一个虚拟的 IP 地址。但是，外部的网络是无法通过这个虚拟 IP 地址访问容器内的应用的。

> 这个虚拟 IP 地址只提供 Docker 内部各个容器相互通信使用。即通过这个 IP 地址实现了 Docker 容器之间的相互连通。

下面对容器间的虚拟 IP 地址通信方式进行测试。

（1）利用 4.3.1 节中创建的自定义网络 bridge2 创建两个容器，并指定它们的 IP 地址。

```
docker run -it --network=bridge2 --ip=192.168.1.3 busybox
docker run -it --network=bridge2 --ip=192.168.1.4 busybox
```

（2）在其中一个容器内执行"ping"命令，确定是否能够通过该容器的虚拟 IP 地址与另外一个容器进行通信，如图 4-18 所示。

通过容器的虚拟 IP 地址能够实现容器间的相互通信。但是这种通信方式存在一定的局限性，如图 4-19 所示。

```
[root@master ~]# docker run -it --network=bridge2 --ip=192.168.1.3 busybox
/ # ping 192.168.1.4
PING 192.168.1.4 (192.168.1.4): 56 data bytes
64 bytes from 192.168.1.4: seq=0 ttl=64 time=0.168 ms
64 bytes from 192.168.1.4: seq=1 ttl=64 time=0.065 ms
64 bytes from 192.168.1.4: seq=2 ttl=64 time=0.063 ms
64 bytes from 192.168.1.4: seq=3 ttl=64 time=0.063 ms
^C
--- 192.168.1.4 ping statistics ---
4 packets transmitted, 4 packets received, 0% packet loss
round-trip min/avg/max = 0.063/0.089/0.168 ms
/ #

[root@master ~]# docker run -it --network=bridge2 --ip=192.168.1.4 busybox
/ #
```

图 4-18

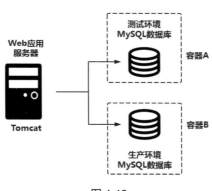

图 4-19

在图 4-19 这个例子中，Web 应用服务器在测试环境中需要连接容器 A 中的 MySQL 数据库；而在生产环境中，它需要连接容器 B 中的 MySQL 数据库。如果要让 Web 应用服务器通过容器 A 的虚拟 IP 地址和容器 B 的虚拟 IP 地址分别对它们进行访问，则需要修改 Web 应用服务器上 Tomcat 中的 IP 地址信息。随着更多新的容器创建，Tomcat 中的 MySQL 地址需要不断进行修改，并且还需要重启 Tomcat。

因此，在实际的使用过程中，通过容器的虚拟 IP 地址进行容器间的相互通信就非常不利于管理和维护了。那有没有解决的办法呢？答案当然是有的——通过 Docker DNS Server 进行通信。

4.4.2 【实战】通过 Docker DNS Server 进行通信

从 Docker 1.10 版本开始，Docker 引擎自带了一个内嵌的 DNS Server。有了它，容器之间可以直接通过容器名称进行通信。这种通信方式的使用方法也非常简单：在启动容器时使用 "--name" 参数为容器命名即可。

下面对容器间的 DNS Server 通信方式进行测试。

（1）利用 4.3.1 节中创建的自定义网络 bridge2 创建两个容器，并指定容器的名称，如：box1

和 box2。

```
docker run -it --network=bridge2 --name box1 busybox
docker run -it --network=bridge2 --name box2 busybox
```

这时容器 box1 和 box2 可以通过容器名称进行相互通信。

（2）在其中一个容器内执行"ping"命令，以确认是否能使用容器名称与对方进行通信，如图 4-20 所示。

```
[root@master ~]# docker run -it --network=bridge2 --name box1 busybox
/ # ping box2
PING box2 (192.168.1.3): 56 data bytes
64 bytes from 192.168.1.3: seq=0 ttl=64 time=0.070 ms
64 bytes from 192.168.1.3: seq=1 ttl=64 time=0.067 ms
64 bytes from 192.168.1.3: seq=2 ttl=64 time=0.232 ms
64 bytes from 192.168.1.3: seq=3 ttl=64 time=0.061 ms

[root@master ~]# docker run -it --network=bridge2 --name box2 busybox
/ #
```

图 4-20

要使用 Docker DNS Server 进行容器间的通信，有一个限制条件：这种方式只能在用户自定义的 bridge 网络中使用，在 Docker 默认的 bridge 网络中是无法使用的。

4.4.3 【实战】通过 Joined 方式进行通信

Joined 是 Docker 引擎提供的一种特殊的容器间通信方式，其本质上使用了 container 模式。因为在 container 模式下，多个容器共享同一个网络环境，也共享网卡的配置。因此，在 container 模式下，容器之间可以直接通过 localhost 或者 127.0.0.1 进行通信。

下面来演示如何使用 Joined 方式实现容器间的通信。

（1）基于 httpd 的镜像创建一个容器，命名为"box1"。

```
docker run -it --name box1 httpd
```

httpd 是 Apache HTTP 服务器，可以把它看成是 Web 服务器。

（2）基于 busybox 的镜像创建一个新的容器"box2"，并通过参数 --network=

container:box1 指定与"box1"容器进行通信。

```
docker run -it --network=container:box1 --name box2 busybox
```

 由于在 box1 和 box2 容器之间使用的是 container 模式，所以它们的 IP 地址等网络环境是完全相同的。

（3）在 box2 容器中，通过"wget 127.0.0.1"命令可以直接访问 box1 容器的 HTTP 服务，如图 4-21 所示。而在 box1 容器中会打印出一条日志，表示从"127.0.0.1"地址接收到了一个 HTTP 的 Get 请求。

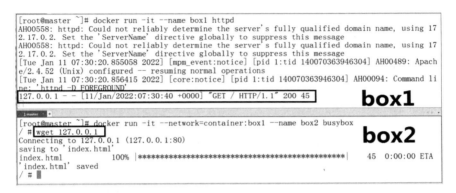

图 4-21

4.4.4 容器间的跨节点通信

在同一台宿主机上，不同的 Docker 容器可以借助 docker0 网桥直接进行通信。而在实际的项目中，一个复杂的系统往往需要部署多个组件，而为了提高组件的运行的效率往往将这些组件部署到不同的主机上。那么在 Docker 中如何实现容器间的跨节点通信呢？

1. 3 种方式

目前主要可以通过 3 种方式来实现 Docker 容器间的跨节点通信：

（1）通过容器在宿主机上的端口映射来实现通信。

使用这种方式实现容器间的跨节点通信，需要宿主机进行转发，所以使用起来非常不方便。

（2）通过 Docker Overlay 网络来实现通信。

这种方式可以直接使用容器本身的虚拟 IP 地址进行相互通信，这与运行在同一台宿主机上的不同容器间的通信方式完全一样。Docker 原生的 Overlay 网络是目前实现容器跨节点通信的主流方案。要使用 Overlay 网络，需要注册中心的支持。

（3）通过第三方网络（如 flannel 网络等）来实现容器间的跨节点通信。

下面着重介绍第（2）种方案。

2. Overlay 网络与注册中心

Overlay 网络是在不改变现有网络的前提下，对 IP 报文进行数据的封装，从而利用 IP 路由协议实现数据的转发功能。在 Overlay 网络中，通过扩展标识位可以支持 16M 的用户。

Docker 的 Swarm 集群便是 Overlay 网络的一个实现，而使用 Overlay 网络需要注册中心的支持。注册中心能够提供服务的注册与发现功能。Docker 支持的注册中心有 ZooKeeper、Consul 和 ETCD。下面以 ZooKeepper 为例来进行介绍。

由于 ZooKeeper 是基于 Java 语言的，因此，在安装 ZooKeeper 之前需要先安装 Java 的 JDK 环境。

下面将在 master 节点上部署一个单节点的 ZooKeeper 实例。

（1）解压缩 JDK 的安装文件，这里使用的版本是 jdk-8u181-linux-x64.tar.gz。

```
tar -zxvf jdk-8u181-linux-x64.tar.gz -C /root/
```

（2）编辑 "/root/.bash_profile" 文件设置 Java 的环境变量：

```
JAVA_HOME=/root/jdk1.8.0_181
export JAVA_HOME

PATH=$JAVA_HOME/bin:$PATH
export PATH
```

（3）生效 Java 的环境变量。

```
source /root/.bash_profile
```

（4）解压缩 ZooKeeper 的安装文件，这里使用的版本是 zookeeper-3.4.10.tar.gz。

```
tar -zxvf zookeeper-3.4.10.tar.gz -C /root
```

（5）进入 ZooKeeper 目录生成 ZooKeeper 的配置文件。

```
cd zookeeper-3.4.10
cp conf/zoo_sample.cfg conf/zoo.cfg
```

（6）启动 ZooKeeper 实例。

```
bin/zkServer.sh start
```

（7）查看 ZooKeeper 实例的状态，如图 4-22 所示。这时 ZooKeeper 是 standalone 状态，说明这是一个单节点的 ZooKeeper 实例。

```
[root@master zookeeper-3.4.10]# bin/zkServer.sh status
ZooKeeper JMX enabled by default
Using config: /root/zookeeper-3.4.10/bin/../conf/zoo.cfg
Mode: standalone
[root@master zookeeper-3.4.10]#
```

图 4-22

3. 实现 Docker 容器间的跨节点通信

有了 ZooKeeper 注册中心的支持，就可以实现 Docker 容器间的跨节点通信了。表 4-2 显示了部署的环境信息。

表 4-2

主机名	IP 地址	已部署的服务
master	192.168.79.11	Docker、ZooKeeper
node1	192.168.79.12	Docker

（1）在 master 节点上修改 "/usr/lib/systemd/system/docker.service" 文件，将参数 "ExecStart" 改为以下内容。

```
ExecStart=/usr/bin/dockerd-current \
  --add-runtime docker-runc=/usr/libexec/docker/docker-runc-current \
  --default-runtime=docker-runc \
  --exec-opt native.cgroupdriver=systemd \
  --userland-proxy-path=/usr/libexec/docker/docker-proxy-current \
  --init-path=/usr/libexec/docker/docker-init-current \
  --seccomp-profile=/etc/docker/seccomp.json \
  -H tcp://0.0.0.0:2375 -H unix:///var/run/docker.sock \
  --cluster-store zk://192.168.79.11:2181 \
  --cluster-advertise 192.168.79.11:2375 \
  $OPTIONS \
  $DOCKER_STORAGE_OPTIONS \
  $DOCKER_NETWORK_OPTIONS \
  $ADD_REGISTRY \
  $BLOCK_REGISTRY \
  $INSECURE_REGISTRY \
  $REGISTRIES
```

其中增加的主要参数说明如下。

- --cluster-store：指定 ZooKeeper 的 IP 地址和端口号。
- --cluster-advertise：将 Docker 注册到 ZooKeeper 中的地址信息。

（2）重启 master 节点上的 Docker 服务。

```
systemctl daemon-reload
systemctl restart docker
```

（3）在 node1 节点上修改"/usr/lib/systemd/system/docker.service"文件，将参数"ExecStart"改为以下内容。

```
ExecStart=/usr/bin/dockerd-current \
  --add-runtime docker-runc=/usr/libexec/docker/docker-runc-current \
  --default-runtime=docker-runc \
  --exec-opt native.cgroupdriver=systemd \
  --userland-proxy-path=/usr/libexec/docker/docker-proxy-current \
  --init-path=/usr/libexec/docker/docker-init-current \
  --seccomp-profile=/etc/docker/seccomp.json \
  -H tcp://0.0.0.0:2375 -H unix:///var/run/docker.sock \
  --cluster-store zk://192.168.79.11:2181 \
  --cluster-advertise 192.168.79.12:2375 \
  $OPTIONS \
  $DOCKER_STORAGE_OPTIONS \
  $DOCKER_NETWORK_OPTIONS \
  $ADD_REGISTRY \
  $BLOCK_REGISTRY \
  $INSECURE_REGISTRY \
  $REGISTRIES
```

（4）重启 node1 节点上的 Docker 服务。

```
systemctl daemon-reload
systemctl restart docker
```

（5）进入 ZooKeeper 目录，启动 ZooKeeper 命令行客户端。

```
cd /root/zookeeper-3.4.10
bin/zkCli.sh
```

（6）查看 master 和 node1 节点在 ZooKeeper 中的注册信息，如图 4-23 所示。

```
ls /docker/nodes
```

```
[zk: localhost:2181(CONNECTED) 3] ls /docker/nodes
[192.168.79.12:2375, 192.168.79.11:2375]
[zk: localhost:2181(CONNECTED) 4]
```

图 4-23

（7）在任意节点上创建 Overlay 网络，例如在 master 节点上执行以下语句。

```
docker network create -d overlay my_multi_hosts
```

其中的参数-d 用于指定创建网络的类型，这里指定的是 Overlay。

在 node1 节点上此时会自动同步新创建的网络 my_multi_hosts。

（8）在 master 或者 node1 节点上查看 Docker 的网络信息，如图 4-24 所示。这时可以看到新创建的网络 my_multi_hosts。

```
[root@node1 ~]# docker network ls
NETWORK ID          NAME                DRIVER              SCOPE
4318df36fb48        bridge              bridge              local
e2b5ff5ed409        host                host                local
da37c424cb6e        my_multi_hosts      overlay             global
f694bf4640d5        none                null                local
[root@node1 ~]#
```

图 4-24

（9）在 master 节点上，使用 my_multi_hosts 的 Overlay 网络启动一个容器，并确定容器的 IP 地址，如图 4-25 所示，可以看到 box1 的 IP 地址是 "10.0.0.2"。

```
docker run -it --net=my_multi_hosts --name box1 busybox
ifconfig eth0
```

```
[root@master ~]# docker run -it --net=my_multi_hosts --name box1 busybox
/ # ifconfig eth0
eth0      Link encap:Ethernet  HWaddr 02:42:0A:00:00:02
          inet addr:10.0.0.2  Bcast:0.0.0.0  Mask:255.255.255.0
          inet6 addr: fe80::42:aff:fe00:2/64 Scope:Link
          UP BROADCAST RUNNING MULTICAST  MTU:1450  Metric:1
          RX packets:10 errors:0 dropped:0 overruns:0 frame:0
          TX packets:6 errors:0 dropped:0 overruns:0 carrier:0
          collisions:0 txqueuelen:0
          RX bytes:836 (836.0 B)  TX bytes:508 (508.0 B)

/ #
```

图 4-25

（10）在 node1 节点上，使用 my_multi_hosts 的 Overlay 网络启动一个容器，并确定容器的 IP 地址，如图 4-26 所示，可以看到 box2 的 IP 地址是 "10.0.0.3"。

```
docker run -it --net=my_multi_hosts --name box2 busybox
ifconfig eth0
```

```
[root@node1 ~]# docker run -it --net=my_multi_hosts --name box2 busybox
/ # ifconfig eth0
eth0      Link encap:Ethernet  HWaddr 02:42:0A:00:00:03
          inet addr:10.0.0.3  Bcast:0.0.0.0  Mask:255.255.255.0
          inet6 addr: fe80::42:aff:fe00:3/64 Scope:Link
          UP BROADCAST RUNNING MULTICAST  MTU:1450  Metric:1
          RX packets:12 errors:0 dropped:0 overruns:0 frame:0
          TX packets:7 errors:0 dropped:0 overruns:0 carrier:0
          collisions:0 txqueuelen:0
          RX bytes:988 (988.0 B)  TX bytes:578 (578.0 B)
/ #
```

图 4-26

（11）现在 box1 和 box2 既可以通过虚拟 IP 地址进行通信，也可以通过 DNS Server 进行通信，如图 4-27 所示。

图 4-27

（12）使用 ZooKeeper 的图形化工具 ZooInspector 登录 ZooKeeper。这时就可以完整地看到 Overlay 网络在 ZooKeeper 中的注册信息，如图 4-28 所示。

图 4-28

4.5 容器的网络访问控制

从 Docker 的网络通信模式可以看出，在默认情况下，运行在宿主机上的容器可以与宿主机及外部的网络进行通信。即使是在同一个宿主机上，容器也能够通过 bridge 模式的 docker0 网桥进行相互通信。

但是，由于宿主机上的网络与容器内的网络不属于同一个网段，因此仅仅依靠 bridge 模式的虚拟接口无法让宿主机以外的系统访问宿主机上的容器。为了解决这个问题，Docker 采用端口绑定的方式让外部系统可以访问宿主机上容器的内部，即利用 Linux 的 iptable 表将宿主机上端口映射到容器内的端口。

下面将介绍容器内的应用如何访问外部网络，以及从外部网络如何访问容器内的应用。

4.5.1 容器内的应用访问外部网络

在默认情况下，容器内的应用访问外部网络是通过宿主机上的 docker0 网桥来实现的。当容器内的应用需要访问外部网络时，需要宿主机进行转发。

1. 开启 IP 数据包的转发功能

执行以下指令可以确定宿主机的 Linux 是否开启了 IP 数据包转发功能。

```
sysctl net.ipv4.ip_forward
```

- 如果返回值是 1，则表示已经开启了 Linux 的 IP 数据包转发功能。
- 如果返回值是 0，则表示没有开启此功能。

可以通过执行以下语句开启此功能。

```
sysctl -w net.ipv4.ip_forward=1
```

开启此功能的另一种方式是，在启动 Docker 服务时指定参数 "--ip-forward=true"。这样 Docker 守护进程会自动将宿主机的 ip_forward 参数设置为 1。

2. 实例

下面对容器内的应用访问外部网络进行测试。

（1）基于 busybox 的镜像创建一个容器。

```
docker run -it busybox
```

（2）在容器内部访问百度的首页，可以看到 ping 命令正常返回信息了，如图 4-29 所示。

```
[root@master ~]# docker run -it busybox
/ # ping www.baidu.com
PING www.baidu.com (110.242.68.4): 56 data bytes
64 bytes from 110.242.68.4: seq=0 ttl=127 time=11.202 ms
64 bytes from 110.242.68.4: seq=1 ttl=127 time=24.384 ms
```

图 4-29

4.5.2 从外部网络访问容器内的应用

运行在宿主机上的容器，允许从外部网络访问其内部的应用，这主要是通过-p 参数来实现的。

通过在创建容器时指定-p 参数，可以将容器内的某个端口与宿主机绑定，来完成宿主机与容器的端口映射。其本质是：在宿主机 iptable 表中添加相应的路由转发规则，对访问外部 IP 地址的数据包进行转换，将其访问的目标地址修改为容器的 IP 地址和容器内的端口。

下面通过一个简单的示例来说明这个过程。

（1）基于 Nginx 的镜像创建一个容器，并将容器的 80 端口映射到宿主机的 1234 端口。

```
docker run -d -p 1234:80 nginx
```

（2）在宿主机上，通过以下命令来查看 iptable 表中的路由转发规则，如图 4-30 所示。可以看到，在 iptable 表中有一条路由规则完成了从宿主机的 1234 端口到 IP 地址 "172.17.0.2:80" 的数据包转发。

```
iptables -t nat -L -n
```

```
[root@master ~]# iptables -t nat -L -n
Chain DOCKER (2 references)
target     prot opt source               destination
RETURN     all  --  0.0.0.0/0            0.0.0.0/0
RETURN     all  --  0.0.0.0/0            0.0.0.0/0
DNAT       tcp  --  0.0.0.0/0            0.0.0.0/0            tcp dpt:1234 to:172.17.0.2:80
```

图 4-30

> 为了方便查看 iptables 的路由规则，可以使用 Linux 的管道符过滤输出结果。以下命令将在所有的路由规则中查找包含 "1234" 的规则信息。
>
> iptables -t nat -L -n | grep 1234

第 5 章
使用 Docker Compose 进行服务编排

在使用 Docker 部署应用时,可以通过定义 Dockerfile 文件来完成对应用服务的描述,之后使用 "docker build" "docker run" 等命令操作容器。

但是,随着应用架构的不断复杂和微服务的应用,通常在一个系统中需要包含多个模块,而一般情况下这些模块都会被部署到不同的 Docker 容器中。如果每一个模块都通过手动方式来完成部署,则效率是非常低,且也不利于系统的维护和扩展。

使用 Docker Compose 可以非常方便地定义和运行复杂应用,它是 Docker 提供的一个服务编排工具。Docker Compose 不是通过 shell 脚本命令,而是通过 YML 描述文件来完成对 Docker 容器的管理。

5.1 配置 Docker Compose

Docker Compose 依赖 Docker 的引擎来完成服务的编排,因此,在安装配置 Docker Compose 之前需要先安装好 Docker 的引擎。

在 Docker 的官方网站上展示了 Docker Compose 的安装要求,如图 5-1 所示。

图 5-1

执行以下步骤完成 Docker Compose 的安装和配置。

```
curl -L "https://github.com/docker/compose/releases/download/1.23.2/docker-compose-$(uname -s)-$(uname -m)" -o /usr/local/bin/docker-compose
chmod +x /usr/local/bin/docker-compose
```

以下语句将检查 Docker Compose 的版本信息。

```
docker-compose --version
```

5.2 进行服务编排

Docker Compose 通过 YML 文件来管理一个复杂系统中的多个容器。在该文件中，所有的容器都是通过 Service 来定义的。Docker Compose 使用 docker-compose 脚本来完成服务的启动、停止、管理和扩容等。因此，Docker Compose 非常适合使用多个容器来应对复杂的应用场景。

本节将通过示例来演示如何使用 Docker Compose。图 5-2 描述了本节示例的架构。在本示例系统中包含两个功能模块：Python Web 模块和 Redis DB 模块。

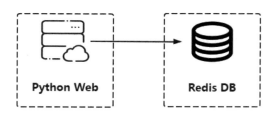

图 5-2

- Redis DB 模块：使用 Redis 来保存用户在 Web 界面上刷新网页的次数。
- Python Web 模块：用于 Web 界面的展示。

利用前面已经掌握的 Docker 知识，完全可以搭建起这样一个系统（通过手动部署的方式来完成）。但是当系统变得越来越复杂时，这样的手动部署的方式是非常不利于管理的。而使用 Docker Compose 则可以非常容易地解决该问题。

下面将分别通过手动部署的方式和使用 Docker Compose 的方式来完成部署和管理，从而体会 Docker Compose 的优点。

5.2.1 【实战】使用手动方式部署应用

使用手动方式部署应用，需要创建 Dockerfile 文件来描述应用。下面是具体的操作步骤。

（1）使用 Python 创建 Web 应用（创建文件"app.py"），文件内容如下：

```
01  from flask import Flask
02  from redis import Redis
03  import os
04  app = Flask(__name__)
05  redis = Redis(host='redis', port=6379)
06
07  @app.route('/')
08  def hello():
09      redis.incr('hits')
10      return 'I have been seen %s times.' % redis.get('hits')
11
12  if __name__ == "__main__":
13      app.run(host="0.0.0.0", debug=True)
```

其中，

- 第 01 行，基于轻量级 Web 框架 Flask 来使用 Python 语言快速实现一个网站或 Web 服务。
- 第 02 行，使用 Python Redis 模块在 Python Web 模块中访问 Redis 数据库，将用户在 Web 界面上的刷新次数保存到 Redis 中。
- 第 05 行，指定 Redis 数据库的地址信息，这里指定了运行 Redis 的主机名和端口号。
- 第 09 行，通过变量"redis"访问 Redis 数据库完成计数功能。
- 第 10 行是 Web 界面上显示的内容，即用户刷新 Web 网页的次数。

（2）创建"requirements.txt"文件，由于在应用中使用了 Flask 和 Redis，因此在文件中输入以下内容：

```
flask
redis
```

（3）创建"Dockerfile"文件，并输入以下内容。

```
FROM python:3.4-alpine
ADD . /code
WORKDIR /code
COPY app.py /code
RUN pip install -r requirements.txt
CMD ["python", "app.py"]
```

（4）使用"docker build"命令编译 Dockerfile 文件。

```
docker build -t myapplication .
```

 "-t"参数用于指定生成的镜像名称为"myapplication"，并将其存放在本地。

（5）由于应用中需要 Redis 的支持，所以先启动一个 Redis 容器。

```
docker run --name myredis -d -p 6379:6379 redis
```

其中，myredis 是容器的名称，在启动 Web 应用时需要用到这个名称。

（6）使用"docker run"命令启动 Web 应用，并使用--link 参数连接到 Redis 容器。

```
docker run --name myapp_using_redis -p 5000:5000 --link myredis:redis -d myapplication
```

其中的参数说明如下。

- -p 5000:5000 表示将宿主机的 5000 端口映射到容器的 5000 端口。
- --link myredis:redis 表示连接运行 Redis 的 myredis 容器，并指定其主机名是"redis"，即 app.py 应用代码中第 05 行所指定的 host 名称。

（7）打开浏览器访问宿主机的 5000 端口并刷新网页，便可以观察到计数器的自增效果，如图 5-3 所示。

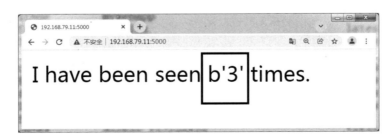

图 5-3

5.2.2 【实战】使用 Docker Compose 部署应用

在 5.2.1 节中,使用手动方式完成了应用的部署和启动,实现了计数器的自增功能。在该应用中只包含两个模块——Python Web 模块和 Redis DB 模块。

随着系统架构的不断复杂和模块的不断增多,这样的手动方式会变得越来越不方便。并且,如果要实现扩容/缩容等操作,则手动方式无法很好地解决这样的问题。

而使用 Docker Compose 则可以非常方便地解决这样的问题,其核心是定义一个 YML 文件来对即将部署的应用进行描述。此时的目录结构如图 5-4 所示。

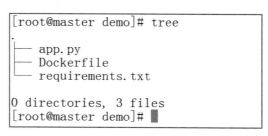

图 5-4

下面演示如何使用 Docker Compose。

(1)在当前目录下创建"docker-compose.yml"文件,输入以下内容。

```
01  version: '3'
02  services:
03    web:
04      build: .
05      ports:
06        - "5000:5000"
07    redis:
08      image: "redis"
```

其中:

- 从第 02 行开始,定义了该应用中所包含的模块。这里定义应用中包含两个模块——第 03 行的 web 模块和第 07 行的 redis 模块。
- 第 03~06 行,定义了 web 模块。通过使用"build"命令在当前目录下编译 Dockerfile 文件,将生成的镜像存放到本地;并且在启动容器时,将宿主机的 5000 端口映射到容器的 5000 端口。
- 第 07 和 08 行定义了 redis 模块。通过参数"image"指定在启动容器时使用"redis"镜像。如果本地没有该镜像,则从镜像仓库中拉取该镜像。

YML 文件的全称是 Yet Another Markup Language，因此也可以称之为 YAML 文件，其设计目的是方便人类读写。这种文件使用缩进表示层级关系，因此特别适合在脚本语言中使用。

（2）通"docker-compose up"命令启动应用。

```
docker-compose up
```

（3）打开浏览器访问访问宿主机的 5000 端口，这时可以观察到与图 5-3 相同的效果。

（4）执行"docker-compose ps"命令检查此时启动的容器信息，如图 5-5 所示。

```
[root@master demo]# docker-compose ps
    Name              Command          State          Ports
-----------------------------------------------------------------------
demo_redis_1   docker-entrypoint.sh redis ...   Up    6379/tcp
demo_web_1     python app.py                    Up    0.0.0.0:5000->5000/tcp
[root@master demo]#
```

图 5-5

可以看出，在该应用中包含两个模块——demo_redis_1 和 demo_web_1，并且这两个模块各自运行在一个 Docker 容器中。

（5）执行"docker-compose images"命令可以查看由 Docker Compose 管理的镜像信息，如图 5-6 所示。

```
[root@master demo]# docker-compose images
 Container      Repository       Tag       Image Id       Size
-----------------------------------------------------------------------
demo_redis_1   docker.io/redis   latest    7614ae9453d1   107 MB
demo_web_1     demo_web          latest    14fbafbf32f4   80.6 MB
[root@master demo]#
```

图 5-6

5.2.3 【实战】使用 Docker Compose 进行服务的在线扩容/缩容

使用 Docker Compose 可以非常方便地实现服务的在线扩容/缩容。图 5-7 展示了"docker-compose scale"命令的帮助信息。

```
[root@master demo]# docker-compose help scale
Set number of containers to run for a service.

Numbers are specified in the form `service=num` as arguments.
For example:

    $ docker-compose scale web=2 worker=3

This command is deprecated. Use the up command with the `--scale` flag
instead.

Usage: scale [options] [SERVICE=NUM...]

Options:
  -t, --timeout TIMEOUT      Specify a shutdown timeout in seconds.
                             (default: 10)
[root@master demo]#
```

图 5-7

1. 扩容

下面以 5.2.3 节中部署的应用为例，来演示如果使用"docker-compose scale"命令完成对服务的在线扩容/缩容。

（1）执行以下命令将服务中的 redis 模块扩展到 3 个容器中。

```
docker-compose scale redis=3
```

（2）在扩容完成后，执行"docker-compose ps"命令查看容器信息，如图 5-8 所示。

```
[root@master demo]# docker-compose ps
     Name                  Command             State         Ports
-------------------------------------------------------------------------
demo_redis_1      docker-entrypoint.sh redis ...   Up    6379/tcp
demo_redis_2      docker-entrypoint.sh redis ...   Up    6379/tcp
demo_redis_3      docker-entrypoint.sh redis ...   Up    6379/tcp
demo_web_1        python app.py                    Up    0.0.0.0:5000->5000/tcp
[root@master demo]#
```

图 5-8

对比图 5-5 与图 5-8 会发现，这时 Docker Compose 启动了 3 个容器来运行 redis 模块，从而实现了负载均衡功能，以达到分摊请求压力的目的。

2. 缩容

当请求量比较少不需要太多的容器时，可以减少容器以达到节省系统资源的目的。

缩容也是执行上面的命令，写入缩容之后的容器数量即可。例如，以下命令把 redis 模块的容器数量减少到 2 个。

```
docker-compose scale redis=2
```

5.2.4 【实战】在 Docker Compose 中控制模块启动和停止的顺序

在 docker-compose.yml 文件中,可以使用"depends_on"参数来控制模块启动和停止的顺序,这样可以满足一些更为复杂的场景需求。

下面来演示如何使用"depends_on"参数。

(1)创建一个新的"docker-compose.yml"文件,在其中输入以下内容:

```yaml
version: '3'
services:
# 定义应用的 web 模块
 web:
   image: centos
   # 指定 web 模块依赖的 db 模块,
   # 在 db 模块启动后,web 模块才可以启动
   depends_on:
     - db
# 定义应用的 db 模块
 db:
   image: centos
   # 在启动 db 模块时,安装 nc 工具用于模拟 MySQL 数据库服务
   command: >
     /bin/bash -c '
     yum install nc -y;
     sleep 5;
     echo "sleep over";
     nc -lk 0.0.0.0 3306;
     '
```

(2)执行"docker-compose up"命令启动应用。通过输出的日志,可以确定确实是"先启动 db 模块,再启动 web 模块"。但是由于 db 模块大约需要 5 秒钟才能完成启动,这会导致 web 模块启动失败,如图 5-9 所示。

```
[root@master demo]# docker-compose up
Starting demo_db_1  ... done
Starting demo_web_1 ... done
Attaching to demo_db_1, demo_web_1
db_1    | Last metadata expiration check: 0:00:17 ago on Wed Jan 19 13:08:05 2022.
demo_web_1 exited with code 0
db_1    | Package nmap-ncat-2:7.70-6.el8.x86_64 is already installed.
db_1    | Dependencies resolved.
db_1    | Nothing to do.
db_1    | Complete!
db_1    | sleep over
```

图 5-9

（3）修改 web 模块的启动脚本——等到 db 模块可以被访问之后再启动 web 模块。代码中的加粗部分用于在启动 web 模块时检查 db 模块是否已经成功启动。

```yaml
version: '3'
services:
  # 定义应用的 web 模块
  web:
    image: centos
    # 指定 web 模块依赖的 db 模块
    # 在 db 模块启动后，web 模块才可以启动
    depends_on:
      - db
    command: >
      /bin/bash -c '
      yum install nc -y;
      while ! nc -z db 3306;
      do
        echo "wait for db";
        sleep 1;
      done;

      echo "db is running!";
      echo "start web service from here";
      '
  # 定义应用的 db 模块
  db:
    image: centos
    # 在启动 db 模块时，安装 nc 工具用于模拟 MySQL 数据库服务
    command: >
      /bin/bash -c '
      yum install nc -y;
      sleep 5;
      echo "sleep over";
      nc -lk 0.0.0.0 3306;
      '
```

（4）重新执行 "docker-compose up" 命令启动应用。通过输出的日志可以看到 "在 db 模块启动完成后，web 模块才启动"，如图 5-10 所示。

```
[root@master demo]# docker-compose up
Starting demo_db_1  ... done
Starting demo_web_1 ... done
Attaching to demo_db_1, demo_web_1
db_1   | Last metadata expiration check: 0:03:15 ago on Wed Jan 19 13:19:47 2022.
db_1   | Package nmap-ncat-2:7.70-6.el8.x86_64 is already installed.
web_1  | Last metadata expiration check: 0:03:10 ago on Wed Jan 19 13:19:53 2022.
db_1   | Dependencies resolved.
db_1   | Nothing to do.
db_1   | Complete!
web_1  | Package nmap-ncat-2:7.70-6.el8.x86_64 is already installed.
web_1  | Dependencies resolved.
web_1  | Nothing to do.
web_1  | Complete!
web_1  | wait for db
web_1  | wait for db
web_1  | wait for db
web_1  | wait for db
web_1  | wait for db
db_1   | sleep over
web_1  | db is running!
web_1  | start web service from here
demo_web_1 exited with code 0
```

图 5-10

 将上面的代码直接嵌入 docker-compose.yml 文件中即可控制模块启动和停止的顺序。但是，这样的方式不利于维护和管理，也容易出现错误。如果有多个依赖或多层依赖，则其维护的复杂度会直线上升。为了解决这个问题，可以单独创建一个名为"wait-for-db.sh"的脚本，并将其包含在 docker-compose.yml 文件中。

（5）创建脚本文件"wait-for-db.sh"，在其中输入以下内容：

```bash
#!/bin/bash
#****************************************
# @file    : wait-for-db.sh
# @author  : zhaoyuqiang
# @date    : 2022-01-19
#****************************************
yum install nc -y;

wait_for() {
    while ! nc -z $1 $2;
    do
      echo "wait for db";
      sleep 1;
    done;
}

#这里将接收两个参数：一个是 db 的主机名；另一个是端口号
host="$1"
port="$2"
```

```
#检查db模块是否已经启动
wait_for $host $port

echo "db is running!";
echo "start web service from here";
```

（6）给"wait-for-db.sh"脚本授予可执行的权限。

```
chmod +x wait-for-db.sh
```

（7）在"docker-compose.yml"文件中引用"wait-for-db.sh"脚本。

```
version: '3'
services:
  # 定义应用的web模块
  web:
    image: centos
    # 指定web模块依赖的db模块
    # 在db模块启动后，web模块才可以启动
    depends_on:
      - db
    volumes:
      - "./wait-for-db.sh:/wait-for-db.sh"
    command: ["/wait-for-db.sh", "db", "3306"]
  # 定义应用的db模块
  db:
    image: centos
    # 在启动db模块时，安装nc工具用于模拟MySQL数据库服务
    command: >
      /bin/bash -c '
      yum install nc -y;
      sleep 5;
      echo "sleep over";
      nc -lk 0.0.0.0 3306;
      '
```

> 在实际使用中，也可以将"wait-for-db.sh"脚本打包到发布的镜像中，这样就不需要配置volumes参数来加载"wait-for-db.sh"脚本了。

（8）执行"docker-compose up"命令重新启动应用，通过输出的日志可以看到"在db模块启动完成后，web模块才开始启动"，如图5-11所示。

```
[root@master demo]# docker-compose up
Starting demo_db_1  ... done
Starting demo_web_1 ... done
Attaching to demo_db_1, demo_web_1
db_1   | Last metadata expiration check: 0:41:00 ago on Wed Jan 19 13:19:47 2022.
db_1   | Package nmap-ncat-2:7.70-6.el8.x86_64 is already installed.
web_1  | Last metadata expiration check: 0:00:12 ago on Wed Jan 19 14:00:36 2022.
db_1   | Dependencies resolved.
db_1   | Nothing to do.
db_1   | Complete!
web_1  | Package nmap-ncat-2:7.70-6.el8.x86_64 is already installed.
web_1  | Dependencies resolved.
web_1  | Nothing to do.
web_1  | Complete!
web_1  | wait for db
web_1  | wait for db
web_1  | wait for db
web_1  | wait for db
web_1  | wait for db
db_1   | sleep over
web_1  | db is running!
web_1  | start web service from here
demo_web_1 exited with code 0
```

图 5-11

5.3 Docker Compose 中的网络

在默认情况下，Docker Compose 会为每一个应用创建单独的网络环境，而该应用中的所有容器在启动时会自动加入其对应的网络环境。当应用启动完成后，该网络环境中的所有容器便可以相互访问。

也可以通过在 docker-compose.yml 文件中定义的名称（即容器的主机名），来实现容器的相互发现。

> 在 Docker Compose 的 1 版本中是不支持网络特性的。

在 Docker Compose 中提供了默认的网络环境，也可以在 docker-compose.yml 文件自定义网络环境。

5.3.1 Docker Compose 中的默认网络环境

以 5.2.2 节中定义的 docker-compose.yml 文件为例，在执行"docker-compose up"命令启动应用时，会对 Docker Compose 网络环境执行以下的操作：

（1）使用 bridge 模式创建一个名为"demo_default"网络，如图 5-12 所示。

```
[root@master demo]# docker-compose up
Creating network "demo_default" with the default driver
Creating demo_redis_1 ... done
Creating demo_web_1   ... done
Attaching to demo_redis_1, demo_web_1
```

图 5-12

（2）使用 web 模块的配置信息创建一个容器，并使用"web"作为容器的主机名加入"demo_default"网络。

（3）使用 redis 模块的配置信息创建一个容器，并使用"redis"作为容器的主机名加入"demo_default"网络。

完成上述步骤后，运行该应用的所有容器，便可以通过主机名"web"和"redis"实现在"demo_default"网络环境中的容器相互访问了。例如，Python Web 应用 app.py 中的第 05 行代码，通过主机名"redis"访问 redis 模块。

```
05    redis = Redis(host='redis', port=6379)
```

> 由于 Docker Compose 会使用 docker-compose.yml 文件中模块的名字作为容器的主机名，因此，在 docker-compose.yml 文件中模块的名字必须唯一。

如果想在 Docker Compose 中覆盖默认的网络配置，则可以在 docker-compose.yml 文件中通过指定参数"networks"的值来实现，如以下代码所示。

```
version: '3'
services:
  web:
    build: .
    ports:
      - "5000:5000"
  redis:
    image: "redis"
networks:
  default:
    driver: bridge2
```

5.3.2　在 Docker Compose 中自定义模块的网络环境

Docker Compose 除提供默认的网络环境外，也允许用户自定义各个模块的网络环境，这样用户可以实现更加复杂的网络。每一个模块也可以通过参数"networks"连接到网络环境中。

以下 docker-compose.yml 是 Docker 官方提供的一个示例。

```yaml
version: "3.9"

services:
  proxy:
    build: ./proxy
    networks:
      - frontend
  app:
    build: ./app
    networks:
      - frontend
      - backend
  db:
    image: postgres
    networks:
      - backend

networks:
  frontend:
    # Use a custom driver
    driver: custom-driver-1
  backend:
    # Use a custom driver which takes special options
    driver: custom-driver-2
    driver_opts:
      foo: "1"
      bar: "2"
```

在这个文件中一共包含 3 个功能模块：proxy、app 和 db。其中，proxy 和 db 模块属于不同的网络环境，即 proxy 模块属于"frontend"网络，而 db 属于"backend"网络。这两个网络彼此隔离，不能进行相互访问。但是，在 app 模块中通过参数 networks 指定了"frontend"网络和"backend"网络，即 app 模块既可以访问 proxy 模块，也可以访问 db 模块。

第 6 章
使用 Docker Machine 进行远程管理

　　Docker Machine 是 Docker 官方提供的一个远程管理工具。通过使用 Docker Machine，可以在远端节点上安装 Docker，以及在远端的虚拟主机上安装虚拟的宿主机并在其中安装 Docker。Docker Machine 还提供了相应的命令以管理这些远端的 Docker 环境和虚拟机。

　　图 6-1 是 Docker Machine 的官方图标，很形象地说明了 Docker Machine 的功能。

图 6-1

6.1 使用 Docker Machine

Docker Machine 可以在多种平台上使用,包括 Linux、Windows 和 MacOS。本书以 Linux 为基础来介绍 Docker Machine 的安装和使用。

Docker Machine 使用 Go 语言开发,并且依赖 Docker。因此在使用 Docker Machine 之前,需要先在本地安装 Docker。

6.1.1 安装 Docker Machine

以 master 节点为例来安装 Docker Machine。

(1)从 GitHub 下载 Docker Machine 的执行脚本"docker-machine"。

```
curl -L
https://github.com/docker/machine/releases/download/v0.16.0/docker-machine-'uname -s'-'uname -m' >/tmp/docker-machine
```

(2)给"docker-machine"脚本添加执行权限。

```
chmod +x /tmp/docker-machine
```

(3)将"docker-machine"脚本复制到"/usr/local/bin/"目录下。

```
cp /tmp/docker-machine /usr/local/bin/docker-machine
```

(4)执行命令验证 Docker Machine,如图 6-2 所示。

```
[root@master ~]# docker-machine -v
docker-machine version 0.16.0, build 702c267f
[root@master ~]#
```

图 6-2

6.1.2 在远端宿主机上安装 Docker

安装好 Docker Machine 后,可以通过"docker-machine"命令来管理远端节点上的 Docker。

参考表 4-2 单独准备一台新的虚拟机 node2,表 6-1 列出了目前使用的 3 台虚拟机的信息。由于已经在 master 节点上安装好了 Docker Machine,所以可以通过"docker-machine"命令远程在 node2 节点上安装 Docker 并管理它。

表 6-1

节点名	IP 地址	已部署的服务
master	192.168.79.11	Docker、ZooKeeper、Docker Machine
node1	192.168.79.12	Docker
node2	192.168.79.13	Linux

由于 Docker Machine 在进行远程管理时，需要免密码登录的支持。因此，在使用"docker-machine"命令之前，需要先配置节点之间的免密码登录。

1. 配置节点之间的免密码登录

免密码登录采用的是不对称加密的认证方式，需要产生密钥对（即一个公钥和一个私钥），其本质就是两个字符串。公钥负责加密；而私钥负责解密。

图 6-3 展示了免密码登录的过程——想从 Server A 免密码登录 Server B。

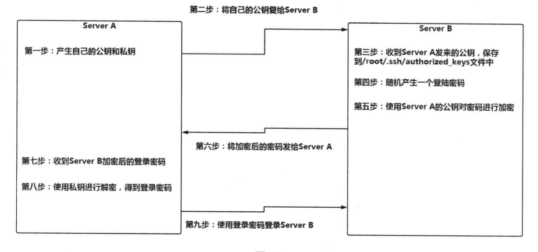

图 6-3

免密码登录是单向的。如果想反过来——从 Server B 免密码登录 Server A，则需要单独进行配置。

在了解了免密码登录的过程后，下面来演示如何配置从 master 节点到 node2 节点的免密码登录。

（1）在 master 节点上生成公钥与私钥：

```
ssh-keygen -t rsa
```

（2）将 master 节点的公钥发送到 node2 节点上：

```
ssh-copy-id -i .ssh/id_rsa.pub root@192.168.79.13
```

由于还没有完成免密码登录的配置，所以此时需要输入 node2 节点的密码。

（3）在 master 节点上输入以下命令验证免密码登录。这时会发现，不需要输入 node2 节点的密码便可以直接从 master 节点登录 node2 节点。

```
ssh 192.168.79.13
```

2. 使用 "docker-machine" 命令在远端节点上安装 Docker

在配置好 master 节点与 node2 节点之间的免密码登录后，便可以在 master 节点上使用 "docker-machine" 命令远程在 node2 节点上安装 Docker。

（1）在 master 节点上执行以下命令（这条命令需要很长的执行时间）。

```
docker-machine create \
-d generic \
--generic-ip-address=192.168.79.13 \
--generic-ssh-user=root \
--generic-ssh-key /root/.ssh/id_rsa node2
```

其中的参数说明如下。

- -d：使用的驱动程序类型，官方支持的驱动程序类型有 amazonec2、azure、digitalocean、exoscale、generic、google、hyperv、none、openstack、rackspace、softlayer、virtualbox、vmwarevcloudair、vmwarefusion、vmwarevsphere。
- --generic-ip-address：远端节点的 IP 地址。
- --generic-ssh-user：远程登录的用户名。
- --generic-ssh-key：免密码登录的私钥文件。
- node2：远端节点的别名。

当 "docker-machine create" 命令执行完成后，通过输出的日志可以看到在远端节点上已经成功安装了 Docker 并且启动成功，如图 6-4 所示。

（2）在 node2 节点上执行以下命令检查 Docker 的版本信息。

```
docker version
```

在执行 "docker-machine create" 命令时，把远端节点的节点名修改为指定的别名。由于这里使用的别名是 "node2"，所以看不出区别。

```
[root@master ~]# docker-machine create \
> -d generic \
> --generic-ip-address=192.168.79.13 \
> --generic-ssh-user=root \
> --generic-ssh-key /root/.ssh/id_rsa node2
Running pre-create checks...
Creating machine...
(remotenode2) Importing SSH key...
Waiting for machine to be running, this may take a few minutes...
Detecting operating system of created instance...
Waiting for SSH to be available...
Detecting the provisioner...
Provisioning with centos...
Copying certs to the local machine directory...
Copying certs to the remote machine...
Setting Docker configuration on the remote daemon...
Checking connection to Docker...
Docker is up and running!
To see how to connect your Docker Client to the Docker Engine running
 on this virtual machine, run: docker-machine env node2
[root@master ~]#
```

图 6-4

（3）在 master 节点上，执行以下命令查看由 Docker Machine 管理的远端节点信息，如图 6-5 所示。

```
[root@master ~]# docker-machine ls
NAME    ACTIVE  DRIVER   STATE    URL
node2    -      generic  Running  tcp://192.168.79.13:2376
[root@master ~]#
```

图 6-5

6.2 Docker Machine 的基本用法

在使用 Docker Machine 创建好远端的 Docker 节点后，便可以使用"docker-machine"命令对远端节点进行管理。

6.2.1 【实战】使用 Docker Machine 的命令

通过执行"docker-machine help"命令可以查看 Docker Machine 提供的命令。表 6-2 列出了这些命令。

表 6-2

命　　令	作　　用
active	查看活跃的 Docker 节点
config	输出连接的配置信息

续表

命令	作用
create	创建一个 Docker 节点
env	显示远端节点的环境变量
inspect	输出远端节点的详细信息
ip	获取远端节点的 IP 地址
kill	停止某个远端节点
ls	列出所有管理的节点
provision	重新设置一个已存在的远端节点
regenerate-certs	为某个远端节点重新生成认证信息
restart	重启某个远端节点
rm	删除某个远端节点
ssh	以 SSH 方式登录远端节点
scp	在节点之间复制文件
mount	挂载远端节点目录到本地
start	启动一个远端节点
status	查看远端节点状态
stop	停止一个远端节点
upgrade	更新远端节点的 Docker 版本为最新
url	获取远端节点的 URL
version	输出 Docker-Machine 的版本信息
help	输出帮助信息

要执行这些命令，只需要在"docker-machine"命令后面加上具体的命令即可。每个命令都提供了帮助信息，可以通过执行"docker-machine 【COMMAND】 --help"命令来查看。

6.2.2 【实战】管理远端的 Docker 节点

了解了"docker-machine"命令的参数后，下面通过几个示例来演示常用命令的使用方法。

1. 显示远端节点的环境变量信息

以下命令将显示远端节点的环境变量信息，如图 6-6 所示。

```
docker-machine env node2
```

这条命令输出的内容可以作为环境变量参数来设置一些 Docker 客户端，从而让本机的 Docker 客户端可以与远端的 Docker 服务器进行通信。其中，node2 是我们之前创建的远端 Docker 服务器的名字。

```
[root@master ~]# docker-machine env node2
export DOCKER_TLS_VERIFY="1"
export DOCKER_HOST="tcp://192.168.79.13:2376"
export DOCKER_CERT_PATH="/root/.docker/machine/machines/node2"
export DOCKER_MACHINE_NAME="node2"
# Run this command to configure your shell:
# eval $(docker-machine env node2)
[root@master ~]#
```

图 6-6

2. 设置本地的环境变量

以下命令将设置本地的环境变量，以操作远端 node2 节点上的 Docker 守护进程。

```
eval $(docker-machine env node2)
```

该命令运行在当前命令行终端中，但在使用该命令后接下来运行的"docker"命令操作的都是远端节点。例如，在 master 节点上执行该命令后尝试拉取一个 Nginx 镜像，则将在远端的 node2 节点上完成 Nginx 镜像的拉取，如图 6-7 所示。

```
[root@master ~]# eval $(docker-machine env node2)
[root@master ~]# docker pull nginx
Using default tag: latest
latest: Pulling from library/nginx
a2abf6c4d29d: Pull complete
a9edb18cadd1: Pull complete
589b7251471a: Pull complete
186b1aaa4aa6: Pull complete
b4df32aa5a72: Pull complete
a0bcbecc962e: Pull complete
Digest: sha256:0d17b565c37bcbd895e9d92315a05c1c3c9a29f762b0
53c9b31
Status: Downloaded newer image for nginx:latest
[root@master ~]#
```

```
[root@node2 ~]# docker images
REPOSITORY    TAG      IMAGE ID       CREATED       SIZE
nginx         latest   605c77e624dd   3 weeks ago   141MB
[root@node2 ~]#
```

图 6-7

3. 启动、停止和重启远端节点上的 Docker 环境

以下命令将启动、停止和重启远端节点上的 Docker 环境。

```
docker-machine start/stop/restart node2
```

generic 驱动类型不支持 stop 命令。

4. 查看远端节点上的 Docker 状态

以下命令将查看远端节点上的 Docker 状态。

```
docker-machine status node2
```

5. 从 master 节点上以 SSH 方式登录远端 node2 节点

Docker Machine 支持 SSH 登录。以下命令将从 master 节点上以 SSH 方式登录远端的 node2 节点。

```
docker-machine ssh node2
```

6.3 Docker Machine 的高级用法

在前面的示例中，Docker Machine 通过远程管理方式管理了远端的宿主机，并可以直接操作运行在其上的 Docker 环境。

但在实际使用中，在远端的宿主机上可能会安装 vSphere 或 VirtualBox 等虚拟机管理软件，这样，远端节点就会是一个虚拟主机。Docker Machine 支持众多类型的虚拟主机。

下面使用 Docker Machine 创建基于 VirtualBox 和 vSphere 的虚拟主机。

6.3.1 【实战】使用 Docker Machine 创建基于 VirtualBox 的虚拟主机

VirtualBox 的全称是 Oracle VM VirtualBox。它不仅具有丰富的特色，而且性能也很优异，使用起来也非常简单。

下面在 master 节点上安装 VirtualBox，并使用 Docker Machine 来管理 VirtualBox。

（1）在 "/etc/yum.repos.d/" 目录下创建 my.repo 的 YUM 源文件，并输入以下内容：

```
[virtualbox]
name=virtualbox
baseurl=http://download.virtualbox.org/virtualbox/rpm/el/$releasever/$basearch
enabled=1
gpgcheck=1
repo_gpgcheck=1
gpgkey=https://www.virtualbox.org/download/oracle_vbox.asc
```

（2）执行命令更新 YUM 源的缓存。

```
yum clean all
yum makecache -y
```

（3）使用 YUM 方式安装 VirtualBox。

```
yum install VirtualBox-6.0 -y
```

（4）在安装完成后启动 VirtualBox。

```
rcvboxdrv setup
```

如果这时出现以下错误信息，则可能是因为宿主机缺少依赖的 RPM 包。

Error:

The distribution packages containing the headers are probably:

kernel-devel kernel-devel-3.10.0-693.el7.x86_64

执行以下命令安装依赖的 RPM 包：

yum install gcc make perl

yum install kernel-devel

yum update kernel -y

yum install kernel-headers

之后执行以下命令重新启动 VirtualBox：

init 6

（5）以下命令将确定 VirtualBox 的版本信息。

```
vboxmanage -v
```

（6）使用 Docker Machine 创建基于 VirtualBox 的虚拟机，并在其中安装 Docker。

```
docker-machine create --driver virtualbox myVirtualBoxLinux
```

这里可能会出现以下错误信息：

Running pre-create checks...

Error with pre-create check: "This computer doesn't have VT-X/AMD-v enabled. Enabling it in the BIOS is mandatory"

这是因为我们在 VMWare WorkStations 上搭建的环境，所以需要打开 VT-X/AMD-v 的虚拟开关。

（7）关闭 master 节点，并在"虚拟机设置"中勾选"虚拟化 Intel VT-x/EPT 或 AMD-V/RVI(V)"复选框，如图 6-8 所示。

图 6-8

（8）重启 master 节点，并重新执行第（6）步的创建命令。在虚拟主机创建成功后，可以通过 "docker-machine" 命令来管理运行在其上的 Docker 了。

在创建 VirtualBox 的虚拟主机时，会从 GitHub 下载 Boot2Docker ISO 文件。它是个基于 Tiny Core Linux 的轻量级发行版，自带 Docker 程序。相关的输出日志如下：
(myVirtualBoxLinux) Downloading /root/.docker/machine/cache/boot2docker.iso from https://github.com/boot2docker/boot2docker/releases/download/v19.03.12/boot2docker.iso...

（9）查看由 Docker Machine 管理的远端节点信息，如图 6-9 所示。

```
[root@master ~]# docker-machine ls
NAME               ACTIVE   DRIVER       STATE     URL                         SWARM   DOCKER
myVirtualBoxLinux  -        virtualbox   Running   tcp://192.168.99.100:2376           v19.03.12
remotenode2        -        generic      Running   tcp://192.168.79.13:2376            v20.10.12
[root@master ~]#
```

图 6-9

（10）通过 Docker Machine 进入 VirtualBox 的虚拟主机。

```
docker-machine ssh myVirtualBoxLinux
```

（11）查看 Boot2Docker Linux 自带的 Docker 版本信息。

```
docker version
```

6.3.2 【实战】使用 Docker Machine 创建基于 vSphere 的虚拟主机

VMware vSphere 是业界领先且最可靠的虚拟化平台，可以将我们的服务器资源整合成资源池进行管理。VMware vSphere 有两个组件——ESXi Server 和 vSphere Client。

ESXi Server 是一款可以独立安装和运行在裸机上的系统。在 ESXi Server 安装好后，可以通过 vSphere Client 远程与 ESXi Server 进行连接，并管理其上的虚拟主机。ESXi Server 从内核就支持硬件虚拟化，运行于其中的虚拟服务器在性能与稳定性方面不逊于普通的硬件服务器，且更易于管理和维护。ESXi Server 也支持通过 Docker Machine 进行远程的管理和操作。

下面使用 Docker Machine 创建基于 VMware vSphere 的虚拟主机。

（1）在 VMWare WorkStations 中安装 ESXi Server，如图 6-10 所示。

> ESXi Server 的安装比较简单，在安装过程中只需要保持默认设置即可，这里就不做详细介绍了。ESXi Server 安装成功的界面如图 6-11 所示。

图 6-10

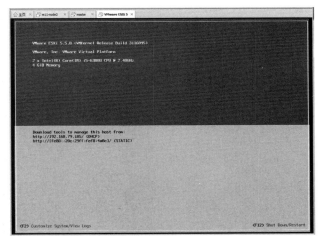

图 6-11

（2）使用 "docker-machine" 命令在 VMware vSphere 上创建 Docker 虚拟机：

```
docker-machine create -d vmwarevsphere \
--vmwarevsphere-password="Welcome_1" \
--vmwarevsphere-vcenter=192.168.79.185 \
--vmwarevsphere-username="root" \
dockervSphereVM
```

其中的参数说明如下。

- -d vmwarevsphere：指定 Docker Machine 的驱动程序是 vmwarevsphere。
- --vmwarevsphere-password：指定 VMware vSphere 的密码。
- --vmwarevsphere-vcenter：指定 VMware vSphere 的 IP 地址或域名。
- --vmwarevsphere-username：指定 VMware vSphere 的用户名。

（3）执行以下命令查看 Docker Machine 管理的所有远端节点信息，如图 6-12 所示。

```
[root@master ~]# docker-machine ls
NAME                ACTIVE   DRIVER           STATE     URL                         SWARM   DOCKER
dockervSphereVM     -        vmwarevsphere    Running   tcp://192.168.79.229:2376           v19.03.12
myVirtualBoxLinux   *        virtualbox       Running   tcp://192.168.99.100:2376           v19.03.12
remotenode2         -        generic          Running   tcp://192.168.79.13:2376            v20.10.12
[root@master ~]#
```

图 6-12

ESXi Server 也可以通过使用其自带的 vSphere Client 进行管理。图 6-13 展示了 vSphere Client 的界面，其中的"dockervSphereVM"虚拟机便是在第（2）步中使用 Docker Machine 创建的 Docker 虚拟机。

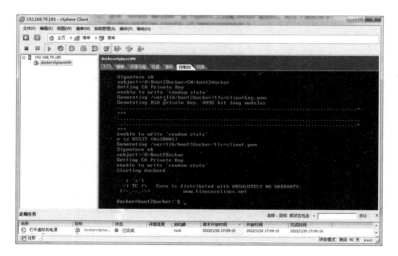

图 6-13

第 7 章
使用 Docker Swarm 构建集群

Docker Swarm 是 Docker 官方提供的一款集群管理工具，其主要作用是把若干台 Docker 宿主机抽象为一个整体，并通过一个入口统一管理这些 Docker 宿主机上的各种 Docker 资源。

Docker Swarm 和 Kubernetes 比较类似，但是它更加轻量级，具有的功能较 Kubernetes 少一些。有了 Docker Swarm 集群，便可以实现应用的负载均衡与失败迁移功能。

7.1 Docker Swarm 集群的体系架构

Docker 从 1.12.0 版本开始将 Docker Swarm 集成进来了。因此，Docker Swarm 不需要单独安装了。由于 Docker Swarm 内置了服务发现功能，因此也不再需要进行服务发现的配置了。

Docker Swarm 与 Docker Compose 类似，都是 Docker 官方提供的服务编排工具。二者所不同的是：Docker Compose 是在单个宿主机上创建多个容器从而进行服务编排的工具；而 Docker Swarm 是在多个服务器或宿主机上创建容器，从而组成集群提供相应的服务的工具。因此，Docker Swarm 的功能比 Docker Compose 更加强大。

图 7-1 展示了 Docker Swarm 的体系架构。Docker Swarm 集群是一个主从架构。其中有一个 Swarm Manager 节点用来管理集群中的容器资源。Swarm Manager 节点对外暴露操作的接口，外部的用户可以通过该接口来实现对集群的管理。用户也可以通过 Swarm Manager 节点向集群发出操作指令。对于较大规模的 Docker 集群，可以将 Swarm Manager 单独部署到一台服务器上，从而提高其性能。Swarm Node 节点从 Swarm Manager 节点接收命令，从而创建相应的容器来运行应用。但是，用户只能笼统地向集群发出指令，而不能具体分配某台服务器干什么（这是

由 Swarm Manager 节点上的 Scheduler 调度器来完成的）。

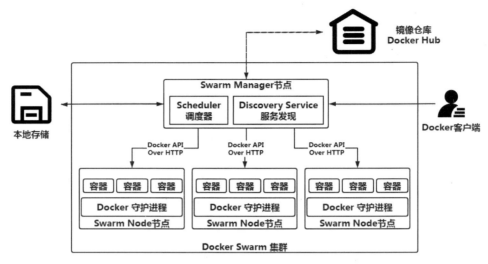

图 7-1

7.2 构建 Docker Swarm 集群

在了解了 Docker Swarm 集群的体系架构后，基于已经部署好 3 台虚拟主机便可以非常容易地构建出 Swarm 集群。表 7-1 列出了这 3 台虚拟机的相关信息。其中，master 节点作为 Swarm Manager 运行；而 node1 和 node2 节点作为 Swarm Node 运行。

表 7-1

节点名	IP 地址	已部署的服务
Master	192.168.79.11	Docker、ZooKeeper、Docker Machine
node1	192.168.79.12	Docker
node2	192.168.79.13	Docker

下面演示如何以 master、node1 和 node2 这 3 个节点来构建一个 Docker Swarm 集群。

（1）在 master 节点上执行以下命令初始化集群。

```
docker swarm init \
--listen-addr 192.168.79.11:8888 \
--advertise-addr 192.168.79.11
```

其中的参数说明如下。

- --listen-addr：指定集群暴露给外界调用的 HTTP API 的 socket 地址。该参数可以省略。
- --advertise-addr：当宿主机有多个网卡时，该参数用于指定绑定的网卡地址。该参数可以省略。

由于在 4.4.4 节中修改了"/usr/lib/systemd/system/docker.service"文件，所以需要删除 master 和 node1 节点上该文件中的以下内容：

-H tcp://0.0.0.0:2375 -H unix:///var/run/docker.sock \
--cluster-store zk://192.168.79.11:2181 \
--cluster-advertise 192.168.79.11:2375 \

否则会出现以下错误信息：

Error response from daemon: --cluster-store and --cluster-advertise daemon configurations are incomp

（2）初始化命令执行成功后，会输出以下信息：

```
Swarm initialized: current node (rzmqa9903qxt1k06lul8bggig) is now a manager.

To add a worker to this swarm, run the following command:

docker swarm join \
  --token SWMTKN-1-16akjyb4q5aqcss3c68rie38fhzogg7g76fge8sus6u4p6j26k-118e95tjq9rdoourp8tk5u922 \
  192.168.79.11:8888

To add a manager to this swarm, run 'docker swarm join-token manager' and follow the instructions.
```

如果不小心忘记了集群初始化成功后的输出信息，则可以在 master 节点上运行以下命令重新输出这部分信息。

docker swarm join-token manager

（3）在 node1 和 node2 节点上分别执行以下命令将其作为 worker 节点加入 Swarm 集群。

```
docker swarm join \
--token SWMTKN-1-16akjyb4q5aqcss3c68rie38fhzogg7g76fge8sus6u4p6j26k-
118e95tjq9rdoourp8tk5u922 \
192.168.79.11:8888
```

(4)在 master 节点上查看集群的节点信息,如图 7-2 所示。

```
[root@master ~]# docker node ls
ID                           HOSTNAME   STATUS   AVAILABILITY   MANAGER STATUS
adnqnbuw2xxvzcwm53pb95bip    node1      Ready    Active
l7m5m9na7ojwgh3xfudlpf2p7    node2      Ready    Active
rzmqa9903qxt1k06lul8bggig *  master     Ready    Active         Leader
[root@master ~]#
```

图 7-2

节点 ID 后带星号表示该节点是当前连接节点。从图 7-2 中可以看出,在这个集群中有 1 个 master 节点和 3 个 worker 节点(master 节点本身也作为 worker 节点加入集群)。

7.3 在 Docker Swarm 集群中部署应用与 HAProxy

在构建完 Docker Swarm 集群后,就可以在上面部署应用了。这里将基于 Nginx 的基础镜像来创建一个 Dockerfile 文件,从而开发自己的应用。为了使 Swarm 集群中的所有节点都能访问镜像,需要将生成的镜像上传到镜像仓库中。由于应用部署在集群中,因此集群也就具有了高可用的功能。另外,可以使用 HAProxy 来实现集群的外部负载均衡功能。

7.3.1 【实战】在集群中部署应用

(1)创建 Dockerfile 文件,在其中输入以下内容。

```
FROM nginx
RUN echo '<h1>Swarm Version 1 <h1>' > /usr/share/nginx/html/index.html
```

(2)使用"docker build"命令进行编译。由于需要把生成的镜像上传到 Docker Hub 镜像仓库中,所以在编译时需要加上 Docker Hub 中的仓库名称前缀。

```
docker build -t collenzhao/swarm_nginx_demo:v1 .
```

这里用"v1"表示这是应用的第 1 个版本。

（3）登录 Docker Hub，输入注册的账号和密码。

```
docker login
```

登录成功后，将看到以下输出信息：

Login with your Docker ID to push and pull images from Docker Hub.

If you don't have a Docker ID, head over to Docker Hub to create one.

Username (collenzhao):

Password:

Login Succeeded

（4）将镜像上传到镜像仓库中。

```
docker push collenzhao/swarm_nginx_demo:v1
```

（5）在 Swarm 集群中部署应用。

```
docker service create \
-p 7788:80 \
--replicas 3 \
--name myswarmdemo \
collenzhao/swarm_nginx_demo:v1
```

其中的参数说明如下。

- -p 7788:80 将容器的 80 端口映射到宿主机的 7788 端口。
- --replicas 3 指定应用的副本数为 3，即启动 3 个容器来运行应用。
- --name myswarmdemo 指定应用在 Swarm 集群中的服务名称。

（6）通过"docker service ls"命令查看部署的服务，如图 7-3 所示，可以看到服务的名称、副本数和镜像的名称等信息。

```
[root@master ~]# docker service ls
ID              NAME          MODE         REPLICAS   IMAGE
qrr9qvm2opcx    myswarmdemo   replicated   3/3        collenzhao/swarm_nginx_demo:v1
[root@master ~]#
```

图 7-3

（7）通过"docker service ps"命令查看服务的详细信息，如图 7-4 所示。

```
[root@master ~]# docker service ps myswarmdemo
NAME           IMAGE                              NODE    CURRENT STATE
myswarmdemo.1  collenzhao/swarm_nginx_demo:v1     node2   Running 6 minutes ago
myswarmdemo.2  collenzhao/swarm_nginx_demo:v1     master  Running 6 minutes ago
myswarmdemo.3  collenzhao/swarm_nginx_demo:v1     node1   Running 6 minutes ago
[root@master ~]#
```

图 7-4

从图 7-4 中可以看出，该应用一共启动了 3 个副本（运行在 3 个容器中，分别在 node2、master 和 node1 节点上），也可以看出每个副本的状态和运行的时间。因此，可以到每个节点上通过"docker ps"命令查看容器的详细信息。

（8）通过 Swarm 集群中任意节点的 7788 端口可以访问部署的应用，如图 7-5 所示。

图 7-5

（9）通过"docker service scale"命令实现服务的动态伸缩功能。例如以下命令设置 myswarmdemo 应用的副本数为 5。

```
docker service scale myswarmdemo=5
```

伸缩命令执行成功后，将输出以下信息。
myswarmdemo scaled to 5
通过执行"docker service ls"命令可以看到 myswarmdemo 应用的副本数变成了 5。

7.3.2 【实战】测试集群的高可用性

集群的一个重特性就是高可用性（High Availablity，HA）。下面将测试部署在 Swarm 集群中的 myswarmdemo 应用的高可用性。

（1）为了测试方便，将 myswarmdemo 应用的副本数设置为 3。

```
docker service scale myswarmdemo=3
```

（2）在 master 节点上，通过"docker service ps"命令确定应用运行的节点信息。

```
docker service ps myswarmdemo
```

输出的信息如下：

```
NAME               NODE     DESIRED STATE   CURRENT STATE
myswarmdemo.1  ...... node2    Running         Running 51 seconds ago
myswarmdemo.2  ...... master   Running         Running 39 seconds ago
myswarmdemo.3  ...... node1    Running         Running 53 seconds ago
```

从输出的信息可以看出，在 node2、master 和 node1 节点上各启动了一个容器来运行应用。

（3）在 node2 节点上，使用 "docker ps" 命令确定容器的 ID，之后使用 "docker stop" 命令停止该容器。这样可以模拟集群中一个容器的异常宕机。

```
docker ps
docker stop 1dd94402b5e5
```

（4）在 master 节点上，重新执行 "docker service ps" 命令确定节点的信息。

```
docker service ps myswarmdemo
```

输出的信息如下：

```
NAME                NODE     DESIRED STATE   CURRENT STATE
myswarmdemo.1       node2    Running         Running less than a second ago
 \_ myswarmdemo.1   node2    Shutdown        Complete 5 seconds ago
myswarmdemo.2       master   Running         Running 5 minutes ago
myswarmdemo.3       node1    Running         Running 5 minutes ago
```

从输出的信息可以看出，尽管在 node2 节点上运行的容器宕机了，但是由于将应用的副本数设为了 3，所以 Swarm 集群将自动在 node2 节点上重新启动一个容器。

（5）直接将 node2 节点的宿主机关机，通过这样的方式来模拟 Docker 宿主机宕机的情况。执行 "docker node ls" 命令确定节点状态。通过输出的信息可以看出，node2 节点的状态变成了 "Down"。

```
docker node ls
```

输出的信息如下：

```
HOSTNAME   STATUS   AVAILABILITY   MANAGER STATUS
node1      Ready    Active
node2      Down     Active
master     Ready    Active         Leader
```

（6）在 master 节点上，再次执行 "docker service ps" 命令确定应用运行的节点信息。

```
docker service ps myswarmdemo
```

输出的信息如下：

```
NAME                       NODE     DESIRED STATE    CURRENT STATE
myswarmdemo.1       ...... node1    Running          Running 2 minutes ago
 \_ myswarmdemo.1   ...... node2    Shutdown         Running 9 minutes ago
 \_ myswarmdemo.1   ...... node2    Shutdown         Complete 9 minutes ago
myswarmdemo.2       ...... master   Running          Running 15 minutes ago
myswarmdemo.3       ...... node1    Running          Running 15 minutes ago
```

可以看到，由于 node2 节点宕机了，因此运行在其上的所有容器的状态都是 "Shutdown"。但是，Swarm 集群会自动在 node1 节点上重新启动一个副本，从而保证应用的高可用性。

7.3.3 【实战】使用 HAProxy 为集群添加外部负载均衡功能

Swarm 集群本身提供了负载均衡功能。但在某些情况下，可以通过给应用添加外部的负载均衡软件以应对高并发访问的情况。HAProxy 是一个免费的负载均衡软件，可以运行于大部分主流的 Linux 操作系统上。它提供了负载均衡功能，并具备丰富的功能和稳定性。

引入 HAProxy 后，系统的整体架构如图 7-6 所示。

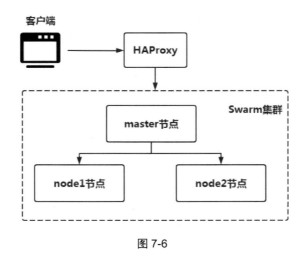

图 7-6

下面是具体的实现步骤。

（1）在任意节点上创建 HAProxy 的配置文件 haproxy.cfg，并在其中输入以下内容：

```
global
    daemon
```

```
defaults
    mode http
frontend http-in
    bind *:8080
    default_backend servers
backend servers
    server server1 192.168.79.11:7788
    server server2 192.168.79.12:7788
    server server3 192.168.79.13:7788
```

这里将 HAProxy 绑定到 8080 端口，并且通过参数"backend servers"指定后端的 Server 地址。

（2）创建 Dockerfile 文件，在其中输入以下内容：

```
FROM haproxy
COPY haproxy.cfg /usr/local/etc/haproxy/haproxy.cfg
```

> 这里基于 HAproxy 的基础镜像创建自己的镜像，并且使用第（1）步中的 haproxy.cfg 文件覆盖 HAProxy 默认的配置文件。

（3）使用 Dockerfile 文件创建镜像：

```
docker build -t myhaproxy .
```

（4）启动 myhaproxy 容器：

```
docker run -d -p 80:8080 myhaproxy
```

（5）通过运行在宿主机上 80 端口的 HAProxy 来代理服务请求，将服务请求转发到 Swarm 集群，结果如图 7-7 所示。

图 7-7

7.3.4 【实战】实现服务的滚动更新

Docker Swarm 可以实现服务的平滑升级，即服务不停机即可更新，客户端无感知。在 7.3.1 节中开发的 Swarm 集群应用的版本是"Version 1"。现在将开发"Version 2"版本，并且利用

Swarm 集群的服务更新实现应用的滚动升级。

（1）更新 Dockerfile 文件，注意版本号变为了 2。

```
FROM nginx
RUN echo '<h1>Swarm Version 2 <h1>' > /usr/share/nginx/html/index.html
```

（2）使用"docker build"命令进行编译。

```
docker build -t collenzhao/swarm_nginx_demo:v2 .
```

（3）登录 Docker Hub，输入注册的账号和密码。

```
docker login
```

（4）使用"docker push"命令将其上传到 Docker Hub 中。

```
docker push collenzhao/swarm_nginx_demo:v2
```

（5）使用"Version 2"版本更新之前在 Swarm 部署的服务。

```
docker service update \
--image collenzhao/swarm_nginx_demo:v2 myswarmdemo
```

（6）重新访问应用，将看到"Version 2"版本，如图 7-8 所示。

图 7-8

7.4　Docker Swarm 集群的数据持久化

与单机环境一样，Docker Swarm 集群中的容器也是无状态的服务。如果在 Swarm 集群中运行了 MySQL 等有状态的服务，若没有将数据挂载到宿主机中，那么一旦容器被销毁，则意味着数据会丢失。

有什么方法可以实现 Swarm 集群中运行的服务的数据持久化？Swarm 集群提供了以下两种方式来解决这个问题。

- volume 模式（默认模式）：将工作节点宿主机的数据同步到容器内。
- NFS 模式：通过网络文件系统实现数据的持久化。

7.4.1 【实战】通过 volume 实现集群的数据持久化

这里说的数据卷（volume）与单机模式下的 Docker 数据卷是完全一样的。Docker 数据卷可以被看成是一个挂载目录，与容器的生命周期无关。因此，Docker 在删除容器时不会自动删除 Docker 数据卷，从而实现了数据的持久化。图 7-9 说明了这种方式的工作原理。

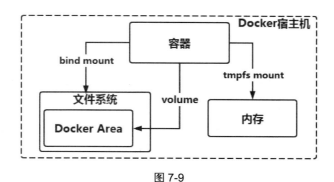

图 7-9

使用 volume 模式时，各个节点的数据不能共享。

下面来演示如何在 Swarm 集群中使用 Docker 数据卷。

（1）执行以下命令在 Swarm 集群中部署应用，并使用 "--mount" 参数挂载数据卷。

```
docker service create \
-p 7788:80 \
--replicas 3 \
--name myswarmdemo \
--mount type=volume,src=myvolumn,dst=/usr/share/nginx/html/ \
collenzhao/swarm_nginx_demo:v1
```

（2）执行以下命令查看数据卷信息。

```
docker volume ls
```

输出的信息如下：

```
DRIVER    VOLUME NAME
local     24015c88a6d4b4186132c0256f0d4
local     3541b10366a68c5571a6555aa8f51
local     57ba94101635ab4c944d59d04c68a
local     5bf278e55f0f8106ce8360f083591
local     5d9b448c513f6f1b75d444bc5e81e
```

```
local    90583a0c1212e15fea0948a4a7f3c
local    9403f5b3e9526266294c3ebad2873
local    d2e8aa4a1784f92f783380d6b1157
local    dd9c3ab4acc23972e4b051b1b7998
local    ddf55402b191e15dddabdfedf2420
local    myvolumn
```

（3）查看数据卷挂载目录的详细信息。

```
docker volume inspect myvolumn
```

输出的信息如下：

```
[
    {
        "Driver": "local",
        "Labels": null,
        "Mountpoint": "/var/lib/docker/volumes/myvolumn/_data",
        "Name": "myvolumn",
        "Options": {},
        "Scope": "local"
    }
]
```

可以看到，通过数据卷挂载的方式将容器的"/usr/share/nginx/html/"目录挂载到了宿主机的"/var/lib/docker/volumes/myvolumn/_data"目录下了。

7.4.2 【实战】通过 NFS 实现集群的数据持久化

通过 Swarm 集群的 volume 模式，可以实现容器与宿主机间的数据持久化，但是无法实现集群中各个节点的数据共享。为了解决这个问题，在 Swarm 集群中更常用的一种方式是，使用 NFS（网络文件系统）来实现数据的共享与持久化。

NFS（网络文件系统）允许计算机之间通过 TCP/IP 网络共享资源。在 NFS 应用中，NFS 客户端可以透明地读写远端 NFS 服务器上的文件，就像访问本地文件一样。

图 7-10 说明了利用 NFS 实现 Docker Swarm 集群数据持久化的工作原理。其中，NFS 可以被看成是 NFS 的服务器端，而 Docker 节点（master 节点、node1 节点、node2 节点）则可以被看成是 NFS 的客户端。因此，整个系统是 Client-Server 结构。

为了方便进行测试，可以将 master 节点作为 NFS Server。但在实际的环境中，一般可以单独搭建一个节点作为 NFS Server。

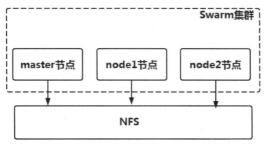

图 7-10

下面先在 master、node1 和 node2 节点上部署 NFS 服务器端和客户端,然后验证 NFS 共享目录在 Swarm 集群中的数据共享功能。

1. 在 Swarm 集群中部署 NFS

(1)执行以下命令在所有节点安装并启动 NFS。

```
yum -y install nfs-utils
systemctl enable nfs
systemctl start nfs
```

(2)在 master 节点上编辑 "/etc/exports" 文件,输入以下配置信息。该节点将作为 NFS 的服务器端。

```
/nfs *(rw,sync,no_root_squash)
```

其中的参数说明如下。

- /nfs:NFS 共享的目录。
- *:表示所有网段可以访问主机网段。
- rw:可读写权限。如果是只读权限,则是 "ro"。
- sync:数据传输采用同步方式。同步方式可以保障数据的安全,但传输速度慢;如果采用异步方式,则这里是 "async"。异步方式的数据传输效率高,但安全性差。
- no_root_squash:NFS 共享目录的属性。如果是 root 用户,则他对于这个目录就具有 root 的权限。

(3)在 master 节点上创建 "/nfs" 目录。

```
mkdir /nfs
```

(4)重启 master 节点上的 NFS 服务。

```
systemctl restart nfs
```

(5)在 node1 节点上启动 NFS 客户端。

```
systemctl start rpcbind
```

(6)执行以下命令在 node1 节点上挂载 NFS 目录进行一个简单的测试。

```
mkdir /node1-nfs
mount -t nfs 192.168.79.11:/nfs /node1-nfs
```

这里先在 node1 节点上创建了一个本地目录"/node1-nfs",然后使用"mount"命令将 NFS Server 共享的目录"/nfs"挂载到了 node1 节点的"/node1-nfs"目录下。

(7)在 node1 节点的"/node1-nfs"目录下随机生成一个文件,该文件将自动同步到 master 节点的"/nfs"目录下。

(8)执行以下命令卸载挂载的 NFS 目录。

```
umount /node1-nfs/
```

(9)启动 node2 节点上的 NFS 客户端。

```
systemctl start rpcbind
```

(10)在 master 节点上执行以下命令在 Docker Swarm 集群中使用 NFS 共享目录。

```
docker service create \
--replicas 3 \
--name myswarmdemo \
-p 7788:80 \
--mount
'type=volume,src=my-nfs-vol,dst=/usr/share/nginx/html,volume-driver=local,vo
lume-nocopy=true,volume-opt=type=nfs,volume-opt=device=192.168.79.11:/nfs,"v
olume-opt=o=addr=192.168.79.11,vers=4,soft,timeo=180,bg,tcp,rw"' \
collenzhao/swarm_nginx_demo:v1
```

其中重要的参数说明如下。

- type=volume:数据存储的类型。
- src=my-nfs-vol:数据卷的名称。
- dst=/usr/share/nginx/html:挂载到容器中的目录。
- volume-opt=type=nfs:数据卷的类型。
- volume-opt=device=192.168.79.11:/nfs:挂载的 NFS 目录。
- volume-opt=o=addr=192.168.79.11:NFS 服务器的地址。

2. 验证 NFS 共享目录在 Swarm 集群中的数据共享功能

(1)在 master 节点上进入 NFS 共享目录下。

```
cd /nfs/
```

(2)创建"a.html"文件,并输入以下代码:

```
<h1>NFS Demo Page</h1>
```

 这时"a.html"将自动同步到运行该应用的所有容器中的"/usr/share/nginx/html"目录下。

(3)访问任意一个节点的"a.html"文件,如图 7-11 所示。

图 7-11

(4)以 node1 节点为例,查看挂载的数据卷目录。

```
docker volume inspect my-nfs-vol
```

输出的信息如下:

```
[
    {
        "Driver": "local",
        "Labels": {},
        "Mountpoint": "/var/lib/docker/volumes/my-nfs-vol/_data",
        "Name": "my-nfs-vol",
        "Options": {
            "device": "192.168.79.11:/nfs",
            "o": "addr=192.168.79.11,vers=4,soft,timeo=180,bg,tcp,rw",
            "type": "nfs"
        },
        "Scope": "local"
    }
]
```

(5)进入 node1 节点的"/var/lib/docker/volumes/my-nfs-vol/_data"目录下,便可以看到从 NFS 服务器同步过来的"a.html"文件。

```
cd /var/lib/docker/volumes/my-nfs-vol/_data
ls
```

输出的信息如下:

```
a.html  a.txt
```

7.5 Docker Swarm 集群的负载均衡

在 7.3.4 节中提到，Swarm 集群本身提供了负载均衡功能。这里先对 Swarm 集群的负载均衡功能进行简单的测试，并以此为基础来介绍 Swarm 集群的负载均衡。

7.5.1 【实战】测试 Docker Swarm 集群的负载均衡

（1）在 Swarm 集群中部署一个简单的 Nginx 服务，并指定副本数为 3。

```
docker service create --name web --replicas 3 -p 80:80 nginx
```

（2）查看部署的 Web 服务运行的节点信息。这里可以看到，Web 服务分别运行在 node1、node2 和 master 节点上了。

```
docker service ps web
```

输出的信息如下：

```
NAME    IMAGE         NODE    DESIRED STATE   CURRENT STATE
web.1   nginx:latest  node1   Running         Running 5 seconds ago
web.2   nginx:latest  node2   Running         Running 5 seconds ago
web.3   nginx:latest  master  Running         Running 7 seconds ago
```

（3）在 master 节点上查看容器的 ID。

```
docker ps --format "table {{.ID}}\t{{.Names}}"
```

输出的信息如下：

```
CONTAINER ID          NAMES
2c8c47cc81a8          web.3.fy8xdgackirpinc0ucmzy67hd
```

（4）进入 master 节点上容器的内部。

```
docker exec -it 2c8c47cc81a8 /bin/bash
```

（5）执行以下命令修改"/usr/share/nginx/html/index.html"的内容。

```
echo "master" > /usr/share/nginx/html/index.html
```

（6）在 node1 节点上完成相同的操作，执行以下命令修改"/usr/share/nginx/html/index.html"的内容。

```
echo "node1" > /usr/share/nginx/html/index.html
```

（7）在 node2 节点上完成相同的操作，执行以下命令修改"/usr/share/nginx/html/index.html"的内容。

```
echo "node2" > /usr/share/nginx/html/index.html
```

（8）在任意节点上，执行以下测试脚本测试负载均衡的效果，如图 7-12 所示。

```
for i in {1..10};
  do curl 192.168.79.11;
done
```

```
[root@node1 ~]# for i in {1..10};
>                do curl 192.168.79.11;
>                done
node1
node2
master
node1
node2
master
node1
node2
master
node1
[root@node1 ~]#
```

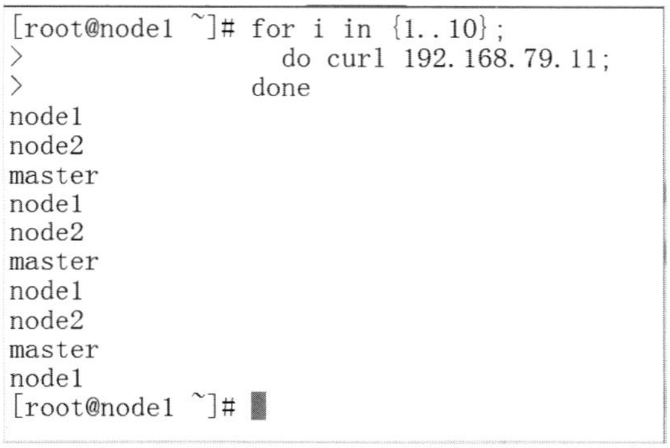

图 7-12

从图 7-12 中可以看到，共请求 Nginx 的 index.html 网页 10 次。第 1 次返回"node1"，第 2 次返回"node2"，第 3 次返回"master"，以此类推。这说明在 Swarm 集群的内部采用轮询的方式进行负载均衡。

7.5.2　选择 Docker Swarm 集群的负载均衡模式

由于 Docker 内置了 DNS 服务，因此 Swarm 集群可以自动为集群中的每个服务分配 DNS 记录。在默认情况下，Swarm 集群采用轮询的方式，根据服务的 DNS 名称给集群内的服务分发请求。而 Swarm 集群在具体实现负载均衡时有两种不同的模式：DNSRR 模式和 VIP 模式。

通过以下命令可以确定部署在 Swarm 集群中的应用的负载均衡模式。这里以 7.5.1 节部署的 Web 应用为例。

```
docker service inspect web | grep Mode
```

输出的信息如下：

```
"Mode": {
   "Mode": "vip",
        "PublishMode": "ingress"
   "Mode": "vip",
```

```
            "PublishMode": "ingress"
        "PublishMode": "ingress"
```

可以看到，Docker Swarm 集群的负载均衡模式默认为 VIP 模式。

1. DNSRR 模式

在 DNSRR 模式下，Docker 内置的 DNS Server 通过解析每个容器内的 IP 地址来实现负载均衡。该模式存在如下问题：如果应用对 DNS 主机名和 IP 地址的映射进行了缓存，则在映射更改后应用会发生超时；如果域名解析记录在 DNS 服务器中的存留时间不为零，则会导致在解析 DNS 信息时发生延迟。

图 7-13 说明了 DNSRR 模式的工作机制。

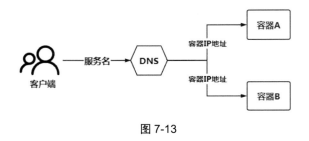

图 7-13

DNSRR 模式不支持容器端口对外映射。使用这种模式可以通过以下语句完成：
```
docker service create --name web1 --endpoint-mode dnsrr --replicas 3 nginx
```
要查看 web1 的负载均衡模式，则执行以下语句：
```
docker service inspect web1 | grep Mode
```
输出的信息如下：
```
"Mode": {
    "Mode": "dnsrr"
    ...
```

2. VIP 模式

VIP 模式解决了 DNSRR 模式的问题，它是 Swarm 集群负载均衡的默认模式，也是 Docker Swarm 推荐的模式。

在 VIP 模式中，每个 Service 的服务都有一个虚拟的 IP 地址（称为 VIP）。该 VIP 地址映射到与该服务关联的多个容器。在这种情况下，即使与该服务关联的容器重新启动了，或容器的 IP 地址发生了变化，与服务关联的服务 VIP 地址也不会发生改变。图 7-14 展示了 VIP 模式的工作机制。

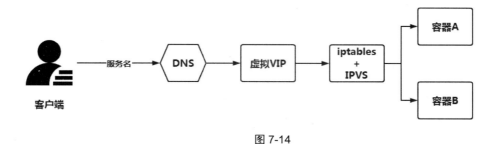

图 7-14

第 8 章
在 Docker 中实现持续集成与持续部署

持续集成（CI/CD）是一种软件开发的经验总结。它用于帮助开发团队和交付团队实现频繁且快速的集成，以及测试他们的工作成果，以尽可能快地发现项目开发和交付工程中的错误。越频繁、越早的项目集成与项目交付，则意味着问题会被越早发现。因此，通过持续集成（CI/CD）可以及时发现和解决代码故障，提高代码质量，减少故障处理成本等。

8.1 什么是持续集成与持续部署（CI/CD）

CI/CD 是一种通过在应用的开发阶段引入自动化来频繁向客户交付应用的方法。CI/CD 包含以下 3 个概念：

- CI（Continuous Integration）：持续集成。
- CD（Continuous Delivery）：持续交付。
- CD（Continuous Deployment）：持续部署。

目前持续集成的生态越来越完善，工具也越来越多，有开源版的也有商业版的。常见持续集成工具如下。

- Jenkins：一个开源的、使用最多的持续集成工具。
- ThoughtWorks GO：ThoughtWorks 公司的一款持续集成和发布的工具，采用 Java 语言开发。

- Bamboo：由澳大利亚公司 Atlassian 开发、开源的持续集成工具。
- Gitlab CI：GitLab 内置的持续集成工具，与 Gitlab 紧密集成。
- Buildbot：基于 Python 语言的、开源的持续集成工具。

下面重点以 Jenkins 为例进行介绍。

8.2 Jenkins 简介与部署

Jenkins 用于监控持续重复的工作，旨在提供一个开放易用的软件平台，使软件项目可以进行持续集成。

其主要功能如下。

（1）实现软件的持续发布与集成。

（2）监控外部调用执行的情况。

8.2.1 Docker 与 Jenkins 集成的体系架构

Docker 作为开源的应用容器引擎，允许开发者打包应用及依赖到 Docker 镜像中，然后发布到任何流行的 Linux 宿主机上，从而实现虚拟化的功能，

而 Jenkins 可以非常方便地实现应用的自动打包与自动部署。

Docker 与 Jenkins 并没有直接的关系。但可以将二者结合起来，从而实现 CI/CD。

图 8-1 展示了 Docker 与 Jenkins 集成实现 CI/CD 的体系架构。

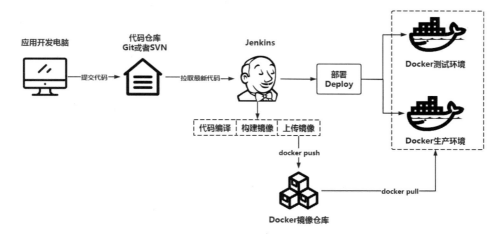

图 8-1

8.2.2 【实战】部署 Jenkins

下面以 master 节点为例来演示如何部署 Jenkins。由于 Jenkins 是基于 Java 语言开发的，因此需要先安装 JDK 环境。在 4.4.4 节中已经在 master 节点上安装了 jdk-8u181-linux-x64.tar.gz，这里可以直接部署 Jenkins。

（1）登录 Jenkins 的官方网站下载 "jenkins.war" 文件，如图 8-2 所示。

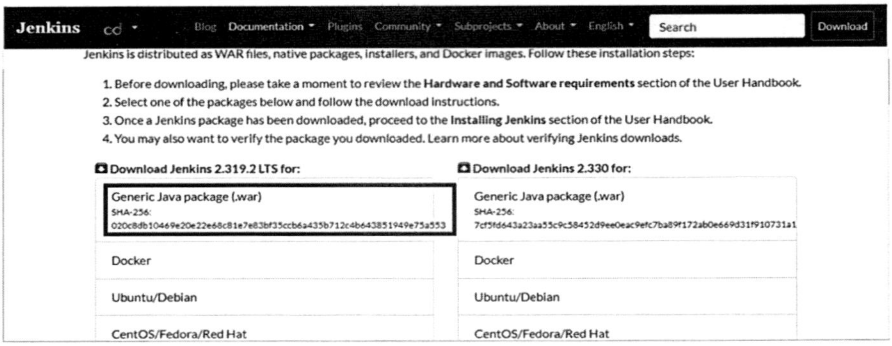

图 8-2

（2）执行以下命令安装 Jenkins。

```
java -jar jenkins.war
```

成功安装 Jenkins 后会输出以下信息：

```
*************************************************************

Jenkins initial setup is required. An admin user has been created and a password generated.
Please use the following password to proceed to installation:

8eb107d3ce39407ebd7db169dc5a3fc4

This may also be found at: /root/.jenkins/secrets/initialAdminPassword

*************************************************************
```

（3）打开浏览器访问 master 节点的 8080 端口，即可打开"解锁 Jenkins"的页面，如图 8-3 所示。

（4）输入第（2）步中输出的管理员密码解锁 Jenkins，并单击"继续"按钮。

（5）在"自定义 Jenkins"页面中单击"安装推荐的插件"链接继续安装 Jenkins。

图 8-3

在安装推荐的插件时需要连接网络，且会耗费较长的时间。

（6）在"创建第一个管理员用户"页面中创建一个管理员，以方便后续的操作。

（7）在"实例配置"页面中直接单击"保存并完成"按钮，之后会展示 Jenkins 的主页，如图 8-4 所示。

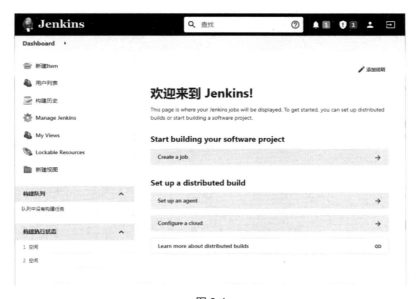

图 8-4

8.2.3 【实战】使用 Jenkins 部署第一个应用

接下来使用 Jenkins 来实现项目的持续集成与持续部署。这里以一个 Java 应用程序为例，用 GitHub 作为代码仓库来介绍 Jenkins 的使用。

（1）在 GitHub 上创建一个项目空间，如 "javacode"。

（2）在 master 节点上安装 Git 客户端。

```
yum install git -y
```

（3）在 master 节点上创建 "javacode" 目录，并初始化 Git 环境。

```
mkdir javacode
cd javacode
git init
```

（3）创建源代码目录 "src"。

```
mkdir src
```

（4）在 src 目录下开发第 1 个版本的 "HelloWorld.java" 程序，代码如下：

```
public class HelloWorld{

  public static void main(String[] args){
    System.out.println("Hello World:Version 1");
  }

}
```

这里输出的信息为 "Hello World:Version 1"，表示是应用的第 1 个版本。

（5）将 src 目录加入 Git。

```
git add src/
```

（6）配置 Git 的用户信息。

```
git config --global user.name "你的用户名"
git config --global user.email "你的邮箱"
```

（7）查看当前 Git 管理的代码状态。

```
git status ./
```

输出的信息如下：

```
# On branch master
#
# Initial commit
#
# Changes to be committed:
#   (use "git rm --cached <file>..." to unstage)
#
# new file:    src/HelloWorld.java
```

（8）提交代码到 GitHub。

```
git commit -m "Version 1"
git branch -M main
git remote add origin git@github.com:collenzhao/javacode.git
git push -u origin main
```

在提交成功后会输出以下信息：

```
Counting objects: 4, done.
Compressing objects: 100% (2/2), done.
Writing objects: 100% (4/4), 365 bytes | 0 bytes/s, done.
Total 4 (delta 0), reused 0 (delta 0)
To git@github.com:collenzhao/javacode.git
 * [new branch]      main -> main
Branch main set up to track remote branch main from origin.
```

（9）在 Jenkins 主页中，选择左上角的"新建 Item"来创建一个项目任务。

（10）输入项目任务名称"javacode"，并选择"Freestyle project"来创建一个自由风格的项目，之后单击"确定"按钮，如图 8-5 所示。

图 8-5

（11）进入"javacode"项目配置界面中的"General"选项卡中，进行如下的配置：

- 在"描述"对话框中输入"My Java Code Demo with Jenkins"。
- 勾选"GitHub 项目"复选框。
- 在"项目 URL"文本框中输入第（1）步中在 GitHub 上创建的"javacode"项目空间的 URL。

（12）在"源码管理"选项卡中进行如下的配置：

- 勾选"Git"单选按钮。
- 在"Repository URL"文本框中输入第（1）步中在 GitHub 上创建的"javacode"项目空间的 URL。
- 单击"Credentials"旁边的"添加"按钮，并使用在 GitHub 上注册的信息添加一个新的 Jenkins 账号信息。
- 在"Branches to build"文本框中设置"指定分支（为空时代表 any）"为"*/main"。

（13）在"构建"的选项卡中，单击"增加构建步骤"并选择"执行 shell"，然后在"命令"对话框中输入以下命令：

```
cd src
javac HelloWorld.java
java HelloWorld
```

（14）单击"保存"按钮。

（15）单击"立即构建"按钮，如图 8-6 所示。

（16）在项目构建完成后，通过 Build History 来查看构建的结果，如图 8-7 所示。

图 8-6

图 8-7

（17）从"控制台输出"的选项中可以看到如图 8-8 所示的输出内容。

图 8-8

（18）开发应用的第 2 个版本——修改"HelloWorld.java"的代码如下。

```
public class HelloWorld{

  public static void main(String[] args){
    System.out.println("Hello World:Version 2");
  }
}
```

这里将输出的信息改为了"Hello World:Version 2"，表示是应用的第 2 个版本。

（19）将代码提交到 GitHub。

```
git commit src/HelloWorld.java -m "Version 2"
git push -u origin main
```

（20）在 Jenkins 中重新单击"立即构建"按钮，Jenkins 会从 GitHub 拉取最新的代码并执行。这时会在控制台中看到"Hello World:Version 2"的信息。

在以上示例中，利用 Jenkins 实现了项目的持续集成与持续部署。每次使用 Jenkins 时，它都从代码仓库中拉取最新的应用代码，并执行预先设置好的脚本代码，从而完成 CI/CD。

8.3 基于 Jenkins 实现 Docker 应用的持续集成与持续部署

在 Docker 应用开发中最常见就是开发 Dockerfile 文件，可以使用代码仓库来管理它。

而在企业私有的开发环境中是无法访问公有代码仓库（如 GitHub）的。这时可以搭建私有代码仓库（如 SVN）来实现代码的管理。Jenkins 支持丰富的代码仓库管理方式。

8.3.1 【实战】准备私有代码仓库 SVN

开发私有代码仓库所需要的 SVN 软件分为服务器端和客户端。

- SVN 的服务器端：VisualSVN-Server-4.3.4-x64.msi。
- SVN 的客户端：TortoiseSVN-1.14.1.29085-x64-svn-1.14.1.msi。

> 这里将 SVN 的服务器端和客户端都安装在 Windows 上。

由于在 Windows 上安装 SVN 代码仓库比较简单，所以下面只对主要步骤的参数进行说明。

（1）安装代码仓库 SVN 的服务器端，如图 8-9 所示。

图 8-9

其中重要的参数说明如下。

- Repositories：指定代码仓库对应的目录。
- Server Port：指定代码仓库服务器端的端口号。

（2）启动"VisualSVN Server"进入主界面后，创建一个新的代码仓库，如图 8-10 所示。

（3）在"Repository Type"界面中单击"下一步"按钮，然后在"Repository Name"界面中输入仓库的名称，如"DockerfileStore"。单击两次"下一步"按钮后，单击"完成"和"Finish"按钮完成代码仓库的创建。

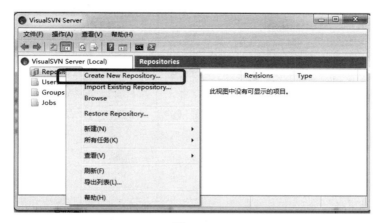

图 8-10

（4）在 Windows 上安装 SVN 的客户端，直接单击"下一步"按钮进行安装即可。

（5）在任意 Windows 目录（如"D:\temp\demo"）上单击鼠标右键，在弹出的菜单中选择"SVN CheckOut"命令，弹出如图 8-11 所示对话框。

图 8-11

> 在执行 CheckOut 动作时，需要输入 SVN 代码仓库的用户名和密码。用户名和密码需要在 SVN 服务器上事先创建。

8.3.2 开发 Dockerfile 文件

在"D:\temp\demo"目录下准备一个 Dockerfile 文件,内容如下:

```
FROM nginx
RUN echo '<h1>Demo Version 1</h1>' > /usr/share/nginx/html/index.html
```

这里基于 Nginx 的基础镜像创建了一个自己的镜像,并修改了 Nginx 的首页。在页面中输出了"Demo Version 1"信息,表示这是应用的第 1 个版本。

将 Dockerfile 文件提交到 SVN 代码仓库的服务器中,如图 8-12 所示。

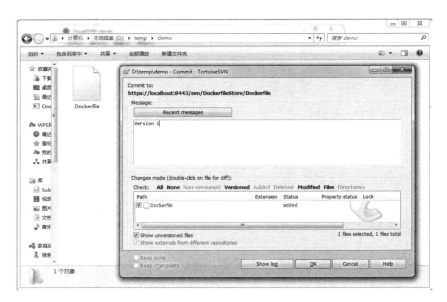

图 8-12

8.3.3 【实战】集成 Jenkins 和 Docker

接下来就可以集成 Jenkins 和 Docker 来实现持续集成与持续部署了。

 由于这里使用的是私有代码仓库 SVN,所以需要在 Jenkins 的"插件管理"页面中安装 SVN 的插件,如 SVN 1.4 Compatibility。

(1)在 Jenkins 中创建一个名为"Freestyle project"的项目空间。

(2)在"源码管理"的选项卡中进行配置,如图 8-13 所示。

- 勾选"Subversion"单选按钮，表示代码由 SVN 进行管理。
- 在"Repository URL"文本框中输入 SVN 的代码仓库地址。
- 单击"Credentials"旁边的"添加"按钮，然后利用 SVN 用户信息添加一个新的 Jenkins 账号信息。

图 8-13

（3）在"构建"选项卡中，单击"增加构建步骤"并选择"执行 shell"，并在"命令"对话框中输入以下命令：

```
docker build -t myjenkinsdemo .
docker run -d -p 7788:80 myjenkinsdemo
```

（4）单击"保存"按钮，并单击"立即构建"链接。

（5）在项目成功构建后，打开浏览器访问 master 节点的 7788 端口即可打开部署的页面，如图 8-14 所示。

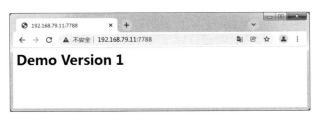

图 8-14

（6）修改第 1 个版本的 Dockerfile 文件，完成第 2 个版本的开发，并提交到 SVN 的代码仓库中。

（7）在 Jenkins 中重新构建项目，然后刷新网页即可看到第 2 个版本的内容。

至此完成了 Jenkins 与 Docker 的集成，从而可以实现应用的 CI/CD。在实际的项目开发中，可能会涉及大量的程序代码，以及依赖的库和环境，但使用 Jenkins 实现 CI/CD 的过程都是一样的。

第 9 章

基于 Consul 实现 Docker 的服务注册与发现

在 7.3.3 节中,给 Docker Swarm 集群添加了 HAProxy 作为其外部负载均衡工具。为了实现这个目的,需要将 Docker Swarm 集群中每个节点上运行的容器地址写入 HAProxy 的配置文件中,这样才能通过 HAProxy 实现请求的路由转发,从而实现负载均衡。

但是,如果 Docker Swarm 集群中的节点启用了新的容器,那 HAProxy 如何才能感知这些新启用的容器呢?另外,由于某些原因导致某些节点上运行的容器异常宕机,那么 HAProxy 又该如何将这些容器从 haproxy.cfg 配置文件中移除呢?

要解决这些问题,都需要实现 Docker 的服务注册与发现功能。

9.1 服务的注册与发现

Docker 提供的 DNS 服务具有服务的注册与发现功能,但服务的注册与发现功能并不是 Docker 特有的。要实现服务的注册与发现则必须依赖注册中心。

9.1.1 什么是服务的注册与发现

服务的注册与发现,包含以下两种功能。

- 服务注册:将服务提供者提供的服务信息(如这个服务的 IP 地址和端口号)保存到一个公共的存储位置。

- 服务发现：服务的使用者从这个公共的存储位置及时发现新注册的服务，并且完成服务的调用。

> 服务的注册和删除都可以自动实现，不需要人为干预。

9.1.2 为什么需要服务的注册与发现

为什么需要服务的注册与发现机制呢？这主要是由于软件架构的不断发展所导致的。软件架构从传统的单体架构，发展到 SOA 架构，再到目前的微服务架构，变得越来越复杂，这就需要一种新型的架构思想来解决软件架构中不同模块的集成和协调。而服务的注册与发现便是这种新型架构思想的核心。

1. 传统的单体架构

图 9-1 展示了传统的单体架构。开发人员会将 Web 系统中的所有功能模块打包到一个 war 压缩包中，然后发布到一个 Web 应用服务器上，并且，所有的功能模块使用同一个数据库来存储数据。单体架构通过 Web UI 或 API 的方式对外提供服务。

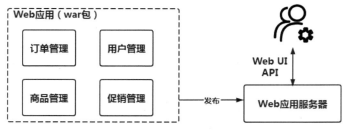

图 9-1

单体架构具有以下优点。

- 开发效率高：模块之间的交互采用本地调用的方式进行，节约了开发的成本。
- 容易测试和部署：所有的软件模块都可以被集成到一个 IDE 开发环境中，所有工作均可在本地完成，大大降低了测试和部署的工作量与成本。

单体架构适用于简单且小型的软件系统。随着软件架构的不断复杂，单体架构展现出了以下不足。

- 可维护性差：所有模块都部署在一起，随着模块的不断增多和复杂度越来越高，系统将变得不可维护。

- 版本升级慢：即使修改了一个很小的代码，也需要将整个应用进行全部编译和重新部署。
- 系统的扩展性差：由于所有模块都部署在一起，所以模块无法实现按需伸缩。

2. SOA 架构

为了解决单体架构的问题，SOA 架构逐步发展起来。图 9-2 展示了从单体架构到 SOA 架构的转换过程。

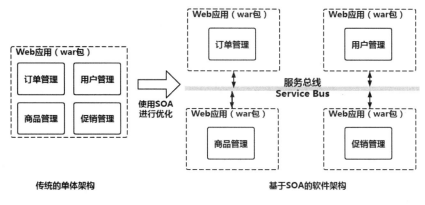

图 9-2

SOA（Service-Oriented Architecture，面向服务的架构）的核心思想是：将系统拆分为不同的服务单元，并通过服务总线 Service Bus 进行相互的通信与调用。这里的服务单元将完成一个特定功能（如图 9-2 中的订单管理、用户管理、商品管理和促销管理）。从整体上看，这些服务单元集成在一起组成完整的应用。

SOA 架构中有两个重要的角色：服务提供者（Provider）和服务使用者（Consumer）。同一个服务单元既可以是服务提供者，也可以是服务使用者。

SOA 架构具有以下优点。

- 更容易维护：服务单元之间是松散耦合关系。改变某一个业务单元时，其他业务单元不受影响，只需要在服务总线上调整服务单元的调用流程或进行某些修改即可，整个应用更容易维护。
- 更高的可用性：服务提供者和服务使用者都无须了解对方的具体实现细节，从而提高了系统的可用性。
- 更好的扩展性：服务提供者和服务使用者彼此独立，都可以根据业务的需要单独实现服务单元的伸缩，以满足新的服务需求。

SOA 架构也有不足。从图 9-2 可以看出，SOA 架构强烈依赖服务总线 Service Bus 来实现整个系统的集成。这就增加了服务单元之间远程通信的成本，增加了开发接口的工作量。

3. 微服务架构

要解决 SOA 架构存在的不足，便产生微服务架构。图 9-3 展示了一个典型的微服务架构。

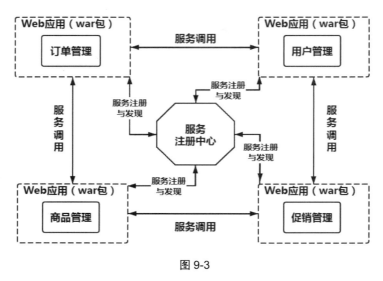

图 9-3

在微服务架构中，系统被拆分成多个独立运行的服务单元。它们将各自提供的服务存储在服务注册中心中（即完成服务的注册）。

- 服务注册中心通过服务发现功能将这些注册的服务暴露给外部，让其他的服务单元能够在服务注册中心中发现所需要的服务。
- 服务单元在发现所需要的服务后，先从服务注册中心获取该服务的相关信息（如服务提供者的主机地址和端口等），然后调用该服务完成相应的业务需求。

微服务架构具有以下优点。

- 开发的敏捷性：每个开发团队相互独立，从而可以更独立、更快速地工作，缩短了开发周期。
- 系统的扩展性：由于系统是由独立的服务单元组成的，因此可以独立扩展各个服务单元以满足应用功能的需求。
- 部署的灵活性：微服务架构支持 CI/CD，从而降低成本。

- 技术的自由性：各个服务单元可以采用不同的技术独立实现，只需要遵循统一的标准即可。
- 代码的重用性：将复杂的系统划分为小型且有明确定义的服务单元，提高了代码的重用性，降低了维护的成本。
- 服务的独立性：如果某个服务单元出现了故障，不会影响其他服务单元的使用。

9.1.3 常见的服务注册中心

微服务架构的核心是服务的注册与发现，因此在架构中需要一个公共的服务注册中心来实现这项功能。常见的服务注册中心有以下几种。

- ZooKeeper：属于 Hadoop 生态体系，是一个分布式的、开放源码的分布式应用协调服务。其提供的功能包括数据同步、节点选举、心跳监听等。
- Consul：Google 基于 Go 语言开发的一个服务发现和配置管理的软件。其提供的功能包括服务注册与发现、分布一致性协议实现、健康检查及键值存储等。
- Eureka：Netflix 开源的服务发现框架，其提供了完整的服务注册与发现功能。它是 Spring Cloud 体系中最核心的组件。

下面将重点介绍如何使用 Consul 实现 Docker 服务的注册与发现。

9.2 注册中心 Consul 的基本使用

在具体使用 Consul 前，先来了解一下 Consul 的调用过程。在服务注册与发现的体系架构中有两种角色：服务生产者和服务消费者。服务生产者提供具体的服务，而服务消费者消费服务生产者的服务，如图 9-4 所示。

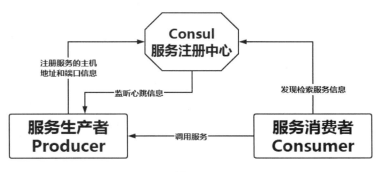

图 9-4

9.2.1 Consul 的安装与启动

下面演示如何在 master 主机上安装和启动 Consul。

（1）登录 Consul 的官方网站，单击页面上的"Linux"链接，在下方选择下载 Consul，如图 9-5 所示。

图 9-5

（2）解压缩 Consul 的压缩包。

```
unzip consul_1.4.0_linux_amd64.zip
```

（3）将解压缩后的二进制文件"consul"复制到"/usr/local/bin"目录下。

```
cp consul /usr/local/bin
```

（4）用以下命令验证 Consul 的版本信息。

```
consul version
```

（5）启动 Consul。

```
consul agent -dev -ui -client=0.0.0.0
```

其中的参数说明如下。

- -dev：以开发模式启动 Consul。
- -ui：启动 Consul 的 Web 界面。
- -client：指定允许连接的客户端地址。0.0.0.0 表示所有 IP 地址的客户端都可以连接。

（6）通过浏览器访问 master 节点的 8500 端口，打开 Consul 的 Web 界面，如图 9-6 所示。

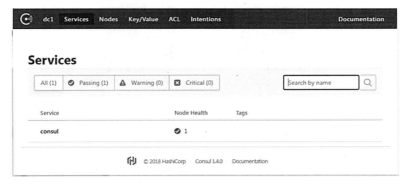

图 9-6

在 Consul 启动成功后，主要通过两种方式将服务信息注册到 Consul 中。这两种方式是：使用 JSON 文件注册和使用 API 注册。

9.2.2 【实战】使用 JSON 文件在 Consul 中注册服务

下面将编写一个 JSON 文件来将一个 Student 的信息注册到 Consul 中，并使用 curl 命令来获取注册的服务信息。

（1）创建 "/etc/consul.d" 目录，用于保存编写的 JSON 文件。

```
mkdir /etc/consul.d
```

（2）执行以下命令在 "/etc/consul.d" 目录下创建一个 "student.json" 文件。

```
echo '{"service":{"name":"student","tags":["student"],"port":80}}' \
>/etc/consul.d/student.json
```

> 这里注册的服务是 student，标签是 student，服务注册的端口是 80。但是，这里注册的 student 服务并不提供任何的实际功能。

（3）使用以下命令重新启动 Consul。

```
consul agent -dev -client 0.0.0.0 -config-dir=/etc/consul.d
```

（4）刷新 Consul Web 界面，查看注册的 student 信息，如图 9-7 所示。

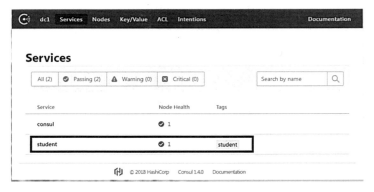

图 9-7

（5）使用 curl 命令获取注册的 student 服务信息。

```
curl http://localhost:8500/v1/catalog/service/student
```

输出的信息如下：

```
[
    {
        "ID": "5bb26463-eb0a-d528-d1e0-cc3f55bf6fff",
        "Node": "master",
        "Address": "127.0.0.1",
        "Datacenter": "dc1",
        "TaggedAddresses": {
            "lan": "127.0.0.1",
            "wan": "127.0.0.1"
        },
        "NodeMeta": {
            "consul-network-segment": ""
        },
        "ServiceKind": "",
        "ServiceID": "student",
        "ServiceName": "student",
        "ServiceTags": [
            "student"
        ],
        "ServiceAddress": "",
        "ServiceWeights": {
            "Passing": 1,
            "Warning": 1
        },
        "ServiceMeta": {},
        "ServicePort": 80,
        "ServiceEnableTagOverride": false,
```

```
        "ServiceProxyDestination": "",
        "ServiceProxy": {},
        "ServiceConnect": {},
        "CreateIndex": 10,
        "ModifyIndex": 10
    }
]
```

9.2.3 【实战】使用 API 在 Consul 中注册服务

下面通过 Consul 的 Java API 将 Nginx 的首页注册到 Consul 中。

(1) 启动一个容器来运行 Nginx 服务。

```
docker run -d -p 1234:80 nginx
```

(2) 使用 Java IDE 开发工具 Eclipse 创建一个 Maven 工程,并在 pom.xml 文件中添加以下依赖信息。

```xml
<dependency>
    <groupId>com.orbitz.consul</groupId>
    <artifactId>consul-client</artifactId>
    <version>0.12.3</version>
</dependency>
```

(3) 通过以下 Java 代码完成 Nginx 首页的注册。

```java
@Test
public void registerService() {
    // 创建 Consul 的客户端
    Consul consul = Consul.builder()
                    .withHostAndPort
                    (HostAndPort.fromString("192.168.79.11:8500"))
                    .build();
    AgentClient client = consul.agentClient();

    // 设置健康检查策略,并将一个 Nginx 首页注册到 Consul 中
    ImmutableRegCheck check = ImmutableRegCheck
                    .builder()
                    .interval("5s")
                    .http("http://192.168.79.11:1234/")
                    .build();

    //设置服务的 ID、标签信息,并添加监控检查策略
    ImmutableRegistration.Builder builder =
                    ImmutableRegistration.builder();
```

```
builder.id("nginx_id")
    .name("nginx")
    .addTags("v1")
    .addChecks(check);

//完成注册
client.register(builder.build());
System.out.println("服务注册完成");
}
```

（4）在 registerService()方法成功执行后刷新 Consul 的 Web 界面，将看到注册的 Nginx 信息，如图 9-8 所示。

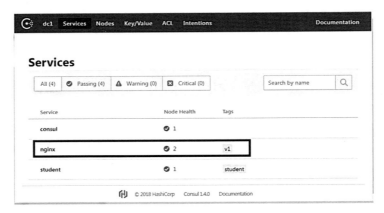

图 9-8

（5）以下 Java 代码将获取在 Consul 中注册的 Nginx 信息。

```
@Test
public void getService() {
    // 创建 Consul 的客户端
    Consul consul = Consul.builder()
                    .withHostAndPort
                    (HostAndPort.fromString("192.168.79.11:8500"))
                    .build();
    HealthClient client = consul.healthClient();

    //输出注册的信息
    client.getHealthyServiceInstances("nginx")
        .getResponse().forEach((response) -> {
      System.out.println(response);
    });
}
```

（6）在 getService() 方法成功执行后，将输出以下信息。

```
ServiceHealth{node=Node{node=master, address=127.0.0.1,
taggedAddresses=TaggedAddresses{wan=127.0.0.1}}, service=Service{id=nginx_id,
service=nginx, tags=[v1], address=, port=0}, checks=[HealthCheck{node=master,
checkId=serfHealth, name=Serf Health Status, status=passing, notes=,
output=Agent alive and reachable, serviceId=, serviceName=},
HealthCheck{node=master, checkId=service:nginx_id, name=Service 'nginx' check,
status=passing, notes=, output=HTTP GET http://192.168.79.11:1234/: 200 OK
Output: <!DOCTYPE html>
<html>
<head>
<title>Welcome to nginx!</title>
<style>
html { color-scheme: light dark; }
body { width: 35em; margin: 0 auto;
font-family: Tahoma, Verdana, Arial, sans-serif; }
</style>
</head>
<body>
<h1>Welcome to nginx!</h1>
<p>If you see this page, the nginx web server is successfully installed and
working. Further configuration is required.</p>

<p>For online documentation and support please refer to
<a href="http://nginx.org/">nginx.org</a>.<br/>
Commercial support is available at
<a href="http://nginx.com/">nginx.com</a>.</p>

<p><em>Thank you for using nginx.</em></p>
</body>
</html>
, serviceId=nginx_id, serviceName=nginx}]}
```

（7）通过执行 curl 命令获取服务的注册信息。

```
curl http://localhost:8500/v1/catalog/service/nginx
```

输出的信息如下：

```
[
  {
    "ID": "5bb26463-eb0a-d528-d1e0-cc3f55bf6fff",
    "Node": "master",
    "Address": "127.0.0.1",
    "Datacenter": "dc1",
    "TaggedAddresses": {
```

```
            "lan": "127.0.0.1",
            "wan": "127.0.0.1"
        },
        "NodeMeta": {
            "consul-network-segment": ""
        },
        "ServiceKind": "",
        "ServiceID": "nginx_id",
        "ServiceName": "nginx",
        "ServiceTags": [
            "v1"
        ],
        "ServiceAddress": "",
        "ServiceWeights": {
            "Passing": 1,
            "Warning": 1
        },
        "ServiceMeta": {},
        "ServicePort": 0,
        "ServiceEnableTagOverride": false,
        "ServiceProxyDestination": "",
        "ServiceProxy": {},
        "ServiceConnect": {},
        "CreateIndex": 61,
        "ModifyIndex": 61
    }
]
```

9.3 集成 Consul 与 Docker

当有新的 Docker 容器被创建或有 Docker 容器被销毁时，可以通过 Consul 感知这个变化，从而动态修改 HAProxy 的配置以达到外部负载均衡的目的。

9.3.1 Docker 服务注册与发现的体系架构

图 9-9 展示了基于 Consul 实现 Docker 服务注册与发现的体系架构。

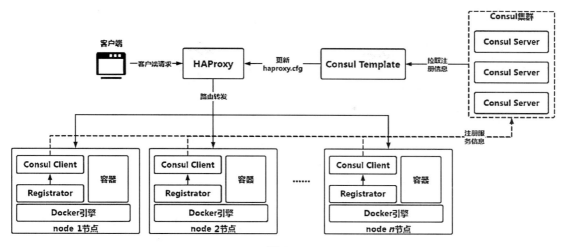

图 9-9

在这个体系架构中，除用到 Consul 外还用到以下两个工具。

- Consul-Template：模板工具，将作为 Docker 守护程序来运行。它用于实时查询 Consul 中注册的信息数据，并更新文件系统中任意数量的指定模板，然后根据这些模板生成对应的配置文件，并执行更新。
- Registrator：Docker 的镜像。它将作为 Consul 的客户端来运行，用于实时监控宿主机上容器的运行状态，并完成自动注册工作。目前 Registrator 支持的服务注册中心有 Consul、ETCD、SkyDNS2 等。

表 9-1 展示了每个节点上部署的组件。

表 9-1

节点名	IP 地址	已部署的服务
master	192.168.79.11	Docker、HAProxy、Consul、Consul-Template
node1	192.168.79.12	Docker、Registrator
node2	192.168.79.13	Docker、Registrator

9.3.2 【实战】使用 Registrator 镜像实现 Docker 服务的注册

（1）在 master 节点上重新启动 Consul。

```
consul agent -dev -ui -client=0.0.0.0
```

这时在 Consul 注册中心中只有一个 Consul 服务。

（2）在 node1 节点上启动 Registrator。

```
docker run -d \
--name=registrator \
-v /var/run/docker.sock:/tmp/docker.sock \
--restart=always \
gliderlabs/registrator:latest \
-ip=192.168.79.12 \
consul://192.168.79.11:8500
```

其中的参数说明如下。

- -ip：指定注册时使用的 IP 地址，一般是宿主机的 IP 地址。
- Consul：指定 Consul 注册中心的地址。

（3）在 node1 节点上随便启动一个新的容器，如运行一个 Nginx 服务的容器。

```
docker run -d -p 1234:80 nginx
```

（4）刷新 master 节点上的 Consul Web 界面，即可看到注册在 Consul 中的 Nginx 信息，如图 9-10 所示。

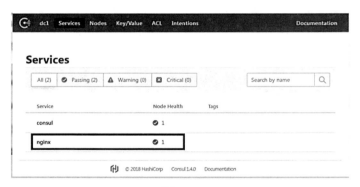

图 9-10

（5）在 node2 节点上启动 Registrator。

```
docker run -d \
--name=registrator \
-v /var/run/docker.sock:/tmp/docker.sock \
--restart=always \
gliderlabs/registrator:latest \
-ip=192.168.79.13 \
consul://192.168.79.11:8500
```

（6）在 node2 节点上随便启动一个新的容器，如运行一个 Tomcat 服务的容器。

```
docker run -d -p 8080:8080 tomcat
```

（7）刷新 master 节点上的 Consul Web 界面，即可看到在 Consul 中注册的 Nginx 和 Tomcat 信息，如图 9-11 所示。

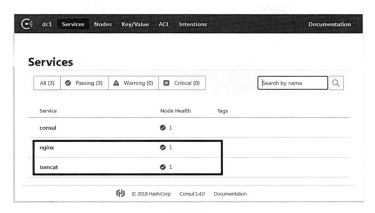

图 9-11

（8）单击服务名称"nginx"，进入 nginx 注册信息的详细页面，如图 9-12 所示。

图 9-12

从图 9-12 中可以看出，在 master 节点的 Consul 注册中心中注册了一个 Nginx 服务。并且该服务只有一个运行的示例，其 IP 地址是"192.168.79.12:1234"。

（9）在 node2 节点上再启动一个 Nginx 服务。

```
docker run -d -p 1234:80 nginx
```

（10）刷新 Consul 的 Web 界面，这时将看到 Nginx 服务有两个运行的实例，如图 9-13 所示。

图 9-13

9.3.3 【实战】使用 Consul-Template 实现 Docker 服务的发现

在完成 Docker 服务的注册后，便可以使用 Consul-Template 实时查询 Consul 中注册的信息，并更新 HAProxy 配置文件中的信息。

（1）在 master 节点上安装 HAProxy，并启动它。

```
yum install -y haproxy
systemctl start haproxy
systemctl enable haproxy
```

这里将 HAProxy 直接安装在宿主机上，也可以根据需要运行在容器中。

（2）在 master 节点上解压缩 Consul-Template 安装包。

```
unzip consul-template_0.19.3_linux_amd64.zip
```

（3）将解压缩后的"consul-template"文件复制到"/usr/bin/"目录下。

```
cp consul-template /usr/bin/
```

（4）在"/root/template/"目录下创建 Consul-Template 的模板文件 haproxy.ctmpl。

```
global
    daemon
defaults
    mode http
frontend http-in
    bind *:80
```

```
        default_backend servers
backend servers
    {{range service "nginx"}}
    server server{{ .Address }}-{{ .Port }} {{ .Address }}:{{ .Port }};
    {{ end }}
```

（5）执行以下命令进行本地测试，检查生成的 HAProxy 的配置文件 "haproxy.cfg"。

```
consul-template \
-consul-addr 192.168.79.11:8500 \
-template "/root/template/haproxy.ctmpl:/root/template/haproxy.cfg" \
-log-level=info
```

（6）查看 "/root/template" 目录下生成的 haproxy.cfg 文件的内容。

```
global
    daemon
defaults
    mode http
frontend http-in
    bind *:80
    default_backend servers
backend servers
    server server192.168.79.12-1234 192.168.79.12:1234;
    server server192.168.79.13-1234 192.168.79.13:1234;
```

可以看出，Consul-Template 实时查询了在 Consul 中注册的 Nginx 服务的信息，并根据预先的模板文件生成了 haproxy.cfg 配置文件，还在 "backend servers" 中自动生成了在 node1 和 node2 节点启动 Nginx 服务的信息。

（7）在 node1 节点上再启动一个 Nginx 服务。

```
docker run -d -p 2233:80 nginx
```

（8）Consul-Template 将重新根据 Consul 注册中心的信息来更新配置文件 haproxy.cfg，如下所示。

```
global
    daemon
defaults
    mode http
frontend http-in
    bind *:80
    default_backend servers
backend servers
    server server192.168.79.12-2233 192.168.79.12:2233;
    server server192.168.79.12-1234 192.168.79.12:1234;
server server192.168.79.13-1234 192.168.79.13:1234;
```

（9）在完成本地测试后，用生成的 HAProxy 模板文件去覆盖 HAProxy 的配置文件，并重新加载该配置文件。修改 consul-template 的命令如下：

```
consul-template \
-consul-addr 192.168.79.11:8500 \
-template "/root/template/haproxy.ctmpl:/etc/haproxy/haproxy.cfg:haproxy \
        -f /etc/haproxy/haproxy.cfg \
        -p /var/run/haproxy.pid -sf $(cat /var/run/haproxy.pid)" \
-log-level=info
```

> 上面的代码中带有深灰底色的部分的作用是：结束原来的 HAProxy 进程，重新加载配置文件，并重新启动 HAProxy。

（10）打开浏览器访问 master 节点的 80 端口，便可以打开 Nginx 的首页，如图 9-14 所示。

图 9-14

（11）刷新 Consul 的 Web 界面，可以看到 Nginx 服务有 3 个实例，如图 9-15 所示。

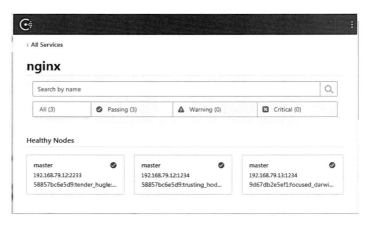

图 9-15

第 10 章
利用图形工具管理 Docker

Docker 提供了命令行工具来管理 Docker 的镜像和运行 Docker 的容器。我们也可以使用图形工具来管理 Docker。目前，主流的 Docker 图形工具有 Docker UI、Portainer 和 Shipyard。

10.1 单机环境中的 Docker 图形工具：Docker UI

Docker UI 是最简单的单机管理 Docker 的工具，适合初学者和小型 Docker 项目的管理。

Docker UI 具有以下的优点：

- 支持容器的管理。
- 运行的稳定性高。
- 可动态显示容器之间的关系图。

10.1.1 部署 Docker UI

下面在 master 节点上安装 Docker UI。

（1）在 Docker Hub 镜像仓库中搜索 "ui-for-docker"。

```
docker search ui-for-docker
```

（2）从镜像仓库拉取 Docker UI 的镜像。

```
docker pull uifd/ui-for-docker
```

（3）启动 Docker UI。

```
docker run -it -d --name docker-web -p 9000:9000 \
```

```
-v /var/run/docker.sock:/var/run/docker.sock \
docker.io/uifd/ui-for-docker
```

 这里需要挂载"/var/run/docker.sock"文件,它是 Docker 守护进程默认监听的 UNIX 域套接字文件。容器中的进程可以通过它与 Docker 守护进程进行通信。

(4)打开浏览器访问 master 节点的 9000 端口,即可打开 Docker 的 Web 界面,如图 10-1 所示。

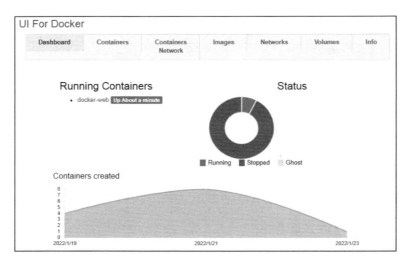

图 10-1

10.1.2 【实战】使用 Docker UI 管理镜像与容器

下面使用 Docker UI 部署一个 Nginx 容器,并在浏览器中访问 Nginx 的首页。

(1)在 Docker UI 的首页中的上方位置单击"Images"标签,进入镜像的管理界面,将看到 master 节点上所有已下载的镜像。

(2)在镜像页面中右边的"Filter"对话框中输入"nginx",过滤出所有与"nginx"相关的镜像,如图 10-2 所示。

(3)单击"docker.io/nginx:latest"左侧的 ID"sha256:605c77e624...",进入 Nginx 镜像的详细信息页面。

(4)单击"Start Container"按钮,在"Create And Start Container From Image"页面上配置启动容器的相关参数。

图 10-2

（5）选择"HostConfig options"标签页，添加一个"PortBindings"的端口映射，如图 10-3 所示。

图 10-3

（6）单击"Create"按钮创建容器。

（7）在 master 节点上查看容器信息。

```
docker ps --format "table {{.ID}}\t{{.Image}}\t{{.Ports}}"
```

输出的信息如下：

```
CONTAINER ID    IMAGE                         PORTS
9aae51834243    605c77e624dd                  0.0.0.0:1122->80/tcp
289abe37543f    docker.io/uifd/ui-for-docker  0.0.0.0:9000->9000/tcp
```

（8）打开浏览器访问宿主机的 1122 端口，即可打开 Nginx 的首页。

Docker UI 存在以下不足：
（1）没有登录验证。
（2）无法分配某个容器给某个用户。
（3）不支持多主机和 Swarm 集群。
（4）不支持控制台命令。

10.2 轻量级的 Docker 图形工具：Portainer

Portainer 是一个轻量级的图形工具。使用 Portainer 可以轻松管理不同的 Docker 环境。Portainer 的部署和使用都非常简单，它由一个可以运行在任何 Docker 引擎上的容器组成。Portainer 可以管理 Docker 的镜像、容器、数据卷和网络等。

Portainer 既可以管理独立运行 Docker 单机环境，也可以管理 Docker Swarm 集群。

Portainer 官方提供了一个部署好的示例（Portainer），初学者可以通过这个示例来了解 Portainer 的功能。图 10-4 展示了 Portainer 的登录界面。使用 Portainer 官方提供的用户名和密码进行登录（用户是"admin"，密码是"tryportainer"）。

图 10-4

Portainer 的首页如图 10-5 所示。可以看出，Portainer 管理了两套 Docker 环境：单机版的 Docker 和 Swarm 集群环境。

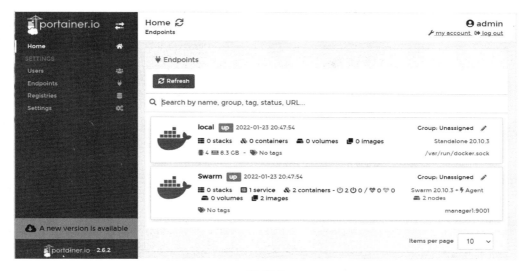

图 10-5

10.2.1 在单机环境中部署 Portainer

在单机环境中 Portainer 的安装和配置比较简单，这里以 master 节点为例来演示。

（1）从镜像仓库中拉取 Portainer 的镜像。

```
docker pull portainer/portainer
```

（2）创建数据卷。

```
docker volume create portainer_db
```

（3）启动 Partainer 容器。

```
docker run -d -p 9000:9000 \
--name portainer \
--restart always \
-v /var/run/docker.sock:/var/run/docker.sock \
-v portainer_db:/data \
portainer/portainer
```

（4）打开浏览器访问 9000 端口。第 1 次访问 Partainer 时需要为"admin"用户设置密码，如图 10-6 所示。

图 10-6

（5）在 Portainer 的运行模式页面上，可以配置 Portainer 运行在本地模式还是远程模式。这里选择的是"Local"，即本地模式，因为这时 Portainer 与 master 节点上的 Docker 运行在同一台主机上，如图 10-7 所示。

图 10-7

（6）单击"Connect"按钮进入 Portainer 的主页，如图 10-8 所示。

图 10-8

10.2.2 【实战】使用 Portainer 管理 Docker 的镜像与容器

下面演示使用 Portainer 管理 Docker 的镜像和容器。将基于 Nginx 的镜像来启动一个容器。

（1）单击图 10-8 中的"local"节点，进入 master 节点的管理主页，如图 10-9 所示。

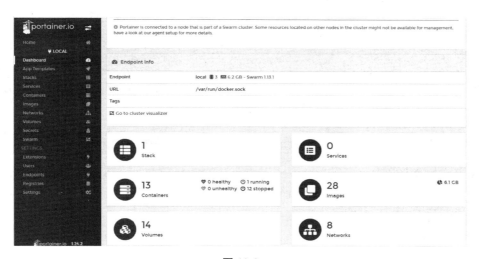

图 10-9

在这里可以看到 master 节点上的 Docker 环境，例如：master 节点上有 13 个容器、28 个镜像、14 个数据卷和 8 个网络配置等。

（2）选择左侧的"Containers"，然后单击"Container list"中的"Add container"按钮添加一个新的容器。

（3）在页面上输入容器的名称、镜像的名称和端口的映射等信息，如图 10-10 所示。

图 10-10

（4）单击 "Deploy the container" 按钮部署容器。

（5）在 Docker 的命令行中确认容器的信息。

```
docker ps --format "table {{.ID}}\t{{.Image}}\t{{.Names}}\t{{.Ports}}"
```

输出的信息如下：

```
CONTAINER ID  IMAGE                NAMES        PORTS
d6026c9725dc  nginx:latest         mycontainer  0.0.0.0:1234->80/tcp
40c8a4ca0d80  portainer/portainer  portainer    0.0.0.0:9000->9000/tcp
```

（6）通过浏览器访问 1234 端口，即可打开 Nginx 的首页面。

10.2.3 【实战】使用 Portainer 管理远端主机上的 Docker

在 10.2.1 节中已经完成了在 master 节点上部署 Portainer。那可以通过 master 节点上的 Portainer 管理其他主机上的 Docker 吗？答案是肯定的。这里将以 master 和 node1 节点为例，来演示如何使用 Portainer 进行远程管理。

（1）在 node1 节点上编辑 "/etc/docker/daemon.json" 文件以启用 2375 端口。

```
{
  "hosts": ["tcp://192.168.79.12:2375", "unix:///var/run/docker.sock"]
}
```

（2）重启 node1 节点上的 Docker 服务。

```
systemctl daemon-reload
systemctl restart docker
```

（3）确定 node1 节点上的 2375 端口已经启用。

```
ss -unlpt | grep 2375
```

输出的信息如下：

```
tcp LISTEN 0 128 192.168.79.12:2375
```

（4）在 Portainer 主页上，单击左侧的"Endpoints"链接，并单击"Add endpoints"。

（5）在"Create endpoint"页面上的"Environment type"中选择"Docker"，并输入"Name"和"Endpoint URL"信息，如图 10-11 所示。

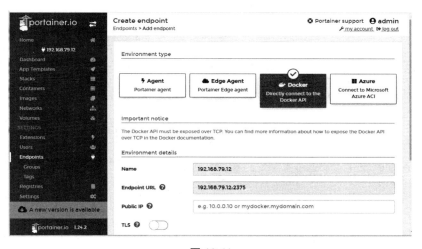

图 10-11

（6）单击"Add endpoint"按钮，这时将在 Portainer 中成功添加 node1 节点，如图 10-12 所示。

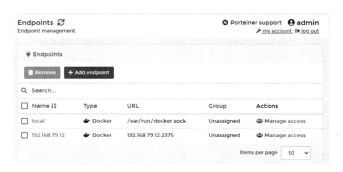

图 10-12

（7）回到 Portainer 的主页，这时可以看到 Portainer 管理了两台 Docker 主机：master 的

local 和 192.168.79.12 的 node1，如图 10-13 所示。

图 10-13

10.2.4　在 Docker Swarm 集群中部署 Portainer

使用 Portainer 管理 Docker 的优势之一就是它支持集群。下面将演示如何在 Docker Swarm 集群中部署 Portainer。

（1）根据 7.2 节的步骤在 master、node1 和 node2 节点上部署 Docker Swarm 集群。

（2）在 master 节点上执行以下命令启动 Portainer 的容器。

```
docker service create \
--name portainer \
--publish 9000:9000 \
--constraint 'node.role == manager' \
--mount type=bind,src=//var/run/docker.sock,dst=/var/run/docker.sock \
portainer/portainer \
-H unix:///var/run/docker.sock
```

（3）第一次登录时设置"admin"用户的密码。

（4）登录 Portainer 后，显示的首页面如图 10-14 所示。

图 10-14

（5）单击左侧的"Swarm"将显示 Docker Swarm 集群中节点的信息，如图 10-15 所示。

图 10-15

（6）单击左侧的"Services"链接，并在打开的页面中单击"Add service"，在 Docker Swarm 集群中部署一个服务。

（7）在"Create service"页面上输入 Service 的名称、镜像、冗余度及端口映射，然后单击"Create the service"按钮，如图 10-16 所示。

图 10-16

（8）图 10-17 展示了 Service 部署成功后的页面。

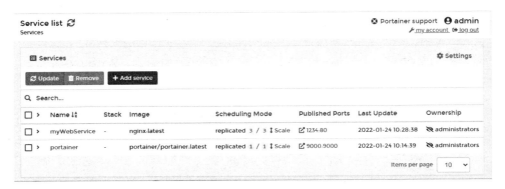

图 10-17

（9）打开浏览器通过访问 1234 端口会打开 Nginx 的首页。

10.3 开源的 Docker 图形工具——Shipyard

Shipyard 是一个集成管理 Docker 镜像、容器和镜像仓库的系统，支持跨多个主机的 Docker 容器的集群。Shipyard 提供的主要功能包括：

- 镜像管理、容器管理、节点管理等。
- 节点的动态扩展。
- 容器监控的可视化管理。
- 在线的命令行终端。

10.3.1 Shipyard 的组件

Shipyard 有两个核心的组件，见表 10-1。

表 10-1

组　件	作　用
engine	监听 Docker 守护进程。Shipyard 通过 engine 调用 Docker API 来管理 Docker 集群。Shipyard 还可以对每个 engine 进行资源限制
rethinkdb	Shipyard 项目中的一个镜像，主要用于存放账号、服务密钥、元数据等信息，但不会存储任何有关容器或镜像的内容

10.3.2 部署 Shipyard

下面演示如何部署 Shipyard。

> 官方提供了 Shipyard 的一键部署脚本 shipyard-deploy。

（1）在 master 节点上拉取 Shipyard 的基础镜像。这里使用的是网易蜂巢的镜像。

```
docker pull hub.c.163.com/library/alpine:latest
docker pull hub.c.163.com/library/rethinkdb:latest
docker pull hub.c.163.com/longjuxu/microbox/etcd:latest
docker pull hub.c.163.com/wangjiaen/shipyard/docker.io/shipyard/docker-proxy:latest
docker pull hub.c.163.com/library/swarm:latest
docker pull hub.c.163.com/wangjiaen/shipyard/docker.io/shipyard/shipyard:latest
```

（2）给拉取的镜像添加一个标签。

```
docker tag 7328f6f8b418 alpine
docker tag 4a511141860c rethinkdb
docker tag 6aef84b9ec5a microbox/etcd
docker tag cfee14e5d6f2 shipyard/docker-proxy
docker tag 0198d9ac25d1 swarm
docker tag 36fb3dc0907d shipyard/shipyard
```

（3）给官方提供的一键部署的脚本 shipyard-deploy 添加执行权限。

```
chmod 755 shipyard-deploy
```

> 由于在 10.2.1 节中部署了 Portainer，所以需要修改"/etc/docker/daemon.json"文件并重启 Docker。在"/etc/docker/daemon.json"中输入以下内容：
> {}
> 然后重启 Dockder：
> systemctl daemon-reload
> systemctl restart docker

（4）在 master 节点上执行脚本 shipyard-deploy。

```
sh shipyard-deploy | bash -s
```

（5）打开浏览器访问 8080 端口，输入用户名"admin"和密码"shipyard"进行登录，如图 10-18 所示。

图 10-18

（6）单击主页上的"节点管理"，可以看到有 Shipyard 管理的 Docker 节点，如图 10-19 所示。

图 10-19

（7）在 node1 节点上重复第（1）（2）（3）步，并执行以下命令将 node1 节点加入 Shipyard 的管理中。

```
cat shipyard-deploy |ACTION=node \
DISCOVERY=etcd://192.168.79.11:4001 bash -s
```

 这里通过参数 DISCOVERY=etcd://192.168.79.11:4001 指定了 Shipyard 的管理地址。

（8）刷新 Shipyard 页面即可观察到新加入的 node1 节点，如图 10-20 所示。

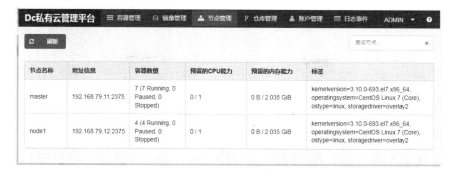

图 10-20

10.3.3 【实战】使用 Shipyard 创建容器

下面将基于 Nginx 的镜像来创建一个容器。

（1）单击 Shipyard 页面上方的"容器管理"，并在容器管理页面上单击"部署容器"按钮。

（2）在"容器构建部署"页面上输入容器的相关参数，见表 10-2。

表 10-2

参数名称	参数值
镜像名称	nginx
容器端口	80
主机端口	1234

（3）单击"部署"按钮。容器启动成功后的界面如图 10-21 所示，可以看到容器被创建在 node1 节点上。

图 10-21

第 11 章
Docker 应用实战

本章将通过两个示例来演示在 Docker 中部署应用。

11.1 Docker 与数据库

在 2.2.1 节中,已经在 Docker 的容器中部署了一个 MySQL 数据库,但是没有通过数据卷对容器中 MySQL 数据库的数据进行持久化。如果这时运行 MySQL 的容器被销毁,则数据将丢失。因此,在 Docker 中部署数据库时,一定要考虑数据持久化的问题。

11.1.1 在 Docker 容器中部署 MySQL

下面来演示如何在 Docker 容器中部署 MySQL,并通过挂载数据卷实现数据的持久化。

(1)从镜像仓库中拉取 MySQL 镜像。

```
docker pull mysql:5.7
```

(2)在宿主机上创建 MySQL 数据持久化目录。

```
mkdir -p /data/mysql/conf
mkdir -p /data/mysql/data
```

(3)执行以下命令启动 MySQL 容器。

```
docker run -d -p 3306:3306 \
-v /data/mysql/conf:/etc/mysql \
-v /data/mysql/data:/var/lib/mysql \
-e MYSQL_ROOT_PASSWORD=Welcome_1 \
```

```
--name mysqldemo \
mysql:5.7
```

> 以上命令挂载了两个数据卷，即以下两行配置参数：
> `-v /data/mysql/conf:/etc/mysql`
> `-v /data/mysql/data:/var/lib/mysql`
> 通过这两个数据卷，可以将容器中的 MySQL 配置文件和数据文件挂载到了宿主机上。

（4）在宿主机上执行以下命令查看"/data/mysql"目录。

```
tree -d -L 2 /data/mysql/
```

输出的信息如下：

```
/data/mysql/
├── conf
└── data
    ├── mysql
    ├── performance_schema
    └── sys
```

11.1.2 数据库不适合 Docker 容器化的原因

由于 Docker 的容器提供的是无状态的服务，因此，不适合将有状态的服务（例如数据库服务）部署到 Docker 的容器中。这主要是因为以下几个原因。

1. 数据的安全性

尽管 Docker 可以通过数据卷的方式将容器中的数据持久化到宿主机上，但仍不能保证不丢失数据。如果容器崩溃了，且数据库未被正确关闭，则可能会丢失数据。

2. 硬件资源的争用

通常在一台 Docker 的宿主机上会启动多个容器，如果将数据库的容器与其他应用的容器部署在同一个宿主机上，由于它们对硬件资源的要求是不同的，则必然会造成资源争用的问题。

3. 网络带宽的占用

Docker 的网络都是虚拟网络，通过宿主机上的 docker0 网桥进行转发。而数据库通常对网络带宽的要求是比较高的。因此，将数据库的容器与其他应用的容器部署在同一个宿主机上，则网络带宽必然会成为数据库性能的瓶颈。

4. 数据额外的隔离

将数据库部署到容器中，毫无疑问会增加对容器的隔离，不利于数据库的水平扩展。

 使用 Docker 是为了更容易地构建新环境和重新部署应用。而在实际情况中，数据库一旦部署完成，则很少会对数据库进行升级或重新部署。因此，从这个角度来看数据库也不适合 Docker 容器化。

11.2 【实战】Docker 与 Python

Python 目前在数据分析中应用得非常广泛。下面将在 CentOS 的容器中部署一套 Python 开发环境，以及 Python 的交互式开发工具 Jupyter Notebook。

（1）基于 CentOS 的镜像启动容器。

```
docker run -it centos bash
```

（2）在容器中添加相关的依赖。

```
yum -y groupinstall "Development tools"
yum -y install zlib-devel bzip2-devel
yum -y install openssl-devel ncurses-devel
yum -y install sqlite-devel readline-devel
yum -y install tk-devel gdbm-devel db4-devel
yum -y install libpcap-devel xz-devel libffi-devel
yum install wget
```

（3）在容器中获取 Python 3 的安装文件。

```
wget "https://www.python.org/ftp/python/3.6.2/Python-3.6.2.tgz"
tar -zxvf Python-3.6.2.tgz -C /tmp
cd /tmp/Python-3.6.2/
```

（4）在容器中安装和配置 Python 3。

```
./configure --prefix=/usr/local/python3
make
make install
```

（5）在容器内中建立软连接。

```
ln -s /usr/local/python3/bin/python3.6 /usr/bin/python3
```

（6）在容器内中验证 Python 3 环境，如图 11-1 所示。

```
python3
```

```
[root@db032a1b3bac /]# python3
Python 3.6.2 (default, Jan 24 2022, 08:22:48)
[GCC 8.5.0 20210514 (Red Hat 8.5.0-4)] on linux
Type "help", "copyright", "credits" or "license" for more information.
>>>
```

图 11-1

（7）新建一个"hello.py"文件，在其中输入以下内容：

```
print("Hello World")
```

（8）执行"hello.py"将输出"Hello World"字符。

```
python3 hello.py
```

（9）使用"docker commit"命令将容器保存为镜像。

```
docker commit ac7753a18311 mycentos_with_python3
```

其中的参数 ac7753a18311 指容器的 ID。

> 上面的步骤使用"docker commit"命令来构建 Python 镜像，也可以使用 Dockerfile 文件进行构建。下面给出了 Dockerfile 文件的完整内容：
>
> FROM centos:7
> MAINTAINER zhaoyuqiang collen7788@126.com
> RUN yum -y groupinstall "Development tools"
> RUN yum -y install zlib-devel bzip2-devel
> RUN yum -y openssl-devel ncurses-devel sqlite-devel
> RUN yum -y readline-devel tk-devel gdbm-devel db4-devel
> RUN yum -y libpcap-devel xz-devel libffi-devel
> RUN mkdir /root/tools
> ADD https://www.python.org/ftp/python/3.6.2/Python-3.6.2.tgz /root/tools
> RUN tar -zxvf　/root/tools/Python-3.6.2.tgz -C /tmp
> WORKDIR /tmp/Python-3.6.2
> RUN ./configure --prefix=/usr/local/python3
> RUN make
> RUN make install
> RUN ln -s /usr/local/python3/bin/python3.6 /usr/bin/python3

（10）使用 mycentos_with_python3 镜像创建一个容器，并开放 8888 端口。

```
docker run -it -p 8888:8888 mycentos_with_python3 bash
```

（11）在容器中安装 Python 包管理工具 pip。

```
python3 -m pip install --upgrade pip
```

（12）安装 Jupyter Notebook。

```
python3 -m pip install jupyter notebook
```

（13）生成 Jupyter Notebook 的配置文件。

```
cd /usr/local/python3/bin/
./jupyter-notebook --generate-config
```

（14）进入 Python 命令行中生成登录密码。

```
>>> from notebook.auth import passwd
>>> passwd()
```

（15）修改 Jupyter Notebook 的配置文件，加入以下内容：

```
c.NotebookApp.ip= '0.0.0.0'
c.NotebookApp.password = u'上一步生成的密码'
c.NotebookApp.open_browser = False
c.NotebookApp.port = 8888
```

（16）启动 Jupyter Notebook。

```
./jupyter-notebook --allow-root
```

（17）打开浏览器访问 8888 端口，即可打开 Jupyter Notebook 的首页，如图 11-2 所示。

图 11-2

11.3 【实战】Docker 与 PHP

PHP 的全名是 Hypertext Preprocessor，即"超文本预处理器"。它是在服务器端执行的脚本语言，尤其适用于 Web 开发并可嵌入 HTML 中。下面使用"docker commit"命令基于 CentOS 的基础镜像来创建自己的镜像，并在镜像中部署 PHP 的运行环境。

（1）从镜像仓库拉取 CentOS 的镜像。

```
docker pull centos:6
```

由于 CentOS 7 在 Docker 上有一个 Bug，所以这里建议下载 CentOS 6 的镜像。

（2）运行 CentOS 6 容器。

```
docker run -it -p 1234:80 5bf9684f4720 /bin/bash
```

其中的参数 5bf9684f4720 指 CentOS 6 镜像的 ID。

（3）在容器中安装 Apache Server。

```
yum install httpd -y
```

（4）设置 Apache Server 开机自启。

```
chkconfig httpd on
service httpd start
```

（5）打开浏览器访问宿主机的 1234 端口，即打开 Apache Server 的首页，如图 11-3 所示。

图 11-3

（6）执行以下命令在容器中安装 PHP：

```
yum install -y php
yum install -y php-mysql
yum install -y php-gd
yum install -y php-imap
yum install -y php-ldap
yum install -y php-odbc
yum install -y php-pear
yum install -y php-xml
```

```
yum install -y php-xmlrpc
```

（7）在"/var/www/html/"目录下创建测试代码文件 info.php，并在其中输入以下内容：

```
<?php
    phpinfo();
?>
```

（8）通过地址"http://IP:1234/info.php"访问页面，这时将无法正常访问，如图 11-4 所示。

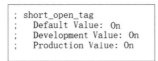

图 11-4

（9）修改"/etc/php.ini"文件，打开"short_open_tags"开关，如图 11-5 所示。

```
; short_open_tag
;   Default Value: On
;   Development Value: On
;   Production Value: On
```

图 11-5

（10）重启 Apache Server，并刷新"info.php"页面。这时即可看到 PHP 的网页，如图 11-6 所示。

```
service httpd restart
```

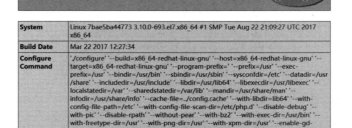

图 11-6

（11）使用"docker commit"命令将容器保存为镜像。

```
docker commit 3364a88765e5 centos_php
```

其中的参数 3364a88765e5 为容器的 ID。

下篇
Kubernetes 从原理到实战

第 12 章
Kubernetes 体系架构

Docker 容器技术将应用及其依赖打包到镜像中,从而很好地解决了应用部署与集成的问题。但在现实中却很少通过 Docker 将应用进行大规模的部署。这主要是因为,Docker 本质上是一种单一的容器技术(或者说是一种工具),并不能很好地将应用组织起来,难以独立地支撑起生产环境中应用的大规模容器化部署。而采用 Kubernetes 则可以很好地解决这个问题。

> Kubernetes 并不能够取代 Docker,它们之间是平台与组件的关系:Kubernetes 可以将 Docker 作为运行时组件,但并不完全依赖 Docker。因此严格地说,Kubernetes 应该被称为容器编排技术,而不是容器技术。

12.1 什么是 Kubernetes

Kubernetes 简称 K8s,它最初源于谷歌内部的 Borg 系统,提供的功能包括:应用的服务编排、容器集群的部署和集群的管理。通过它能够非常方便地进行集群的扩容与缩容。

> Borg 系统是一个集群管理器,它管理着 Google 内部很多个应用集群,而每个集群都有成千上万台机器。Borg 系统通过准入控制、高效的任务打包、超额的资源分配,以及进程级隔离的机器共享,实现了超高的资源利用率。

Kubernetes 通过一个抽象的逻辑单元 Pod 将应用的容器组合在一起,从而让服务更容易被发现和管理。Kubernetes 中沉淀了 Google 多年的生产环境运行经验,目前已经形成了一个完善的生态圈。Google 在 2014 年将 Kubernetes 开源。

Kubernetes 提供了以下 7 个重要功能。

（1）自动发布和回滚。

Kubernetes 通过持久化存储来保存应用发布时的相关配置信息，从而在部署过程中发生问题时能够执行回滚操作。

（2）自动化装箱。

Kubernetes 按照应用对资源的要求将容器进行自动部署，从而提高了资源的利用率，节省了资源。

（3）水平扩容。

Kubernetes 根据应用在运行过程中对 CPU、内存的使用情况，通过简单的命令即可对应用进行扩容和缩容。

（4）配置管理。

Kubernetes 将集群和应用的配置信息进行了持久化存储，可以在不重新构建镜像的情况下更新应用的配置信息。

（5）自愈能力。

Kubernetes 实现了容器的高可用。当节点上运行的容器失败后，Kubernetes 会对容器进行重启。即使节点出现宕机，Kubernetes 也会对容器进行重新部署和重新调度，容器能够正常运行后才会对外提供服务。

（6）服务发现和负载均衡。

Kubernetes 内置了服务发现机制和负载均衡功能，不需要使用额外的服务。

（7）存储编排。

Kubernetes 利用持久卷和持久卷声明完成存储系统的自动挂载，同时支持多种存储系统（如本地存储、云存储和网络存储等）。

12.2 Kubernetes 集群

有必要先了解一下 Kubernetes 集群的体系架构及组件，这对于后续部署及使用 Kubernetes 集群都非常重要。

12.2.1 集群的架构体系

Kubernetes 的体系架构如图 12-1 所示。

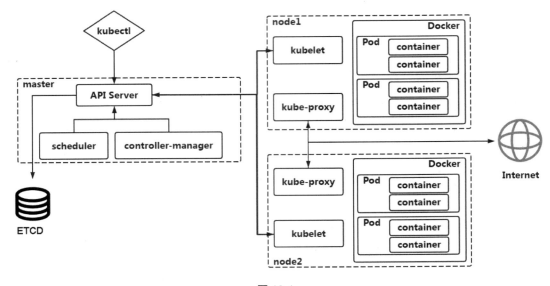

图 12-1

Kubernetes 属于主从分布式架构，主要包括 master 节点（主节点）和 node 节点（工作节点）。实际的生产环境中，至少有两个 node 节点。

- master 节点：控制着整个集群，并对集群进行管理和调度。master 节点上运行着 API Server、scheduler、controller-manager 等服务组件，并且 master 节点还管理着集群的持久化存储 ETCD。
- node 节点：实际运行业务应用容器的节点。node 节点上运行着 kubelet、kube-proxy 和 Docker 容器的守护进程。node 节点通过 kubelet 与 master 节点上的 API Server 进行交互，从而在集群中对各种资源执行增、删除、修改和查询等操作。

除 master 节点和 node 节点外，在 Kubernetes 体系架构中还包括客户端命令行工具 kubectl 和其他一些附加组件。

12.2.2 Kubernetes 的核心组件

下面对 Kubernetes 的核心组件及其作用进行说明。

1. API Server

它提供了操作 Kubernetes 集群的唯一入口，通过它可以访问集群中的所有资源。它也提供了用户的认证及授权功能，并支持访问控制的管理和服务的注册与发现等机制。API Server 也可以与 ETCD 存储器进行交互，从而将集群的信息持久化保存。

2. scheduler

它负责管理和调度集群资源。Kubernetes 允许用户预先定义集群资源的调度策略，通过 scheduler 将 Pod 调度到相应的 node 节点上。

3. controller-manager

它负责维护集群的状态，如故障检测、自动扩展、滚动更新等。

4. kubelet

它负责管理 node 节点上运行的 Pod，包括 Pod 的创建、修改、删除、重启及健康状态监控等。它还负责与 master 节点上的 API Server 进行交互，定时将 node 节点的状态发送给 API Server，并接收 API Server 下发的指令。

5. kube-proxy

它会根据 ETCD 中存储的应用配置信息在 node 节点上启动一个监听进程，从而将外部请求路由转发到后端正确的容器中。另外，kube-proxy 还解决了服务端口冲突的问题，并为部署在 Kubernetes 集群中的应用提供访问外部网络的能力。kube-porxy 支持随机和轮询这两种负载均衡的路由算法。

6. Docker 容器的守护进程

它负责 Docker 的镜像管理，以及 Pod 和容器的真正运行，是 Kubernetes 真正的执行引擎。

> Docker 是 Kubernetes 的执行引擎，但 Kubernetes 也可以使用其他的容器技术作为执行引擎。

12.2.3 Kubernetes 的常用附加组件

表 12-1 中列举了 Kubernetes 的常用附加组件（这些附加组件不是必需的）。利用这些组件可以增强 Kubernetes 的功能。

表 12-1

附加组件名称	组件的功能
kube-dns	为整个 Kubernetes 集群提供 DNS 服务
Ingress	为集群中的服务提供最佳的外部访问方式
Heapster	监控 Kubernetes 集群的资源

续表

附加组件名称	组件的功能
Dashboard	Kubernetes 的 Web UI
Federation	提供跨可用区的 Kubernetes 集群
Fluentd-elasticsearch	提供 Kubernetes 集群日志的采集、存储与查询

12.3 Kubernetes 的对象

Kubernetes 将所有的内容（如 Pod、Service、PV 和 PVC 等）都抽象为"资源"。"资源"的实例是"对象"，而对象可以被持久化。Kubernetes 使用对象去代表整个集群。对象描述了如下信息：

- 集群中有哪些应用，它们运行在哪些节点上。
- 应用可以使用哪些资源或对象。

12.3.1 对象的管理

由于 master 节点上的 API Server 是操作集群资源的唯一入口，因此，对象的管理都是通过 API Server 来完成的。Kubernetes 提供的这些 API 是 RESTFul API，所以，即使使用命令行工具 kubectl 来操作集群，实际上也是调用 API Server 中提供的接口来完成客户端的请求的。

Kubernetes 使用配置文件来描述和管理对象，配置文件格式可以是 JSON 或 YAML 格式，更常用的是 YAML 格式。

下面是 Kubernetes 官方提供的一个 YAML 示例文件，它展示了 Kubernetes Deployment 对象的属性的必需字段和规约。

```yaml
apiVersion: apps/v1
kind: Deployment
metadata:
  name: nginx-deployment
spec:
  selector:
    matchLabels:
      app: nginx
  replicas: 2     # 设置 Deployment 的副本数为 2
  template:
    metadata:
      labels:
        app: nginx
    spec:
```

```
    containers:
    - name: nginx
      image: nginx:1.14.2
      ports:
      - containerPort: 80
```

 "规约"描述了对象被期望的状态，或者对象应该具备的特征。它通过关键字"spec"进行定义。在创建 Kubernetes 对象时，必须提供对象的规约。

在使用 YAML 文件管理对象时，有些字段是必需的。表 12-2 列出了这些必需的字段。

表 12-2

字段的名称	字段的作用
apiVersion	创建该对象所使用的 Kubernetes API 的版本
kind	指定对象的类型
metadata	设置对象的元信息，包括对象的名称、ID 及命名空间等
spec	设定对象的规约信息

12.3.2 对象与命名空间

Kubernetes 支持多个虚拟集群，它们的底层依赖同一个物理集群。通过命名空间可以将这些虚拟集群从逻辑上进行隔离。同一个命名空间中的对象名称必须唯一，并且不能跨越命名空间。每个 Kubernetes 的对象只能属于一个命名空间。不同的命名空间不能互相嵌套。

以下命令将列出集群中现存的命名空间：

```
kubectl get namespaces
```

输出的信息如下：

```
NAME              STATUS    AGE
default           Active    312d
kube-node-lease   Active    312d
kube-public       Active    312d
kube-system       Active    312d
```

在部署完 Kubernetes 集群后，会自动创建 4 个命名空间。

- default：在创建 Kubernetes 对象时，如果没有指定所属的命名空间，则对象将创建在 default 命名空间中。
- kube-node-lease：该命名空间与所有 node 节点上的对象相关联，并接收 node 节点上 kubelet 发送的心跳信息，由此检测 node 节点的状态。

- kube-public：该命名空间中的对象具有公开的属性，即 Kubernetes 中的所有用户都可以读取该命名空间中的对象。
- kube-system：Kubernetes 系统的命名空间，是 Kubernetes 保留的命名空间。其中的对象都是供 Kubernetes 系统使用的。用户不应该使用该命名空间。

如果要获取某个命名空间（如"kube-system"命名空间）中的所有资源对象，则执行以下语句。

```
kubectl get all -n kube-system
```

而以下语句只会获取"kube-system"命名空间中的 Pod 类型的资源。

```
kubectl get pods -n kube-system
```

输出的信息如下：

```
NAME                                  READY   STATUS    RESTARTS   AGE
coredns-bccdc95cf-68dgw               1/1     Running   3          312d
coredns-bccdc95cf-7dp9n               1/1     Running   3          312d
etcd-master                           1/1     Running   0          312d
kube-apiserver-master                 1/1     Running   1          312d
kube-controller-manager-master        1/1     Running   1          312d
kube-flannel-ds-amd64-7v5jq           1/1     Running   1          312d
kube-flannel-ds-amd64-dwpkm           1/1     Running   0          312d
kube-flannel-ds-amd64-pgmt7           1/1     Running   1          312d
kube-proxy-hjkc8                      1/1     Running   0          312d
kube-proxy-l7b4b                      1/1     Running   0          312d
kube-proxy-xkt9j                      1/1     Running   0          312d
kube-scheduler-master                 1/1     Running   2          312d
```

12.3.3　对象的标签

标签（Labels）是附加到 Kubernetes 对象上的键值对。通过使用标签，用户可以为对象指定有意义且相关的标识属性。标签用于组织和选择对象。可以在创建对象时创建标签，也可以在对象创建成功后随时添加标签。可以为一个对象添加多个标签，但是同一个对象上的标签不能重复。

标签的定义可以在"metadata"字段中通过"labels"关键字进行定义。下面是一个示例：

```
"metadata": {
  "labels": {
    "key1" : "value1",
    "key2" : "value2"
  }
}
```

对象的标签可以与"selector"选择器配合使用，并用表达式对条件加以限制，从而实现更精确、更灵活的资源查找。

> 标签是 Kubernetes 中非常强大的一个功能。所有的 Kubernetes 资源都可以使用标签，例如：在 node 节点上增加标签，然后利用 Pod 的标签选择器将 Pod 分配到不同类型的 node 节点上。

第 13 章

部署 Kubernetes 集群

由于 Kubernetes 本身不提供容器引擎,因此在部署 Kubernetes 前需要安装好 Docker 环境。Kubernetes 有几种部署方式:kubeadmin 方式、YUM 方式、二进制包方式和 minikubei 工具方式。

13.1　Kubernetes 的部署方式

Kubernetes 是主从式架构,分为 master 节点和 node 节点,因此不管使用哪种方式进行部署,其核心都是部署这两部分。

13.1.1　使用 kubeadmin 部署 Kubernetes 集群

kubeadmin 是 Kubernetes 提供的一个部署工具,用来简化 Kubernetes 的部署过程。下面以 7.2 节中的 master、node1 和 node2 这 3 个节点为例,来演示如何使用 kubeadmin 部署 Kubernetes 集群。

(1)由于 Kubernetes 集群要求运行的宿主机至少有 2 个 CPU,因此需要修改 master、node1 和 node2 这 3 个节点的 CPU 核数,如图 13-1 所示。

> 如果只有一个 CPU,则在部署 Kubernetes 时会出现以下错误:
> [ERROR NumCPU]: the number of available CPUs 1 is less than the required 2

图 13-1

（2）在 master、node1 和 node2 节点上编辑"/etc/hosts"文件，设置主机名和 IP 地址的映射关系。

```
192.168.79.11 master
192.168.79.12 node1
192.168.79.13 node2
```

Kubernetes 集群需要保证所有节点的时间是同步的，否则 node 节点无法加入集群。

（3）在 master、node1 和 node2 节点上关闭防火墙和 SELinux，并且禁用 Swap。

```
systemctl stop firewalld.service
systemctl disable firewalld.service
swapoff -a
sed -i 's/enforcing/disabled/' /etc/selinux/config
setenforce 0
```

（4）在 master、node1 和 node2 节点上配置阿里云的 Kubernetes 源。

```
cat > /etc/yum.repos.d/kubernetes.repo << EOF
[kubernetes]
name=Kubernetes
baseurl=https://mirrors.aliyun.com/kubernetes/yum/repos/kubernetes-el7-x86_64
   enabled=1
   gpgcheck=0
   repo_gpgcheck=0
   gpgkey=https://mirrors.aliyun.com/kubernetes/yum/doc/yum-key.gpg
https://mirrors.aliyun.com/kubernetes/yum/doc/rpm-package-key.gpg
   EOF
```

（5）在 master、node1 和 node2 节点上分别安装 kubeadm、kubelet 和 kubectl。

```
yum install -y kubelet-1.15.0 kubeadm-1.15.0 kubectl-1.15.0
```

由于 Kubernetes 版本更新频繁，所以这里要指定 Kubernetes 的版本号。

（6）在 master、node1 和 node2 节点上设置 kubelet 开机自启。

```
systemctl enable kubelet
systemctl start kubelet
```

（7）在 master 节点上对 Kubernetes 的 master 节点进行初始化。

```
kubeadm init \
--apiserver-advertise-address=192.168.79.11 \
--image-repository registry.aliyuncs.com/google_containers \
--kubernetes-version v1.15.0 \
--service-cidr=10.1.0.0/16 \
--pod-network-cidr=10.244.0.0/16
```

由于 kubeadm 拉取镜像的默认地址从国内无法访问，所以这里通过参数 "--image-repository" 指定使用阿里云镜像仓库地址。

初始化成功后将输出以下信息：

```
Then you can join any number of worker nodes by running the following on each as root:
kubeadm join 192.168.79.11:6443 \
--token kje09p.q0qkle6fl66uhmn9 \
--discovery-token-ca-cert-hash \
sha256:ce122287eb530991dadab0886956157716db8929d52fd41e1ebe78a6e9b02105
```

（8）在 master 节点上配置 kubectl 命令行工具。

```
mkdir -p $HOME/.kube
sudo cp -i /etc/kubernetes/admin.conf $HOME/.kube/config
sudo chown $(id -u):$(id -g) $HOME/.kube/config
```

（9）在 master 节点上下载 flannel 网络的配置文件。

```
wget https://raw.githubusercontent.com/coreos/flannel/a70459be0084506e4ec919aa1c114638878db11b/Documentation/kube-flannel.yml
```

> 如果无法访问 kube-flannel.yml 文件中的 quay.io/coreos/flannel:v0.11.0-amd64 镜像,则可以使用笔者提供的镜像 collenzhao/flannel:v0.11.0-amd64 代替它。

(10) 在 master 节点上应用 flannel 网络的配置文件。

```
kubectl apply -f kube-flannel.yml
```

命令执行成功后将输出以下信息:

```
clusterrole.rbac.authorization.k8s.io/flannel created
clusterrolebinding.rbac.authorization.k8s.io/flannel created
serviceaccount/flannel created
configmap/kube-flannel-cfg created
daemonset.extensions/kube-flannel-ds-amd64 created
daemonset.extensions/kube-flannel-ds-arm64 created
daemonset.extensions/kube-flannel-ds-arm created
daemonset.extensions/kube-flannel-ds-ppc64le created
daemonset.extensions/kube-flannel-ds-s390x created
```

(11) 在 node1 和 node2 节点上拉取 quay.io/coreos/flannel:v0.11.0-amd64 镜像。

```
docker pull quay.io/coreos/flannel:v0.11.0-amd64
```

> 如果无法拉取 quay.io/coreos/flannel:v0.11.0-amd64 镜像,则可以使用笔者提供的镜像 collenzhao/flannel:v0.11.0-amd64 代替它。使用命令以下:
>
> docker pull collenzhao/flannel:v0.11.0-amd64

(12) 在 node1 和 node2 节点上分别执行以下命令将这两个节点加入 Kubernetes 集群。

```
kubeadm join 192.168.79.11:6443 \
--token kje09p.q0qkle6fl66uhmn9 \
--discovery-token-ca-cert-hash \
sha256:ce122287eb530991dadab0886956157716db8929d52fd41e1ebe78a6e9b02105
```

> 这里的命令会在第 (7) 步 master 节点初始化成功后打印在屏幕上。
> 如果在单节点上执行 join 操作时出现错误,则可以加上参数 "--ignore-preflight-errors=all" 来忽略检查时的错误信息。

在 node 节点成功加入 Kubernetes 集群后,会输出以下信息:

```
This node has joined the cluster:
```

```
    * Certificate signing request was sent to apiserver and a response was
received.
    * The Kubelet was informed of the new secure connection details.

Run 'kubectl get nodes' on the control-plane to see this node join the cluster.
```

（13）在 mater 节点上执行以下命令查看集群的节点信息，如图 13-2 所示。

```
kubectl get nodes
```

```
[root@master ~]# kubectl get nodes
NAME     STATUS   ROLES    AGE     VERSION
master   Ready    master   5m48s   v1.15.0
node1    Ready    <none>   3m59s   v1.15.0
node2    Ready    <none>   3m56s   v1.15.0
[root@master ~]#
```

图 13-2

通过以下命令查看 Kubernetes 集群中各个节点的详细信息。

```
kubectl get nodes -o wide
```

13.1.2　使用 YUM 方式部署 Kubernetes 集群

下面来演示使用 YUM 方式部署包含 3 个节点的 Kubernetes 集群。

> 这里依然以 7.2 节中的 master、node1 和 node2 这 3 个节点为例，并根据 13.1.1 节中的前 3 步修改每个节点的配置信息。

（1）在 master 节点上安装 ETCD 服务。

```
yum -y install etcd
```

（2）修改 "/etc/etcd/etcd.conf" 文件，在其中输入以下内容：

```
ETCD_NAME=k8s-etcd
ETCD_DATA_DIR="/var/lib/etcd/k8s-etcd"
ETCD_LISTEN_PEER_URLS="http://192.168.79.11:2380"
ETCD_LISTEN_CLIENT_URLS="http://127.0.0.1:2379,http://192.168.79.11:2379"
ETCD_INITIAL_ADVERTISE_PEER_URLS="http://192.168.79.11:2380"
ETCD_INITIAL_CLUSTER="k8s-etcd=http://192.168.79.11:2380"
ETCD_INITIAL_CLUSTER_STATE="new"
ETCD_INITIAL_CLUSTER_TOKEN="etcd-test"
ETCD_ADVERTISE_CLIENT_URLS="http://192.168.79.11:2379"
```

（3）启动 ETCD 服务。

```
systemctl start etcd
systemctl enable etcd
```

（4）检查 ETCD Cluster 的状态。

```
etcdctl cluster-health
```

输出的信息如下：

```
member fd4d0bd2446259d9 is healthy:
got healthy result from http://192.168.79.11:2379
cluster is healthy
```

"cluster is healthy"表示 ETCD 服务已经正常启动。

（5）检查 ETCD 服务的成员列表。

```
etcdctl member list
```

输出的信息如下：

```
fd4d0bd2446259d9: name=k8s-etcd
peerURLs=http://192.168.79.11:2380
clientURLs=http://192.168.79.11:2379 isLeader=true
```

由于这里只在 master 节点上部署了一个单节点的 ETCD 服务，所以在成员列表中只有一个成员。

（6）在 master 节点上部署"kubernetes-master"。

```
yum -y install kubernetes-master
```

（7）修改 Kubernetes 的配置文件"/etc/kubernetes/config"，输入以下内容：

```
KUBE_LOGTOSTDERR="--logtostderr=true"
KUBE_LOG_LEVEL="--v=0"
KUBE_ALLOW_PRIV="--allow-privileged=false"
KUBE_MASTER="--master=http://192.168.79.11:8080"
```

（8）修改 API Server 的配置文件"/etc/kubernetes/apiserver"，输入以下内容：

```
KUBE_API_ADDRESS="--address=0.0.0.0"
KUBE_API_PORT="--port=8080"
KUBELET_PORT="--kubelet-port=10250"
KUBE_ETCD_SERVERS="--etcd-servers=http://192.168.79.11:2379"
```

```
    KUBE_SERVICE_ADDRESSES="--service-cluster-ip-range=10.254.0.0/16"
    KUBE_API_ARGS="--service_account_key_file=/etc/kubernetes/serviceaccount
.key"
    KUBE_ADMISSION_CONTROL="--admission-control=NamespaceLifecycle,Namespace
Exists,LimitRanger,SecurityContextDeny,ResourceQuota"
    KUBE_API_ARGS="--service_account_key_file=/etc/kubernetes/serviceaccount
.key"
```

（9）修改 controller-manager 的配置文件"/etc/kubernetes/controller-manager"，输入以下内容：

```
    KUBE_CONTROLLER_MANAGER_ARGS="--service_account_private_key_file=/etc/ku
bernetes/serviceaccount.key"
```

（10）修改 scheduler 的配置文件"/etc/kubernetes/scheduler"，输入以下内容：

```
# Add your own!
KUBE_SCHEDULER_ARGS=""
```

（11）执行以下命令在 master 节点上生成密钥 serviceaccount.key。

```
openssl genrsa -out /etc/kubernetes/serviceaccount.key 2048
```

（12）在 master 节点上配置 ETCD 服务的网络。

```
etcdctl set /k8s/network/config '{"Network": "10.255.0.0/16"}'
```

（13）查看 master 节点上 ETCD 服务的网络配置信息。

```
etcdctl get /k8s/network/config
```

输出的信息如下：

```
{"Network": "10.255.0.0/16"}
```

（14）在 master、node1 和 node2 节点上安装 flannel 网络控件。

```
yum -y install kubernetes-node flannel
```

（15）在 master、node1 和 node2 节点上修改 flannel 网络的配置文件"/etc/sysconfig/flanneld"。

```
FLANNEL_ETCD_ENDPOINTS="http://192.168.79.11:2379"
FLANNEL_ETCD_PREFIX="/k8s/network"
```

> 这里指定了 master 节点上的 ETCD 服务地址，而"/k8s/network"就是在 ETCD 服务中创建的数据目录。

（16）在 node1 和 node2 节点上修改配置文件"/etc/kubernetes/config"。

```
KUBE_MASTER="--master=http://192.168.79.11:8080"
```

这里指定的是 master 节点上的 KUBE_MASTER 地址。

（17）在 node1 和 node2 节点上修改配置文件 "/etc/kubernetes/kubelet"。

```
KUBELET_ADDRESS="--address=192.168.79.12"
KUBELET_PORT="--port=10250"
KUBELET_HOSTNAME="--hostname-override=192.168.79.12"
KUBELET_API_SERVER="--api-servers=http://192.168.79.11:8080"
KUBELET_POD_INFRA_CONTAINER="--pod-infra-container-image=docker.io/colle
nzhao/pod-infrastructure"
```

其中的参数说明如下。

- KUBELET_ADDRESS 和 KUBELET_HOSTNAME：都是 node 节点的地址，即 192.168.79.12 和 192.168.79.13。
- KUBELET_POD_INFRA_CONTAINER：Pod 的基础镜像，这里使用了笔者提供的镜像，即：

```
--pod-infra-container-image=docker.io/collenzhao/pod-infrastructure
```

（18）在 node1 和 node2 节点上修改配置文件 "/etc/kubernetes/proxy"。

```
KUBE_PROXY_ARGS="0.0.0.0"
```

（19）在 master 节点上按以下顺序启动服务。

```
systemctl start kube-apiserver \
                kube-controller-manager \
                kube-scheduler \
                flanneld \
                kube-proxy
```

（20）在 node1 和 node2 节点上都按以下顺序启动服务。

```
systemctl start flanneld \
                kubelet \
                kube-proxy
```

（21）在 master 节点上查看 Kubernetes 集群中 node 节点的状态。

```
kubectl get nodes
```

输出的信息如下：

```
NAME            STATUS      AGE
```

```
192.168.79.12    Ready    24s
192.168.79.13    Ready    21s
```

（22）编辑"k8s-yum-demo.yaml"文件在 Kubernetes 集群中部署一个应用。

```
apiVersion: extensions/v1beta1
kind: Deployment
metadata:
  labels:
    app: nginx
  name: nginx
spec:
  replicas: 3
  selector:
    matchLabels:
      app: nginx
  template:
    metadata:
      labels:
        app: nginx
    spec:
      containers:
      - image: nginx
        name: nginx
        imagePullPolicy: IfNotPresent
```

（23）执行以下命令部署应用。

```
kubectl apply -f k8s-yum-demo.yaml
```

（24）查看部署的应用信息。

```
kubectl get pod,deploy
```

输出的信息如下：

```
NAME                             READY    STATUS     RESTARTS   AGE
po/nginx-3086523004-25tcx        1/1      Running    0          1m
po/nginx-3086523004-bbc66        1/1      Running    0          1m
po/nginx-3086523004-j258l        1/1      Running    0          1m

NAME           DESIRED   CURRENT   UP-TO-DATE   AVAILABLE   AGE
deploy/nginx   3         3         3            3           1m
```

13.1.3　使用二进制包部署 Kubernetes 集群

在企业的私有环境中可能无法连接外部的网络。如果要在这样的环境中部署 Kubernetes 集群，则需要采用 Kubernetes 离线安装的方式（即使用二进制安装包部署 Kubernetes 集群），采用的版本是 Kubernetes v1.18.20。

下面来演示使用二进制包部署包含 3 个节点的 Kubernetes 集群。

这里依然以 7.2 节中的 master、node1 和 node2 这 3 个节点为例，并根据 13.1.1 节中的前 3 步修改每个节点的配置信息。表 13-1 列举了每个节点上的 Kubernetes 服务信息。

表 13-1

节点名称	IP 地址	部署的服务
master	192.168.79.11	kube-apiserver kube-controller-manager kube-scheduler ETCD
node1	192.168.79.12	kubelet kube-proxy flannel
node2	192.168.79.13	kube-apiserver kube-controller-manager kube-scheduler

1. 部署 ETCD 服务

（1）从 GitHub 下载 ETCD 服务的二进制安装包 "etcd-v3.3.27-linux-amd64.tar.gz"。

（2）从 cfssl 官方网站下载 cfssl 安装包并安装。

```
chmod +x cfssl_linux-amd64 cfssljson_linux-amd64
mv cfssl_linux-amd64 /usr/local/bin/cfssl
mv cfssljson_linux-amd64 /usr/local/bin/cfssljson
```

cfssl 是一个命令行工具包，其中包含运行一个认证中心所需要的功能。

（3）创建用于生成 CA 证书和私钥的配置文件。

```
mkdir -p /opt/ssl/etcd
cd /opt/ssl/etcd
cfssl print-defaults config > config.json
cfssl print-defaults csr > csr.json

cat > config.json <<EOF
```

```
{
  "signing": {
    "default": {
      "expiry": "87600h"
    },
    "profiles": {
      "kubernetes": {
        "usages": [
            "signing",
            "key encipherment",
            "server auth",
            "client auth"
        ],
        "expiry": "87600h"
      }
    }
  }
}
EOF

cat > csr.json <<EOF
{
  "CN": "etcd",
  "key": {
    "algo": "rsa",
    "size": 2048
  },
  "names": [{
    "C": "CN",
    "ST": "BeiJing",
    "L": "BeiJing",
    "O": "k8s",
    "OU": "System"
  }]
}
EOF
```

（4）生成 CA 证书和私钥。

```
cfssl gencert -initca csr.json | cfssljson -bare etcd
```

（5）在"/opt/ssl/etcd"目录下添加"etcd-csr.json"文件，该文件用于生成 ETCD 服务的证书和私钥，文件内容如下：

```
cat > etcd-csr.json <<EOF
{
  "CN": "etcd",
```

```
    "hosts": [
      "192.168.79.11"
    ],
    "key": {
      "algo": "rsa",
      "size": 2048
    },
    "names": [
      {
        "C": "CN",
        "ST": "BeiJing",
        "L": "BeiJing",
        "O": "etcd",
        "OU": "Etcd Security"
      }
    ]
}
EOF
```

 这里只部署了一个 ETCD 节点。如果部署 ETCD 集群，则需要修改"hosts"字段以添加多个 ETCD 节点。

（6）安装 ETCD。

```
tar -zxvf etcd-v3.3.27-linux-amd64.tar.gz
cd etcd-v3.3.27-linux-amd64
cp etcd* /usr/local/bin
mkdir -p /opt/platform/etcd/
```

（7）编辑"/opt/platform/etcd/etcd.conf"文件添加 ETCD 服务的配置信息，内容如下：

```
ETCD_NAME=k8s-etcd
ETCD_DATA_DIR="/var/lib/etcd/k8s-etcd"
ETCD_LISTEN_PEER_URLS="http://192.168.79.11:2380"
ETCD_LISTEN_CLIENT_URLS="http://127.0.0.1:2379,http://192.168.79.11:2379"
ETCD_INITIAL_ADVERTISE_PEER_URLS="http://192.168.79.11:2380"
ETCD_INITIAL_CLUSTER="k8s-etcd=http://192.168.79.11:2380"
ETCD_INITIAL_CLUSTER_STATE="new"
ETCD_INITIAL_CLUSTER_TOKEN="etcd-test"
ETCD_ADVERTISE_CLIENT_URLS="http://192.168.79.11:2379"
```

（8）将 ETCD 服务加入系统服务，编辑"/usr/lib/systemd/system/etcd.service"文件的内容如下：

```
[Unit]
Description=Etcd Server
After=network.target
After=network-online.target
Wants=network-online.target

[Service]
Type=notify
EnvironmentFile=/opt/platform/etcd/etcd.conf
ExecStart=/usr/local/bin/etcd \
--cert-file=/opt/ssl/etcd/etcd.pem \
--key-file=/opt/ssl/etcd/etcd-key.pem \
--peer-cert-file=/opt/ssl/etcd/etcd.pem \
--peer-key-file=/opt/ssl/etcd/etcd-key.pem \
--trusted-ca-file=/opt/ssl/etcd/etcd.pem \
--peer-trusted-ca-file=/opt/ssl/etcd/etcd.pem
Restart=on-failure
LimitNOFILE=65536

[Install]
WantedBy=multi-user.target
```

（9）创建 ETCD 服务的数据存储目录，然后启动 ETCD 服务。

```
mkdir -p /opt/platform/etcd/data
chmod 755 /opt/platform/etcd/data
systemctl daemon-reload
systemctl enable etcd.service
systemctl start etcd.service
```

（10）验证 ETCD 服务的状态。

```
etcdctl cluster-health
```

输出的信息如下：

```
member fd4d0bd2446259d9 is healthy:
got healthy result from http://192.168.79.11:2379
cluster is healthy
```

（11）查看 ETCD 服务的成员列表。

```
etcdctl member list
```

输出的信息如下：

```
fd4d0bd2446259d9: name=k8s-etcd peerURLs=http://192.168.79.11:2380
clientURLs=http://192.168.79.11:2379 isLeader=true
```

由于是单节点的 ETCD，因此这里只有一个成员信息。

(12) 将 ETCD 服务的证书文件复制到 node1 和 node2 节点上。

```
cd /opt
scp -r ssl/ root@node1:/opt
scp -r ssl/ root@node2:/opt
```

2. 部署 flannel 网络

（1）在 master 节点上将分配的子网段写入 ETCD 服务中供 flannel 网络使用：

```
etcdctl set /coreos.com/network/config '{ "Network": "172.17.0.0/16", "Backend": {"Type": "vxlan"}}'
```

（2）在 master 节点上查看写 flannel 子网的信息：

```
etcdctl get /coreos.com/network/config
```

输出的信息如下：

```
{ "Network": "172.17.0.0/16", "Backend": {"Type": "vxlan"}}
```

（3）在 node1 节点上解压缩 flannel-v0.10.0-linux-amd64.tar.gz 安装包：

```
tar -zxvf flannel-v0.10.0-linux-amd64.tar.gz
```

（4）在 node1 节点上创建 Kubernetes 的工作目录。

```
mkdir -p /opt/kubernetes/{cfg,bin,ssl}
mv mk-docker-opts.sh flanneld /opt/kubernetes/bin/
```

（5）在 node1 节点上创建定义 flannel 网络的脚本文件 "flannel.sh"：

```
#!/bin/bash
ETCD_ENDPOINTS=${1}

cat <<EOF >/opt/kubernetes/cfg/flanneld
FLANNEL_OPTIONS="--etcd-endpoints=${ETCD_ENDPOINTS} \
-etcd-cafile=/opt/ssl/etcd/etcd.pem \
-etcd-certfile=/opt/ssl/etcd/etcd.pem \
-etcd-keyfile=/opt/ssl/etcd/etcd-key.pem"
EOF

cat <<EOF >/usr/lib/systemd/system/flanneld.service
[Unit]
Description=Flanneld overlay address etcd agent
```

```
After=network-online.target network.target
Before=docker.service

[Service]
Type=notify
EnvironmentFile=/opt/kubernetes/cfg/flanneld
ExecStart=/opt/kubernetes/bin/flanneld --ip-masq \$FLANNEL_OPTIONS
ExecStartPost=/opt/kubernetes/bin/mk-docker-opts.sh -k
DOCKER_NETWORK_OPTIONS -d /run/flannel/subnet.env
Restart=on-failure

[Install]
WantedBy=multi-user.target
EOF

systemctl daemon-reload
systemctl enable flanneld
systemctl restart flanneld
```

（6）在 node1 节点上启用 flannel 网络：

```
bash flannel.sh http://192.168.79.11:2379
```

这里指定了在 master 节点上部署的 ETCD 服务地址。

（7）在 node1 节点上查看 flannel 网络的状态：

```
systemctl status flanneld
```

输出的信息如下：

```
flanneld.service - Flanneld overlay address etcd agent
   Loaded: loaded (/usr/lib/systemd/system/flanneld.service; enabled; vendor preset: disabled)
   Active: active (running) since Tue 2022-02-08 22:30:46 CST; 6s ago
```

（8）在 node1 节点上修改"/usr/lib/systemd/system/docker.service"文件以配置运行在 node1 节点上的 Docker 连接 flannel 网络——在文件中增加以下这一行：

```
EnvironmentFile=/run/flannel/subnet.env
```

（9）在 node1 节点上重启 Docker 服务。

```
systemctl daemon-reload
systemctl restart docker.service
```

（10）执行以下命令查看 node1 节点上的 flannel 网络信息，如图 13-3 所示。

```
ifconfig
```

```
docker0: flags=4099<UP,BROADCAST,MULTICAST>  mtu 1500
        inet 172.17.89.1  netmask 255.255.255.0  broadcast 0.0.0.0
        ether 02:42:d3:82:b9:a5  txqueuelen 0  (Ethernet)
        RX packets 0  bytes 0 (0.0 B)
        RX errors 0  dropped 0  overruns 0  frame 0
        TX packets 0  bytes 0 (0.0 B)
        TX errors 0  dropped 0  overruns 0  carrier 0  collisions 0
flannel.1: flags=4163<UP,BROADCAST,RUNNING,MULTICAST>  mtu 1450
        inet 172.17.89.0  netmask 255.255.255.255  broadcast 0.0.0.0
        inet6 fe80::a842:e8ff:fe5d:ac19  prefixlen 64  scopeid 0x20<link>
        ether aa:42:e8:5d:ac:19  txqueuelen 0  (Ethernet)
        RX packets 0  bytes 0 (0.0 B)
        RX errors 0  dropped 0  overruns 0  frame 0
        TX packets 0  bytes 0 (0.0 B)
        TX errors 0  dropped 8  overruns 0  carrier 0  collisions 0
```

图 13-3

（11）在 node2 节点上配置 flannel 网络，重复第（3）~（10）步。

3. 部署 master 节点

（1）创建 Kubernetes 集群证书目录。

```
mkdir -p /opt/ssl/k8s
cd /opt/ssl/k8s
```

（2）创建脚本文件"k8s-cert.sh"用于生成 Kubernetes 集群的证书——在脚本中输入以下内容：

```
cat > ca-config.json <<EOF
{
  "signing": {
    "default": {
      "expiry": "87600h"
    },
    "profiles": {
      "kubernetes": {
        "usages": [
            "signing",
            "key encipherment",
            "server auth",
            "client auth"
        ],
        "expiry": "87600h"
      }
    }
```

```
    }
  }
EOF

cat > ca-csr.json <<EOF
{
  "CN": "kubernetes",
  "key": {
    "algo": "rsa",
    "size": 2048
  },
  "names": [{
    "C": "CN",
    "ST": "BeiJing",
    "L": "BeiJing",
    "O": "k8s",
    "OU": "System"
  }]
}
EOF

cfssl gencert -initca ca-csr.json | cfssljson -bare ca

cat >server-csr.json<<EOF
{
  "CN": "kubernetes",
  "hosts": [
    "192.168.79.11",
    "127.0.0.1",
    "kubernetes",
    "kubernetes.default",
    "kubernetes.default.svc",
    "kubernetes.default.svc.cluster",
    "kubernetes.default.svc.cluster.local"
  ],
  "key": {
    "algo": "rsa",
    "size": 2048
  },
  "names": [
    {
      "C": "CN",
      "L": "BeiJing",
      "ST": "BeiJing",
      "O": "system:masters",
```

```
      "OU": "System"
    }
  ]
}
EOF

cfssl gencert -ca=ca.pem -ca-key=ca-key.pem \
-config=ca-config.json -profile=kubernetes \
server-csr.json | cfssljson -bare server

cat >admin-csr.json <<EOF
{
  "CN": "admin",
  "hosts": [],
  "key": {
    "algo": "rsa",
    "size": 2048
  },
  "names": [
    {
      "C": "CN",
      "L": "BeiJing",
      "ST": "BeiJing",
      "O": "system:masters",
      "OU": "System"
    }
  ]
}
EOF
cfssl gencert -ca=ca.pem -ca-key=ca-key.pem \
-config=ca-config.json -profile=kubernetes \
admin-csr.json | cfssljson -bare admin

cat > kube-proxy-csr.json <<EOF
{
  "CN": "system:kube-proxy",
  "hosts": [],
  "key": {
    "algo": "rsa",
    "size": 2048
  },
  "names": [
    {
      "C": "CN",
      "L": "BeiJing",
```

```
      "ST": "BeiJing",
      "O": "k8s",
      "OU": "System"
    }
  ]
}
EOF

cfssl gencert -ca=ca.pem -ca-key=ca-key.pem \
-config=ca-config.json -profile=kubernetes \
kube-proxy-csr.json | cfssljson -bare kube-proxy
```

（3）执行脚本文件"k8s-cert.sh"。

```
bash k8s-cert.sh
```

（4）复制证书。

```
mkdir -p /opt/kubernetes/ssl/
mkdir -p /opt/kubernetes/logs/
cp ca*pem server*pem /opt/kubernetes/ssl/
```

（5）解压缩 kubernetes 压缩包。

```
tar -zxvf kubernetes-server-linux-amd64.tar.gz
```

（6）复制关键命令文件。

```
mkdir -p /opt/kubernetes/bin/
cd kubernetes/server/bin/
cp kube-apiserver kube-scheduler kube-controller-manager \
   /opt/kubernetes/bin
cp kubectl /usr/local/bin/
```

（7）随机生成序列号。

```
mkdir -p /opt/kubernetes/cfg
head -c 16 /dev/urandom | od -An -t x | tr -d ' '
```

输出内容如下：

```
05cd8031b0c415de2f062503b0cd4ee6
```

（8）创建"/opt/kubernetes/cfg/token.csv"文件，在其中输入以下内容：

```
05cd8031b0c415de2f062503b0cd4ee6,kubelet-bootstrap,10001,"system:node-bo
otstrapper"
```

（9）创建 API Server 的配置文件"/opt/kubernetes/cfg/kube-apiserver.conf"，在其中输入以下内容：

```
KUBE_APISERVER_OPTS="--logtostderr=false \
```

```
--v=2 \
--log-dir=/opt/kubernetes/logs \
--etcd-servers=http://192.168.79.11:2379 \
--bind-address=192.168.79.11 \
--secure-port=6443 \
--advertise-address=192.168.79.11 \
--allow-privileged=true \
--service-cluster-ip-range=10.0.0.0/24 \
--enable-admission-plugins=NamespaceLifecycle,LimitRanger,ServiceAccount
,ResourceQuota,NodeRestriction \
--authorization-mode=RBAC,Node \
--enable-bootstrap-token-auth=true \
--token-auth-file=/opt/kubernetes/cfg/token.csv \
--service-node-port-range=30000-32767 \
--kubelet-client-certificate=/opt/kubernetes/ssl/server.pem \
--kubelet-client-key=/opt/kubernetes/ssl/server-key.pem \
--tls-cert-file=/opt/kubernetes/ssl/server.pem \
--tls-private-key-file=/opt/kubernetes/ssl/server-key.pem \
--client-ca-file=/opt/kubernetes/ssl/ca.pem \
--service-account-key-file=/opt/kubernetes/ssl/ca-key.pem \
--etcd-cafile=/opt/ssl/etcd/etcd.pem \
--etcd-certfile=/opt/ssl/etcd/etcd.pem \
--etcd-keyfile=/opt/ssl/etcd/etcd-key.pem \
--audit-log-maxage=30 \
--audit-log-maxbackup=3 \
--audit-log-maxsize=100 \
--audit-log-path=/opt/kubernetes/logs/k8s-audit.log"
```

（10）执行以下命令使用系统的 systemd 来管理 API Server：

```
cat > /usr/lib/systemd/system/kube-apiserver.service << EOF
[Unit]
Description=Kubernetes API Server
Documentation=https://github.com/kubernetes/kubernetes
[Service]
EnvironmentFile=/opt/kubernetes/cfg/kube-apiserver.conf
ExecStart=/opt/kubernetes/bin/kube-apiserver \$KUBE_APISERVER_OPTS
Restart=on-failure
[Install]
WantedBy=multi-user.target
EOF
```

（11）启动 API Server。

```
systemctl daemon-reload
systemctl start kube-apiserver
systemctl enable kube-apiserver
```

（12）查看 API Server 的状态。

```
systemctl status kube-apiserver.service
```

输出的信息如下：

```
kube-apiserver.service - Kubernetes API Server
   Loaded: loaded (/usr/lib/systemd/system/kube-apiserver.service; enabled; vendor preset: disabled)
   Active: active (running) since Tue 2022-02-08 21:11:47 CST; 24min ago
```

（13）查看监听的 6433 和 8080 端口的信息，如图 13-4 所示。

```
netstat -ntap | grep 6443
netstat -ntap | grep 8080
```

```
[root@master ~]# netstat -ntap | grep 6443
tcp        0      0 192.168.79.11:6443       0.0.0.0:*               LISTEN      4818/kube-apiserver
tcp        0      0 192.168.79.11:36644      192.168.79.11:6443      ESTABLISHED 4818/kube-apiserver
tcp        0      0 192.168.79.11:6443       192.168.79.11:36644     ESTABLISHED 4818/kube-apiserver
[root@master ~]# netstat -ntap | grep 8080
tcp        0      0 127.0.0.1:8080           0.0.0.0:*               LISTEN      4818/kube-apiserver
```

图 13-4

（14）授权 kubelet-bootstrap 用户可以请求证书。

```
kubectl create clusterrolebinding kubelet-bootstrap \
--clusterrole=system:node-bootstrapper \
--user=kubelet-bootstrap
```

（15）执行以下命令创建 kube-controller-manager 的配置文件：

```
cat > /opt/kubernetes/cfg/kube-controller-manager.conf << EOF
KUBE_CONTROLLER_MANAGER_OPTS="--logtostderr=false \
--v=2 \
--log-dir=/opt/kubernetes/logs \
--leader-elect=true \
--master=127.0.0.1:8080 \
--bind-address=127.0.0.1 \
--allocate-node-cidrs=true \
--cluster-cidr=10.244.0.0/16 \
--service-cluster-ip-range=10.0.0.0/24 \
--cluster-signing-cert-file=/opt/kubernetes/ssl/ca.pem \
--cluster-signing-key-file=/opt/kubernetes/ssl/ca-key.pem \
--root-ca-file=/opt/kubernetes/ssl/ca.pem \
--service-account-private-key-file=/opt/kubernetes/ssl/ca-key.pem \
--experimental-cluster-signing-duration=87600h0m0s"
EOF
```

（16）执行以下命令使用 systemd 服务来管理 kube-controller-manager。

```
cat > /usr/lib/systemd/system/kube-controller-manager.service << EOF
[Unit]
Description=Kubernetes Controller Manager
Documentation=https://github.com/kubernetes/kubernetes
[Service]
EnvironmentFile=/opt/kubernetes/cfg/kube-controller-manager.conf
ExecStart=/opt/kubernetes/bin/kube-controller-manager \$KUBE_CONTROLLER_MANAGER_OPTS
Restart=on-failure
[Install]
WantedBy=multi-user.target
EOF
```

（17）启动 kube-controller-manager。

```
systemctl daemon-reload
systemctl start kube-controller-manager
systemctl enable kube-controller-manager
```

（18）查看 kube-controller-manager 的状态。

```
systemctl status kube-controller-manager
```

输出的信息如下：

```
kube-controller-manager.service - Kubernetes Controller Manager
   Loaded: loaded (/usr/lib/systemd/system/kube-controller-manager.service; enabled; vendor preset: disabled)
   Active: active (running) since Tue 2022-02-08 20:42:08 CST; 1h 2min ago
```

（19）执行以下命令创建 kube-scheduler 的配置文件。

```
cat > /opt/kubernetes/cfg/kube-scheduler.conf << EOF
KUBE_SCHEDULER_OPTS="--logtostderr=false \
--v=2 \
--log-dir=/opt/kubernetes/logs \
--leader-elect \
--master=127.0.0.1:8080 \
--bind-address=127.0.0.1"
EOF
```

（20）执行以下命令使用 systemd 服务来管理 kube-scheduler。

```
cat > /usr/lib/systemd/system/kube-scheduler.service << EOF
[Unit]
Description=Kubernetes Scheduler
Documentation=https://github.com/kubernetes/kubernetes
[Service]
```

```
EnvironmentFile=/opt/kubernetes/cfg/kube-scheduler.conf
ExecStart=/opt/kubernetes/bin/kube-scheduler \$KUBE_SCHEDULER_OPTS
Restart=on-failure
[Install]
WantedBy=multi-user.target
EOF
```

(21) 启动 kube-scheduler。

```
systemctl daemon-reload
systemctl start kube-scheduler
systemctl enable kube-scheduler
```

(22) 查看 kube-scheduler 的状态。

```
systemctl status kube-scheduler.service
```

输出的信息如下：

```
kube-scheduler.service - Kubernetes Scheduler
  Loaded: loaded (/usr/lib/systemd/system/kube-scheduler.service; enabled;
vendor preset: disabled)
  Active: active (running) since Tue 2022-02-08 20:43:01 CST; 1h 8min ago
```

(23) 查看 master 节点的状态信息。

```
kubectl get cs
```

输出的信息如下：

```
NAME                 STATUS    MESSAGE             ERROR
etcd-0               Healthy   {"health":"true"}
controller-manager   Healthy   ok
scheduler            Healthy   ok
```

4. 部署 node 节点

(1) 在 master 节点上创建脚本文件"kubeconfig"，在其中输入以下内容：

```
APISERVER=${1}
SSL_DIR=${2}

# 创建 kubelet bootstrapping kubeconfig
export KUBE_APISERVER="https://$APISERVER:6443"

# 设置集群参数
kubectl config set-cluster kubernetes \
  --certificate-authority=$SSL_DIR/ca.pem \
  --embed-certs=true \
  --server=${KUBE_APISERVER} \
  --kubeconfig=bootstrap.kubeconfig
```

```
# 设置客户端认证参数
# 注意，这里的 token ID 需要与 token.csv 文件中的 ID 一致
kubectl config set-credentials kubelet-bootstrap \
  --token=05cd8031b0c415de2f062503b0cd4ee6 \
  --kubeconfig=bootstrap.kubeconfig

# 设置上下文参数
kubectl config set-context default \
  --cluster=kubernetes \
  --user=kubelet-bootstrap \
  --kubeconfig=bootstrap.kubeconfig

# 设置默认上下文
kubectl config use-context default --kubeconfig=bootstrap.kubeconfig

#----------------------

# 创建 kube-proxy kubeconfig 文件

kubectl config set-cluster kubernetes \
  --certificate-authority=$SSL_DIR/ca.pem \
  --embed-certs=true \
  --server=${KUBE_APISERVER} \
  --kubeconfig=kube-proxy.kubeconfig

kubectl config set-credentials kube-proxy \
  --client-certificate=$SSL_DIR/kube-proxy.pem \
  --client-key=$SSL_DIR/kube-proxy-key.pem \
  --embed-certs=true \
  --kubeconfig=kube-proxy.kubeconfig

kubectl config set-context default \
  --cluster=kubernetes \
  --user=kube-proxy \
  --kubeconfig=kube-proxy.kubeconfig

kubectl config use-context default --kubeconfig=kube-proxy.kubeconfig
```

（2）执行脚本文件 "kubeconfig"。

```
bash kubeconfig 192.168.79.11 /opt/ssl/k8s/
```

输出的信息如下：

```
Cluster "kubernetes" set.
User "kubelet-bootstrap" set.
```

```
Context "default" created.
Switched to context "default".
Cluster "kubernetes" set.
User "kube-proxy" set.
Context "default" created.
Switched to context "default".
```

(3)将 master 节点上生成的配置文件复制到 node1 和 node2 节点上。

```
scp bootstrap.kubeconfig kube-proxy.kubeconfig \
root@node1:/opt/kubernetes/cfg/

scp bootstrap.kubeconfig kube-proxy.kubeconfig \
root@node2:/opt/kubernetes/cfg/
```

(4)在 node1 节点上解压缩 kubernetes-node-linux-amd64.tar.gz 文件。

```
tar -zxvf kubernetes-node-linux-amd64.tar.gz
```

(5)在 node1 节点上将 kubelet 和 kube-proxy 复制到目录"/opt/kubernetes/bin/"下。

```
cd kubernetes/node/bin/
cp kubelet kube-proxy /opt/kubernetes/bin/
```

(6)在 node1 节点上创建脚本文件"kubelet.sh",在其中输入以下内容:

```
#!/bin/bash

NODE_ADDRESS=$1
DNS_SERVER_IP=${2:-"10.0.0.2"}

cat <<EOF >/opt/kubernetes/cfg/kubelet

KUBELET_OPTS="--logtostderr=true \\
--v=4 \\
--hostname-override=${NODE_ADDRESS} \\
--kubeconfig=/opt/kubernetes/cfg/kubelet.kubeconfig \\
--bootstrap-kubeconfig=/opt/kubernetes/cfg/bootstrap.kubeconfig \\
--config=/opt/kubernetes/cfg/kubelet.config \\
--cert-dir=/opt/kubernetes/ssl \\
--pod-infra-container-image=registry.cn-hangzhou.aliyuncs.com/google-containers/pause-amd64:3.0"

EOF

cat <<EOF >/opt/kubernetes/cfg/kubelet.config

kind: KubeletConfiguration
apiVersion: kubelet.config.k8s.io/v1beta1
```

```
  address: ${NODE_ADDRESS}
  port: 10250
  readOnlyPort: 10255
  cgroupDriver: systemd
  clusterDNS:
  - ${DNS_SERVER_IP}
  clusterDomain: cluster.local.
  failSwapOn: false
  authentication:
    anonymous:
      enabled: true
EOF

cat <<EOF >/usr/lib/systemd/system/kubelet.service
[Unit]
Description=Kubernetes Kubelet
After=docker.service
Requires=docker.service

[Service]
EnvironmentFile=/opt/kubernetes/cfg/kubelet
ExecStart=/opt/kubernetes/bin/kubelet \$KUBELET_OPTS
Restart=on-failure
KillMode=process

[Install]
WantedBy=multi-user.target
EOF

systemctl daemon-reload
systemctl enable kubelet
systemctl restart kubelet
```

（7）在 node1 节点上执行脚本文件"kubelet.sh"。

```
bash kubelet.sh 192.168.79.12
```

这里指定的是 node1 节点的 IP 地址。

（8）在 node1 节点上查看 Kubelet 的状态。

```
systemctl status kubelet
```

输出的信息如下：

```
kubelet.service - Kubernetes Kubelet
   Loaded: loaded (/usr/lib/systemd/system/kubelet.service; enabled; vendor
preset: disabled)
   Active: active (running) since Tue 2022-02-08 23:23:52 CST; 3min 18s ago
```

(9)在 node1 节点上创建脚本文件"proxy.sh",在其中输入以下内容

```
#!/bin/bash

NODE_ADDRESS=$1

cat <<EOF >/opt/kubernetes/cfg/kube-proxy

KUBE_PROXY_OPTS="--logtostderr=true \\
--v=4 \\
--hostname-override=${NODE_ADDRESS} \\
--cluster-cidr=10.0.0.0/24 \\
--proxy-mode=ipvs \\
--kubeconfig=/opt/kubernetes/cfg/kube-proxy.kubeconfig"

EOF

cat <<EOF >/usr/lib/systemd/system/kube-proxy.service
[Unit]
Description=Kubernetes Proxy
After=network.target

[Service]
EnvironmentFile=-/opt/kubernetes/cfg/kube-proxy
ExecStart=/opt/kubernetes/bin/kube-proxy \$KUBE_PROXY_OPTS
Restart=on-failure

[Install]
WantedBy=multi-user.target
EOF

systemctl daemon-reload
systemctl enable kube-proxy
systemctl restart kube-proxy
```

(10)在 node1 节点上执行脚本文件"proxy.sh"。

```
bash proxy.sh 192.168.79.12
```

(11)在 node1 节点上查看 kube-proxy 的状态。

```
systemctl status kube-proxy.service
```

输出的信息如下：

```
kube-proxy.service - Kubernetes Proxy
  Loaded: loaded (/usr/lib/systemd/system/kube-proxy.service; enabled; vendor preset: disabled)
  Active: active (running) since Tue 2022-02-08 23:30:51 CST; 9s ago
```

（12）在 master 节点上检查 node1 节点加入集群的请求信息：

```
kubectl get csr
```

输出的信息如下：

```
NAME                                                    ... CONDITION
node-csr-Qc2wKIo6AIWh6AXKW6tNwAvUqpxEIXFPHkkIe1jzSBE ... Pending
```

（13）在 master 节点上批准 node1 节点的请求：

```
kubectl certificate approve \
node-csr-Qc2wKIo6AIWh6AXKW6tNwAvUqpxEIXFPHkkIe1jzSBE
```

（14）在 master 节点上查看 Kubernetes 集群中的节点信息：

```
kubectl get node
```

输出的信息如下：

```
NAME             STATUS   ROLES    AGE   VERSION
192.168.79.12    Ready    <none>   85s   v1.18.20
```

> 这时 node1 节点已经成功加入 Kubernetes 集群。

（15）在 node2 节点上重复前面的第（4）~（14）步把 node2 节点加入集群。

（16）在 master 节点上查看 Kubernetes 集群中的节点信息：

```
kubectl get node
```

输出的信息如下：

```
NAME             STATUS   ROLES    AGE     VERSION
192.168.79.12    Ready    <none>   5m47s   v1.18.20
192.168.79.13    Ready    <none>   11s     v1.18.20
```

至此，成功使用二进制包部署了包含 3 个节点的 Kubernetes 集群。

13.1.4 使用 minikube 工具部署 Kubernetes 单机版集群

minikube 是一个本地运行 Kubernetes 集群的简易工具。利用 minikube 可以轻松搭建单机版

Kubernetes 集群,并使用它进行日常的开发与测试。

下面在单机环境中使用 minikube 部署 Kubernetes 集群,并部署一个简单的应用。

(1)根据表 13-2 的要求,按照 1.4.1 节的步骤单独准备一个虚拟机,并安装和启动 Docker。

表 13-2

虚拟机参数	参考值
CPU 核数	2 个及以上
虚拟机内存大小	2GB 及以上
虚拟机硬盘大小	20GB 及以上

 默认情况下 minikube 只能在本地访问。为了方便使用 minikube,建议在安装 CentOS Linux 时勾选"Server with GUI",这样就有了 CentOS 的图形界面。

(2)安装 kubectl。

```
curl -LO \
"https://dl.k8s.io/release/$(curl -L -s https://dl.k8s.io/release/stable.txt)/bin/linux/amd64/kubectl"
sudo install -o root -g root -m 0755 kubectl /usr/local/bin/kubectl
kubectl version --client
```

(3)下载 minikube 的安装文件并安装。

```
curl \-LO \
https://storage.googleapis.com/minikube/releases/latest/minikube-linux-amd64
sudo install minikube-linux-amd64 /usr/local/bin/minikube
```

(4)由于 minikube 不能在"root"用户下执行,因此需要创建一个用户并设置该用户的密码。

```
useradd minikube
passwd minikube
```

(5)在文件"/etc/sudoers"文件中增加"minikube"用户。

```
minikube ALL=(ALL)        ALL
```

(6)切换到 minikube 用户。

```
su - minikube
```

(7)创建 docker 组。

```
sudo groupadd docker
```

（8）将 minikube 用户添加到 docker 组中。

```
sudo usermod -aG docker $USER && newgrp docker
```

（9）重启 Docker 服务。

```
sudo systemctl restart docker.service
```

（10）启动 minikube。

```
minikube start
```

　　如果是第一次启动 minikube，则 minikube 会自动下载所需要的镜像信息，如图 13-5 所示。

```
* Starting control plane node minikube in cluster minikube
* Pulling base image ...
* Downloading Kubernetes v1.23.1 preload ...
    > preloaded-images-k8s-v16-v1...:  504.42 MiB / 504.42 MiB  100.00% 912.10 K
    > index.docker.io/kicbase/sta...:  226.51 MiB / 378.98 MiB   59.77% 175.43 Ki
```

图 13-5

minikube 在启动成功后会输出以下信息：

```
* Done! kubectl is now configured to use "minikube" cluster and "default" namespace by default
```

在默认情况下，minikube 会从官方的镜像仓库下载镜像，下载的时间会比较长。可以在启动 minikube 时指定国内的镜像仓库地址，以加速下载过程。命令如下：

```
minikube start \
--registry-mirror=https://registry.docker-cn.com \
--image-repository=registry.cn-hangzhou.aliyuncs.com/google_containers \
--vm-driver=docker \
--alsologtostderr \
-v=8 \
--base-image \
registry.cn-hangzhou.aliyuncs.com/google_containers/kicbase:v0.0.10
```

（11）查看 minikube 的节点数。在默认情况下，minikube 在单节点上实现一个集群环境。

```
minikube node list
```

输出的信息如下：

```
minikube   192.168.49.2
```

（12）向 minikube 集群添加两个节点。

```
minikube node add
minikube node add
```

(13)重新查看 minikube 集群的节点信息。

```
minikube node list
```

输出的信息如下：

```
minikube         192.168.49.2
minikube-m02     192.168.49.3
minikube-m03     192.168.49.4
```

(14)启动 minikube 集群的 Web 可视化界面，如图 13-6 所示。

```
minikube dashboard
```

图 13-6

 如果启动失败，则可以通过以下命令尝试重新启动。

minikube delete

minikube start

在默认情况下，minikube 集群的 Web 可视化界面只能从本地访问，可以通过以下命令从其他主机访问。

kubectl proxy --port=8888 --address='0.0.0.0' --accept-hosts='^.*'

下面是访问的 URL：

http://IP:8888/api/v1/namespaces/kubernetes-dashboard/services/http:kubernetes-dashboard:/proxy/#/node?namespace=default

(15)在命令行中查看 minikube 集群的节点信息。

```
kubectl get nodes
```

输出的信息如下：

```
NAME            STATUS    ROLES                   AGE      VERSION
minikube        Ready     control-plane,master    45m      v1.23.1
minikube-m02    Ready     <none>                  6m28s    v1.23.1
minikube-m03    Ready     <none>                  3m37s    v1.23.1
```

（16）基于 Nginx 镜像创建一个的 Deployment 应用，并暴露其容器的 80 端口。

```
docker pull nginx
kubectl create deployment nginx --image=nginx
kubectl expose deployment nginx --port=80 --type=NodePort
```

（17）获取 Kubernetes 的 Pod 和 Service 信息，如图 13-7 所示。

```
kubectl get pod,svc -o wide
```

图 13-7

从图 13-7 可以看到，部署的 Nginx 服务被调到 minikube-m03 上了，端口号是 30504。

（18）根据第（13）步中"minikube-m03"的 IP 地址，使用浏览器访问 30504 端口，即可打开 Nginx 首页，如图 13-8 所示。

图 13-8

13.1.5 Kubernetes 集群的高可用

Kubernetes 是主从式架构，主节点是 master 节点，从节点是 node 节点。一旦位于中央位置的 master 节点中断服务或发生宕机，则会导致所有 node 节点均不可控，有可能造成严重的事故。这就是主从架构的单点故障问题。

因此在实际的生产环境中，应实现 Kubernetes 集群的 HA（High Availablity，高可用）。图 13-9 展示了 Kubernetes 高可用集群架构。

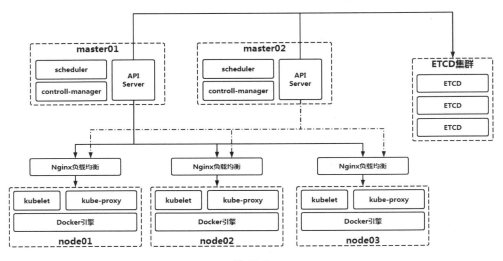

图 13-9

Kubernetes 高可用架构主要包含以下几方面。

（1）ETCD 的高可用。

ETCD 是 Kubernetes 中的持久化存储。虽然单节点的 ETCD 也可以正常运行，但为了保证 ETCD 不存在单点故障，推荐采用 3 个或者 5 个节点组成 ETCD 集群，供 Kubernetes 作为后端存储器使用。

（2）API Server 的高可用。

API Server 是一个无状态服务，它提供了操作集群的唯一入口。推荐在实际的生产环境中，至少在不同的 master 节点上搭建两个以上的 API Server，以实现 API Server 的高可用。

（3）controller-manager 与 scheduler 的高可用。

controller-manager 与 scheduler 也属于 master 的一部分。它们的高可用实现与 API Server 的高可用实现相同——在不同的 master 节点上运行多个实例。

13.2 Kubernetes 的客户端工具

随着 Kubernetes 应用得越来越广泛,出现了许多客户端工具。开发人员和运维人员使用这些客户端工具,可以很方便地与 Kubernetes 进行交互。这里主要介绍两款客户端工具:图形管理工具 DashBoard UI 和命令行管理工具 Kubectl。

13.2.1 Kubernetes 图形管理工具——DashBoard UI

Dashboard UI 是基于 Web 的图形管理工具,它提供了集群中应用的概览信息。通过 Dashboard UI,可以将应用部署到 Kubernetes 中,也可以对容器应用进行编排。另外,Dashboard UI 还可以管理集群的资源,如创建和修改 Deployment、Job、DaemonSet 等。Dashboard UI 还可以展示 Kubernetes 集群中的资源状态信息,以及所有的报错信息。

下面在 Kubernetes 集群中部署 Dashboard UI。

(1) 获取 DashBoard UI 的 YAML 配置文件。

```
wget \
https://raw.githubusercontent.com/kubernetes/dashboard/v1.10.1/src/deploy/recommended/kubernetes-dashboard.yaml
```

默认情况下,Kubernetes 将从 Google 网站下载需要的文件,所以,需要对 "kubernetes-dashboard.yaml" 文件进行适当的修改。

(2) 修改 "kubernetes-dashboard.yaml" 文件的第 112 行。

修改前:

```
image: k8s.gcr.io/kubernetes-dashboard-amd64:v1.10.1
```

修改后:

```
image: registry.cn-hangzhou.aliyuncs.com/google_containers/kubernetes-dashboard-amd64:v1.10.1
```

这里改为从阿里云网站下载 DashBoard UI 所需要的文件。

(3)修改"kubernetes-dashboard.yaml"文件中第 149 行以后的内容,增加 Service 的"type"参数和暴露的端口号,如以下加粗加斜部分所示:

```
# ------------------- Dashboard Service ------------------- #

kind: Service
apiVersion: v1
metadata:
  labels:
    k8s-app: kubernetes-dashboard
  name: kubernetes-dashboard
  namespace: kube-system
spec:
  type: NodePort
  ports:
    - port: 443
      targetPort: 8443
      nodePort: 31620
  selector:
    k8s-app: kubernetes-dashboard
```

(4)执行以下命令应用 DashBoard UI 的配置文件。

```
kubectl apply -f kubernetes-dashboard.yaml
```

执行成功后,输出如下信息:

```
secret/kubernetes-dashboard-certs created
serviceaccount/kubernetes-dashboard created
role.rbac.authorization.k8s.io/kubernetes-dashboard-minimal created
rolebinding.rbac.authorization.k8s.io/kubernetes-dashboard-minimal created
deployment.apps/kubernetes-dashboard created
service/kubernetes-dashboard created
```

(5)创建 service account 账号。

```
kubectl create serviceaccount dashboard-admin -n kube-system
```

(6)将创建的 service account 账号与集群管理员角色 cluster-admin 进行绑定。

```
kubectl create clusterrolebinding dashboard-admin \
 --clusterrole=cluster-admin \
 --serviceaccount=kube-system:dashboard-admin
```

(7)执行以下命令查看 DashBoard UI 登录时的 Token,如图 13-10 所示。

```
kubectl describe secrets -n kube-system \
$(kubectl -n kube-system \
get secret | awk '/dashboard-admin/{print $1}')
```

```
[root@master ~]# kubectl describe secrets -n kube-system \
> $(kubectl -n kube-system \
> get secret | awk '/dashboard-admin/{print $1}')
Name:         dashboard-admin-token-ggq9m
Namespace:    kube-system
Labels:       <none>
Annotations:  kubernetes.io/service-account.name: dashboard-admin
              kubernetes.io/service-account.uid: 5df4bdf8-27a9-431d-8c0b-7d6453ccfccc

Type:  kubernetes.io/service-account-token

Data
====
token:      eyJhbGciOiJSUzI1NiIsImtpZCI6IiJ9.eyJpc3MiOiJrdWJlcm5ldGVzL3NlcnZpY2VhY2NvdW50
VudC9uYW1lc3BhY2UiOiJrdWJlLXN5c3RlbSIsImt1YmVybmV0ZXMuaW8vc2VydmljZWFjY291bnQvc2VjcmV0Lm5
hbWUiOiJkYXNoYm9hcmQtYWRtaW4tdG9rZW4tZ2dxOW0iLCJrdWJlcm5ldGVzLmlvL3NlcnZpY2VhY2NvdW50L3N
lcnZpY2UtYWNjb3VudC5uYW1lIjoiZGFzaGJvYXJkLWFkbWluIiwia3ViZXJuZXRlcy5pby9zZXJ2aWNlYWNjb3V
udC5zZXJ2aWNlLWFjY291bnQudWlkIjoiNWRmNGJkZjgtMjdhOS00MzFkLThjMGItN2Q2NDUzY2NmY2NjIiwic3V
iIjoic3lzdGVtOnNlcnZpY2VhY2NvdW50Omt1YmUtc3lzdGVtOmRhc2hib2FyZC1hZG1pbiJ9.CZAOrVDXEJ45fp1UDzj_f3_znR7sD255W1kdhJCwpNf
IFOk2i_sExJx-GlROTh6SJJLKkVLee75XI9_OYxmwU8x6eN8IjgJ7UoLgB1ZyskkSpIw6jysoZFO6Av_bet2dhq_Y
d2soEj-wyqt1x9tkNQgqNP1xugJuJwKkPCbsVyHLEfzT6MRmvu4ULN9pUJjF7ZQ6kEXf4trGCFj-_B1xho1VYXHZu
3u8bkGA
ca.crt:     1025 bytes
namespace:  11 bytes
[root@master ~]#
```

图 13-10

（8）打开浏览器，使用 HTTPS 访问 master 节点的 31620 端口，即可看到 DashBoard UI 的登录页面，如图 13-11 所示。

图 13-11

（9）选择"令牌"单选按钮，并输入在第（7）步中生成的 Token，单击"登录"按钮。

（10）DashBoard UI 的首页如图 13-12 所示。

图 13-12

13.2.2 Kubernetes 命令行管理工具——kubectl

kubectl 是 Kubernetes 提供的命令行管理工具。通过使用 kubectl，可以管理和操作 Kubernetes。表 13-3 中列出了 kubect 的常用命令。

表 13-3

命令类型	命 令	说 明
基础命令	create	通过文件名或标准输入创建 Kubernetes 的资源
	expose	将 Kubernetes 的资源暴露为一个服务
	run	在集群中运行一个特定的镜像
	set	修改对象的特定功能
	explain	给资源添加文档说明
	get	获取资源信息
	edit	编辑资源的属性
	delete	通过文件名、标准输入、资源名称或标签选择器来删除资源
部署命令	rollout	管理资源的部署状态
	scale	对资源进行扩容/缩容
	autoscale	创建一个能够自动扩容或缩容的资源
集群管理命令	certificate	修改证书的资源
	cluster-info	显示集群信息
	top	监控集群资源的使用
	cordon	将节点标记为不可调度
	uncordon	将节点标记为可调度
	drain	指定维护期间排除的节点
	taint	更新节点上的污点
故障诊断与调试命令	describe	显示特定资源或资源组的详细信息
	logs	显示 Pod 中容器的日志信息
	attach	连接到一个运行的容器
	exec	在容器中执行命令
	port-forward	将本地端口转发到 Pod 中
	proxy	在 Kubernetes API Server 上运行一个 Proxy
	cp	复制文件或目录到容器中
	auth	检查授权
高级命令	diff	对比实时版本和潜在版本
	apply	通过文件名或者标准输入对资源应用进行配置
	patch	使用补丁修改或更新资源
	replace	通过文件名或标准输入替换一个资源

续表

命令类型	命 令	说 明
	convert	在不同的 API 版本之间转换配置文件
	kustomize	从目录或 URL 创建 kustomization 对象
设置命令	label	更新资源上的标签
	annotate	更新资源上的注释
	completion	实现 kubectl 工具的自动补全功能
其他命令	api-resources	输出集群支持的 API 资源
	api-versions	输出集群支持的 API 资源的版本
	config	修改 kubeconfig 文件
	plugin	运行一个命令行插件
	version	输出客户端和服务版本信息

下面来演示 kubectl 常用命令的使用。

（1）显示 "kube-system" 命名空间中的 Pod 详细信息，如图 13-13 所示。

```
kubectl get pod -o wide -n kube-system
```

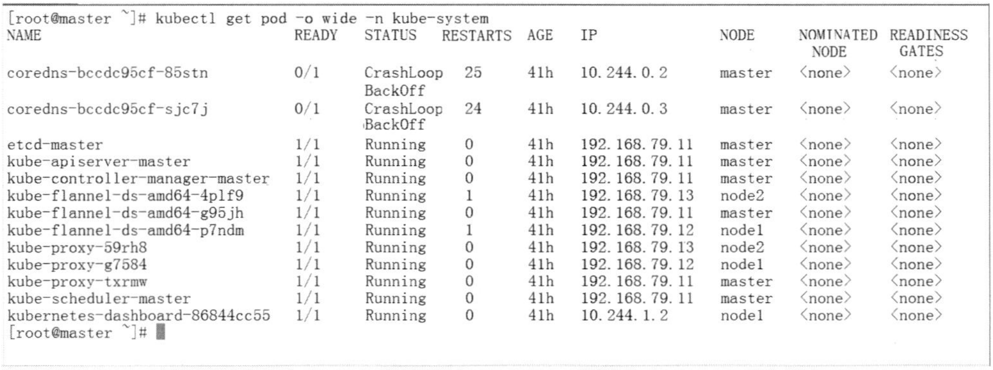

```
[root@master ~]# kubectl get pod -o wide -n kube-system
NAME                                   READY   STATUS             RESTARTS   AGE   IP             NODE     NOMINATED   READINESS
                                                                                                            NODE        GATES
coredns-bccdc95cf-85stn                0/1     CrashLoop          25         41h   10.244.0.2     master   <none>      <none>
                                               BackOff
coredns-bccdc95cf-sjc7j                0/1     CrashLoop          24         41h   10.244.0.3     master   <none>      <none>
                                               BackOff
etcd-master                            1/1     Running            0          41h   192.168.79.11  master   <none>      <none>
kube-apiserver-master                  1/1     Running            0          41h   192.168.79.11  master   <none>      <none>
kube-controller-manager-master         1/1     Running            0          41h   192.168.79.11  master   <none>      <none>
kube-flannel-ds-amd64-4plf9            1/1     Running            1          41h   192.168.79.13  node2    <none>      <none>
kube-flannel-ds-amd64-g95jh            1/1     Running            0          41h   192.168.79.11  master   <none>      <none>
kube-flannel-ds-amd64-p7ndm            1/1     Running            1          41h   192.168.79.12  node1    <none>      <none>
kube-proxy-59rh8                       1/1     Running            0          41h   192.168.79.13  node2    <none>      <none>
kube-proxy-g7584                       1/1     Running            0          41h   192.168.79.12  node1    <none>      <none>
kube-proxy-txrmw                       1/1     Running            0          41h   192.168.79.11  master   <none>      <none>
kube-scheduler-master                  1/1     Running            0          41h   192.168.79.11  master   <none>      <none>
kubernetes-dashboard-86844cc55         1/1     Running            0          41h   10.244.1.2     node1    <none>      <none>
[root@master ~]#
```

图 13-13

（2）使用镜像 "nginx:1.14" 创建一个名为 "nginx" 的 Deployment 资源。该资源有 3 个副本，并且暴露资源的 80 端口。

```
kubectl run nginx --replicas=3 --image=nginx:1.14 --port=80
```

（3）为名为 "nginx" 的 Deployment 资源创建一个 "NodePort" 类型的服务。

```
kubectl expose deployment nginx --port=80 --type=NodePort \
--target-port=80 --name=nginx-service
```

（4）获取 "default" 命名空间中的 Pod、service 和 deployment，如图 13-14 所示。

```
kubectl get pod,service,deployment
```

```
[root@master ~]# kubectl get pod,service,deployment
NAME                            READY   STATUS    RESTARTS   AGE
pod/nginx-65fc77987d-kxzz1      1/1     Running   0          67s
pod/nginx-65fc77987d-q15n8      1/1     Running   0          67s
pod/nginx-65fc77987d-tcb9p      1/1     Running   0          67s

NAME                    TYPE        CLUSTER-IP    EXTERNAL-IP   PORT(S)        AGE
service/kubernetes      ClusterIP   10.1.0.1      <none>        443/TCP        41h
service/nginx-service   NodePort    10.1.202.134  <none>        80:30151/TCP   58s

NAME                         READY   UP-TO-DATE   AVAILABLE   AGE
deployment.extensions/nginx  3/3     3            3           67s
[root@master ~]#
```

图 13-14

（5）使用镜像"nginx:1.15"更新名为"nginx"的 Deployment 资源。

```
kubectl set image deployment/nginx nginx=nginx:1.15
```

（6）查看名为"nginx"的 Deployment 资源的历史版本。

```
kubectl rollout history deployment/nginx
```

（7）回滚名为"nginx"的 Deployment 资源到上一个版本。

```
kubectl rollout undo deployment/nginx
```

（8）删除名为"nginx"的 Deployment 资源。

```
kubectl delete deploy/nginx
```

（9）删除名为"nginx"的 Service 资源。

```
kubectl delete svc/nginx-service
```

13.3 【实战】使用 Kubectl 在 Kubernetes 中部署第一个应用

在 Kubernetes 中部署应用主要分为 5 个步骤，如图 13-15 所示。

第一步 制作镜像	使用Dockerfile制作Docker镜像。
第二步 创建控制器	根据需要创建相应的控制管理Pod，最常用的控制器是Deployment控制器。
第三步 暴露应用	根据需要使用对应的Service类型暴露集群内部的应用。
第四步 发布应用	对外发布集群部署的应用程序，让外部可以进行访问。
第五步 监控应用	搭建集群的监控系统，如ELK，监视集群中应用的运行。

图 13-15

下面以 2.4.4 节中的 "MyDemoWeb.war" Web 应用为例，在 Kubernetes 集群中进行部署。

（1）创建 Dockerfile 文件。

```
FROM tomcat
ADD ./MyDemoWeb.war /usr/local/tomcat/webapps
```

（2）编译 Dockerfile 文件，并将结果保存到 Docker Hub 中。

```
docker build -t collenzhao/k8s-javaweb-demo .
docker push collenzhao/k8s-javaweb-demo
```

这里的 "collenzhao" 是 Docker Hub 上笔者的仓库名称。读者可以根据需要换成自己的仓库。

（3）生成应用部署的 YAML 文件。

```
kubectl create deployment javawebdemo \
--image=collenzhao/k8s-javaweb-demo \
--dry-run -o yaml > javawebdemo.yaml
```

生成的 "javawebdemo.yaml" 文件内容如下：

```
apiVersion: apps/v1
kind: Deployment
metadata:
  creationTimestamp: null
  labels:
    app: javawebdemo
  name: javawebdemo
spec:
  replicas: 1
  selector:
    matchLabels:
      app: javawebdemo
  strategy: {}
  template:
    metadata:
      creationTimestamp: null
      labels:
        app: javawebdemo
    spec:
      containers:
      - image: collenzhao/k8s-javaweb-demo
        name: k8s-javaweb-demo
        resources: {}
```

```
status: {}
```

（4）在"javawebdemo.yaml"文件中修改 Deployment 资源的副本数，改为 3 个副本。

```
replicas: 3
```

（5）将应用部署到 Kubernetes 集群中。

```
kubectl apply -f javawebdemo.yaml
```

（6）执行以下语句查看 expose 语句生成的 YAML 文件。

```
kubectl expose deployment javawebdemo --port=80 \
--target-port=8080 --type=NodePort -o yaml --dry-run \
> javawebdemosvc.yaml
```

生成的"javawebdemosvc.yaml"文件内容如下：

```
apiVersion: v1
kind: Service
metadata:
  creationTimestamp: null
  labels:
    app: javawebdemo
  name: javawebdemo
spec:
  ports:
  - port: 80
    protocol: TCP
    targetPort: 8080
  selector:
    app: javawebdemo
  type: NodePort
status:
  loadBalancer: {}
```

这里采用 NodePort 服务类型将宿主机的 80 端口映射到 Pod 容器内的 8080 端口。

（7）将应用暴露给外部。

```
kubectl apply -f javawebdemosvc.yaml
```

（8）查看 Kubernetes 集群中的应用，如图 13-16 所示。

```
kubectl get pods,svc,deploy
```

```
[root@master ~]# kubectl get pods,svc,deploy
NAME                                     READY   STATUS    RESTARTS   AGE
pod/javawebdemo-64758f5f66-krbq8          1/1     Running   0          5m50s
pod/javawebdemo-64758f5f66-mvs6h          1/1     Running   0          5m50s
pod/javawebdemo-64758f5f66-q2577          1/1     Running   0          5m50s

NAME                   TYPE        CLUSTER-IP    EXTERNAL-IP   PORT(S)         AGE
service/javawebdemo    NodePort    10.1.47.147   <none>        80:31462/TCP    2m50s
service/kubernetes     ClusterIP   10.1.0.1      <none>        443/TCP         43h

NAME                                   READY   UP-TO-DATE   AVAILABLE   AGE
deployment.extensions/javawebdemo      3/3     3            3           5m50s
[root@master ~]#
```

图 13-16

　　由于在 javawebdemo.yaml 中设置了 Deployment 的副本数是 3，所以在这里可以看到在 Pod 中启动了 3 个容器。通过以下命令可以看到这 3 个容器分别运行在哪个节点上。
　　kubectl get pod -o wide

（9）打开浏览器，通过"http://IP:Port/MyDemoWeb/"访问应用，结果如图 13-17 所示。

图 13-17

第 14 章
Kubernetes 中的最小可部署对象 Pod

Kubernetes 通过一个抽象的逻辑单元 Pod 将应用的容器组合在一起，从而使得应用更容易被发现和管理。Pod 是 Kubernetes 的核心。

14.1 什么是 Pod

Pod 是 Kubernetes 中的一个逻辑单位，它代表集群中正在运行的一个进程，是 Kubernetes 集群中的一个应用实例，由一个或者多个容器组成，如图 14-1 所示。在 Pod 中还可以包含数据的持久化存储、网络配置等资源。Pod 支持多种容器的执行环境，而 Docker 则是 Pod 最常见的执行环境。Pod 也支持用其他容器引擎作为执行环境。

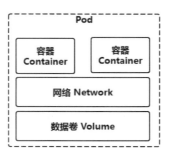

图 14-1

在 Kubernetes 中，主要有两种使用 Pod 的方式。

- 单容器的 Pod：最常见的使用 Pod 的方式，即在 Pod 中只运行一个容器，Kubernetes 通过 Pod 来管理这个容器。
- 多容器的 Pod：当应用需要多个容器一起协同工作时，Pod 中可以运行多个容器，从而满足应用的要求。Kubernetes 的调度器会保证这些容器都运行在同一台物理主机或虚拟主机上，从而达到容器之间的资源共享。

> 由于这些容器运作在同一个宿主机上，因此它们可以通过 localhost 互相访问。不过，除非是具有强耦合关系的容器，否则不推荐使用这种方式。

Kubernetes 使用 Pod 来管理容器具有以下优点：

- Pod 作为最小的逻辑部署单元，简化了应用部署的难度，同时也极大地方便了应用的管理。
- Pod 作为独立运行的服务单元，可以方便地进行部署、水平扩展/收缩、调度管理与资源分配。
- Pod 中的容器共享命名空间和网络地址，方便进行统一的资源管理与资源分配。

14.2 【实战】Pod 的基本使用方法

通过使用 YAML 文件可以描述一个 Pod。

Pod 的配置信息包含以下几部分。其中最重要的部分是 metadata 和 spec。

- apiVersion：创建该对象所使用的 Kubernetes API 的版本。
- kind：指定对象的类型，即 "Pod"。
- metadata：设置 Pod 的元信息，包括对象的名称、ID 及命名空间等。
- spec：设置 Pod 的规约信息。
- status：设置 Pod 运行时的状态。

下面是 Pod 的一个示例文件 "firstpod.yaml"：

```
apiVersion: v1
kind: Pod
metadata:
  name: first-pod
  labels:
    app: demo
    type: bash
```

```
spec:
  replicas: 2
  containers:
  - name: busybox-container
    image: busybox
    command: ['sh', '-c', 'echo Hello Kubernetes Pod! && sleep 1000']
```

下面演示了 Pod 的基本使用方法。

（1）使用 firstpod.yml 方式创建 Pod。

```
kubectl create -f firstpod.yaml
```

（2）查看正在运行的 Pod。

```
kubectl get pod
```

输出的信息如下：

```
NAME         READY   STATUS    RESTARTS   AGE
first-pod    1/1     Running   0          2m4s
```

以下命令将获取 Pod 的详细信息。

kubectl get pod -o wide

（3）查看正在运行的 Pod 的配置信息。

```
kubectl get pod first-pod -o yaml
```

输出的信息如下：

```
apiVersion: v1
kind: Pod
metadata:
  creationTimestamp: "2022-01-27T02:55:29Z"
  labels:
    app: demo
    type: bash
  name: first-pod
  namespace: default
  resourceVersion: "25955"
  selfLink: /api/v1/namespaces/default/pods/first-pod
  uid: 187fed31-ed44-4764-ad14-3f613b220e22
spec:
  containers:
  - command:
    - sh
```

```yaml
    - -c
    - echo Hello Kubernetes Pod! && sleep 1000
      image: busybox
      imagePullPolicy: Always
      name: busybox-container
      resources: {}
      terminationMessagePath: /dev/termination-log
      terminationMessagePolicy: File
      volumeMounts:
      - mountPath: /var/run/secrets/kubernetes.io/serviceaccount
        name: default-token-gp47w
        readOnly: true
    dnsPolicy: ClusterFirst
    enableServiceLinks: true
    nodeName: node2
    priority: 0
    restartPolicy: Always
    schedulerName: default-scheduler
    securityContext: {}
    serviceAccount: default
    serviceAccountName: default
    terminationGracePeriodSeconds: 30
    tolerations:
    - effect: NoExecute
      key: node.kubernetes.io/not-ready
      operator: Exists
      tolerationSeconds: 300
    - effect: NoExecute
      key: node.kubernetes.io/unreachable
      operator: Exists
      tolerationSeconds: 300
    volumes:
    - name: default-token-gp47w
      secret:
        defaultMode: 420
        secretName: default-token-gp47w
status:
  conditions:
  - lastProbeTime: null
    lastTransitionTime: "2022-01-27T02:55:29Z"
    status: "True"
    type: Initialized
  - lastProbeTime: null
    lastTransitionTime: "2022-01-27T02:55:49Z"
    status: "True"
```

```
      type: Ready
    - lastProbeTime: null
      lastTransitionTime: "2022-01-27T02:55:49Z"
      status: "True"
      type: ContainersReady
    - lastProbeTime: null
      lastTransitionTime: "2022-01-27T02:55:29Z"
      status: "True"
      type: PodScheduled
  containerStatuses:
  - containerID: docker://8a535752c063266d839c7aa366fc4e6b953cf9f5dde10921795a1af28d2a6811
      image: docker.io/busybox:latest
      imageID: docker-pullable://docker.io/busybox@sha256:5acba83a746c7608ed544dc1533b87c737a0b0fb730301639a0179f9344b1678
      lastState: {}
      name: busybox-container
      ready: true
      restartCount: 0
      state:
        running:
          startedAt: "2022-01-27T02:55:48Z"
    hostIP: 192.168.79.13
    phase: Running
    podIP: 10.244.2.11
    qosClass: BestEffort
    startTime: "2022-01-27T02:55:29Z"
```

（4）查看 Pod 的标准输出日志。

```
kubectl logs first-pod
```

输出的信息如下：

```
Hello Kubernetes Pod!
```

如果在 Pod 中有多个容器，则在查看某个容器的日志时需要指定容器的名称，例如：kubectl logs pod-name -c container-name。

（5）显示 Pod 的标签。

```
kubectl get pods --show-labels
```

输出的信息如下：

```
NAME        READY   STATUS    RESTARTS   AGE     LABELS
first-pod   1/1     Running   0          6m13s   app=demo,type=bash
```

标签 Label 是 Kubernetes 管理 Pod 的重要属性，我们可以在 YAML 文件的 metadata 配置参数中指定，也可以通过命令行来指定。

（6）根据标签查询 Pod。

```
kubectl get pods -l app=demo --show-labels
```

（7）通过命令行给 Pod 增加标签。

```
kubectl label pod first-pod side=frontend
```

（8）通过命令行修改 Pod 的标签。

```
kubectl label pod first-pod side=unkonwn --overwrite
```

（9）重新显示 Pod 的标签。

```
kubectl get pods --show-labels
```

输出的信息如下：

```
NAME        READY   STATUS    RESTARTS   AGE    LABELS
first-pod   1/1     Running   0          7m7s   app=demo,side=unkonwn,type=bash
```

也可以将标签显示为列，命令如下：

kubectl get pods -L app,type,side

输出的信息如下：

```
NAME        READY   STATUS    RESTARTS   AGE      APP    TYPE   SIDE
first-pod   1/1     Running   0          7m39s    demo   bash   unkonwn
```

14.3 Pod 中的容器

Pod 由一个或者多个容器组成。这里的容器通常指运行应用的业务容器。在 Pod 中，除业务容器外，还有基础容器、初始化容器和临时容器。

14.3.1 基础容器

基础容器（Infrastructure Container）负责维护整个 Pod 的网络空间。这种类型的容器对用户是透明的，用户不能操作这种容器。

在 node 节点上，通过"docker ps"命令可以查看基础容器，如图 14-2 所示。

```
docker ps --format "table {{.ID}}\t{{.Image}}\t{{.Command}}" | \
grep pause
```

```
0a3a5bddd491  registry.aliyuncs.com/google_containers/pause:3.1  "/pause"
7a3718d6aa9a  registry.aliyuncs.com/google_containers/pause:3.1  "/pause"
f1779b235df4  registry.aliyuncs.com/google_containers/pause:3.1  "/pause"
[root@node2 ~]#
```

图 14-2

基础容器将使用 pause 镜像来创建和维护 Pod 的网络环境。

14.3.2 初始化容器

初始化容器晚于基础容器运行，但先于业务容器运行。如果 Pod 的初始化容器运行失败，则默认情况下 Kubernetes 会不断尝试重启 Pod，直到初始化容器运行成功。如果将 Pod 的配置参数 "restartPolicy"设置为"Never"，则 Kubernetes 不会执行重启动作。

如果要将 Pod 中的容器指定为初始化容器，则需要在"spec"中添加"initContainers"字段。一个 Pod 可以指定多个初始化容器，它们会按顺序逐个运行：一个初始化容器运行成功，下一个才能够运行。当所有的初始化容器运行完成后，Kubernetes 才会执行业务容器从而运行应用。

下面是一个初始化容器的示例。

（1）创建"initcontainer.yaml"文件，并在其中输入以下内容。

```
apiVersion: v1
kind: Pod
metadata:
  name: myapp-pod
  labels:
    app: myapp
spec:
  containers:
  - name: myapp-container
    image: busybox:1.28
    command: ['sh', '-c', 'echo The app running! && sleep 5']
  initContainers:
  - name: init-myservice
    image: busybox:1.28
```

```
      command: ['sh', '-c', 'echo The init-myservice running! && sleep 5']
    - name: init-mydb
      image: busybox:1.28
      command: ['sh', '-c', 'echo The init-mydb running! && sleep 5']
```

（2）执行以下语句创建 Pod。

```
kubectl apply -f initcontainer.yaml
```

（3）在 Pod 创建成功后，使用"describe"命令查看输出信息，如图 14-3 所示。

```
kubectl describe -f initcontainer.yaml
```

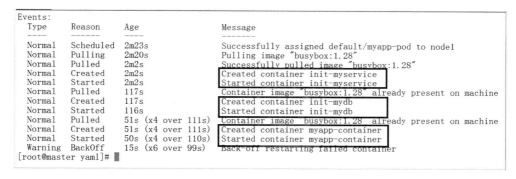

图 14-3

从输出信息可以看到，Kubernetes 先启动了两个初始化容器（init-myservice 和 init-mydb），之后启动了业务容器（myapp-container）。

初始化容器与业务容器是分离的，具有以下优势：

- 在初始化容器中可以提前安装在业务容器中要使用的工具，或者运行一些初始化的脚本。
- 将应用依赖的工具和脚本分离到初始化容器中，可以避免这些工具降低应用镜像的安全性。
- 开发人员可以进行组件镜像的独立创建和部署，而不需要将所有的组件构建成一个大的应用镜像。
- 初始化容器可以独立访问 Kubernetes 中的一些敏感信息，如 Secret。
- 由于初始化容器必须在业务容器之前执行完成，因此，可以利用初始化容器来阻塞或延迟业务容器的启动，从而进行先决条件的检查。

14.3.3　临时容器

临时容器是一种特殊的容器。它在现有的 Pod 中临时运行，以完成用户发起的操作（例如故障排查和性能诊断等）。由于临时容器没有端口要配置，且资源分配是不可变的，所以它不适合用来构建应用。

临时容器的最大用途是调试其他的容器。因为，当 Pod 中的容器异常退出，或者容器镜像不包含调试工具（例如没有 shell）时，会导致 "kubectl exec" 命令无法使用。这时临时容器对于交互式故障排查就很有用了。

下面是 Kubernetes 官方提供的一个临时容器的示例。

（1）使用 "k8s.gcr.io/pause:3.1" 镜像创建一个 Pod。

```
kubectl run ephemeral-demo --image=k8s.gcr.io/pause:3.1 --restart=Never
```

这里使用 "k8s.gcr.io/pause:3.1" 镜像是因为它不包含任何调试程序。

（2）使用 "kubectl exec" 命令创建 shell 进入容器。

```
kubectl exec -it ephemeral-demo -- sh
```

由于该镜像不包含任何调试程序，因此会出现以下错误信息。

OCI runtime exec failed: exec failed: container_linux.go:346: starting container process caused "exec: \"sh\": executable file not found in $PATH": unknown

（3）使用 "kubectl debug" 命令为 "ephemeral-demo" 容器添加一个临时容器，以达到调试的目的。

```
kubectl debug -it ephemeral-demo --image=busybox --target=ephemeral-demo
```

这里使用 busybox 的镜像来创建临时容器，Kubernetes 将自动启动临时容器的控制台。在临时容器启动后，会输出如下信息：

Defaulting debug container name to debugger-8xzrl.

If you don't see a command prompt, try pressing enter.

/ #

14.3.4　业务容器

业务容器（Containers）是实际运行应用的容器，例如 14.2 节中创建的 "busybox-container" 容器。

14.4 Pod 的生命周期

Pod 在运行过程中会遵循预定义的生命周期。该周期从 Pending 阶段开始。

- 如果 Pod 中有一个容器正常启动了，则 Pod 会进入 Running 阶段；
- 如果 Pod 中的所有容器成功运行并结束，则 Pod 进入 Succeeded 阶段，否则进入 Failed 阶段。

在整个生命周期内，Pod 只会被调度一次。一旦 Pod 被调度到某个节点上，则 Pod 会一直在该节点上运行，直到 Pod 停止或者被终止。

14.4.1 Pod 的阶段与容器的状态

1. Pod 的阶段

Pod 的 "status" 字段是一个 PodStatus 对象，其中包含 "phase" 字段。以 14.2 节的 "firstpod.yaml" 文件为例，以下语句可以查看 Pod 的当前阶段。

```
kubectl get pod first-pod -o yaml|grep phase
```

输出的信息如下：

```
phase: Running
```

表 14-1 列举了 "phase" 可能的值（即 Pod 的阶段）。

表 14-1

Pod 的阶段	描述
Pending	Kubernetes 已经开始创建 Pod，但由于 Pod 中的容器还未创建成功，所以 Pod 还处于挂起的状态。这时 Pod 可能在等待被调度，或者在等待下载镜像
Running	Pod 已经被调度到某个节点上了，Pod 中的所有容器都被成功创建，并且至少有一个容器正处于启动、重启、运行这 3 个状态中的 1 个
Succeeded	Pod 中的所有容器都已成功执行完成，并且不会再重启
Failed	Pod 中的所有容器都已停止运行，并且至少有一个容器是因为失败而退出（即容器以非 0 状态退出或者被系统强制终止）
Unknown	因为某些原因导致无法取得 Pod 的状态。这种情况通常是由于网络的造成，例如，Pod 所在主机通信失败等

2. 容器的状态

当 Pod 处于不同阶段时，Kubernetes 会跟踪 Pod 中每个容器的状态。一旦 Pod 被调度到某个具体的节点上，则 Kubelet 会创建 Pod 的容器。容器的状态有 3 种：Waiting、Running 和

Terminated。表 14-2 中列出了这 3 种状态的含义。

表 14-2

容器的状态	含 义
Waiting	处于等待状态的容器仍在运行它完成启动所需要的依赖操作,例如,等待镜像被成功拉取等。当使用 kubectl 来获取等待状态的容器的 Pod 信息时,Kubernetes 会提供一个"Reason"字段,该字段给出了容器处于等待状态的原因
Running	Pod 中的容器正在正常执行
Terminated	处于终止状态的容器可能已经正常结束了运行,或者因某些原因而运行失败。如果使用 kubectl 的命令来获取终止状态的容器的 Pod 信息,则 Kubernetes 会提供容器进入终止状态的原因、退出代码,以及容器执行的起止时间

14.4.2 Pod 中容器的重启策略

重启策略表示在一个 Pod 出现故障时需要做的动作,它适用于 Pod 中的所有容器。

Pod 通过在"spec"中使用"restartPolicy"字段来标识 Pod 采用哪一种重启策略。表 14-3 列出了"restartPolicy"字段支持的值(即具体的重启策略)。

表 14-3

重启策略	描 述
Always	当容器终止退出后,总是重启容器。这是默认的重启策略
OnFailure	当容器异常退出或者退出状态码非 0 时,才重启容器
Never	当容器异常退出或者退出状态码非 0 时,从不重启容器

以 13.3 节中 javawebdemo 为例,在"javawebdemo.yaml"文件中增加"restartPolicy"字段,将其值设置为"Always"。完整的 YAML 文件如下:

```
apiVersion: apps/v1
kind: Deployment
metadata:
  creationTimestamp: null
  labels:
    app: javawebdemo
  name: javawebdemo
spec:
  replicas: 3
  selector:
    matchLabels:
      app: javawebdemo
  strategy: {}
  template:
    metadata:
      creationTimestamp: null
```

```yaml
    labels:
      app: javawebdemo
  spec:
    containers:
    - image: collenzhao/k8s-javaweb-demo
      name: k8s-javaweb-demo
      resources: {}
  restartPolicy: Always
status: {}
```

14.4.3 【实战】Pod 的健康检查

当 Pod 处于运行状态时，kubelet 使用探针（Probe）对容器的健康状态进行检查和诊断。表 14-4 列出了 Kubernetes 支持的 3 种探针。

表 14-4

探针的类型	简称	描述
livenessProbe	存活探针	检查 Pod 中的容器是否正在运行。如果检查失败，则 kubelet 将"杀死"容器，并根据 Pod 的 restartPolicy 字段进行操作。如果容器不提供存活探针，则默认状态为 Success
readinessProbe	就绪探针	检查 Pod 中的容器是否准备好为请求提供服务。如果就绪探针检查失败，则 Kubernetes 会把 Pod 从 Service Endpoint 中剔除，从而让外部无法进行访问。如果容器不提供就绪态探针，则默认状态为 Success
startupProbe	启动探针	检查在 Pod 中的容器中部署的应用是否已经启动。如果启动探针检查失败，则 kubelet 将"杀死"对应的容器，并且根据重启策略进行容器的重启。如果容器没有提供启动探针，则默认状态为 Success。需要注意的是：如果提供了启动探针，则其他类型的探针都会被禁用，直到启动探针成功执行为止

所有探针都支持 3 种检查方式：HTTPGetAction、ExecAction 和 TCPSocketAction。下面分别对这 3 种检查方式进行介绍。

1. HTTPGetAction

对指定容器的 IP 地址和端口执行 HTTP Get 请求，如果返回状态码在（200,400）区间内，则诊断是成功的。

（1）创建"httpgetaction.yaml"文件，在其中输入以下内容。

```yaml
apiVersion: v1
kind: Pod
metadata:
  labels:
    test: liveness
  name: liveness-http
spec:
  containers:
```

```
        - name: liveness-http
          image: nginx
          ports:
          - name: http
            containerPort: 80
          lifecycle:
            postStart:
              exec:
                command:
                - /bin/sh
                - -c
                - 'echo healty > /usr/share/nginx/html/healthz'
          livenessProbe:
            httpGet:
              path: /healthz
              port: http
              scheme: HTTP
            initialDelaySeconds: 10
            periodSeconds: 5
```

> "httpgetaction.yaml"文件使用 Nginx 的镜像创建 Pod。同时在容器 liveness-http 中执行 "echo Healty > /usr/share/nginx/html/healthz" 命令创建了一个 nginx 目录。通过存活探针进行检测，第 1 次监测时间为 Pod 容器启动后的 10 s，以后每隔 5 s 监测一次。

（2）使用 "kubeclt apply" 命令应用 "httpgetaction.yaml" 文件。

```
kubectl apply -f httpgetaction.yaml
```

（3）查看 Pod 的运行信息，如图 14-4 所示。

```
kubectl describe pods liveness-http
```

```
Events:
  Type    Reason     Message
  ----    ------     -------
  Normal  Scheduled  Successfully assigned default/liveness-http to node2
  Normal  Pulling    Pulling image "nginx"
  Normal  Pulled     Successfully pulled image "nginx"
  Normal  Created    Created container liveness-http
  Normal  Started    Started container liveness-http
[root@master yaml]#
```

图 14-4

（4）使用 "kubeclt exec" 命令删除容器中的 "/usr/share/nginx/html/healthz" 目录。

```
kubectl exec liveness-http rm /usr/share/nginx/html/healthz
```

（5）再次查看查看 Pod 的运行信息，如图 14-5 所示。

```
kubectl describe pods liveness-http
```

```
Events:
 Type      Reason     Message
 ----      ------     -------
 Normal    Scheduled  Successfully assigned default/liveness-http to node2
 Normal    Pulled     Successfully pulled image "nginx"
 Normal    Created    Created container liveness-http
 Normal    Started    Started container liveness-http
 Normal    Pulling    Pulling image "nginx"
 Warning   Unhealthy  Liveness probe failed: HTTP probe failed with statuscode: 404
 Normal    Killing    Container liveness-http failed liveness probe, will be restarted
[root@master yaml]#
```

图 14-5

2. ExecAction

在容器内执行 Shell 命令，如果命令退出时返回码为 0，则诊断的结果是成功的。下面来演示如何使用 ExecAction 方式的探针。

（1）创建 "execaction.yaml" 文件，在其中输入以下内容。

```yaml
apiVersion: v1
kind: Pod
metadata:
  labels:
    test: liveness
  name: liveness-exec
spec:
  containers:
  - name: liveness
    image: busybox
    args:
    - /bin/sh
    - -c
    - touch /tmp/healthy; sleep 10; rm -rf /tmp/healthy; sleep 10
    # 这里以存活探针为例
    livenessProbe:
      # 指定存活探针检查方式是 ExecAction()
      exec:
        # 指定在容器内具体执行的命令
        command:
        - cat
        - /tmp/healthy
      # 指定在容器启动 5s 后才进行检查
      initialDelaySeconds: 5
```

```
    # 指定每隔5s检查一次
    periodSeconds: 5
```

 这里使用 ExecAction 方式检测 "/tmp/healthy" 文件。该文件在创建10s后就会被删除。如果使用 ExecAction 方式检测文件已被删除，则存活探针会返回失败，然后根据重启策略进行 Pod 的重启。下次使用 ExecAction 方式检测时再重复这个过程，检测过程将无限循环。

（2）使用 "kubeclt apply" 命令应用 "execaction.yaml" 文件。

```
kubectl apply -f execaction.yaml
```

（3）多执行几次以下命令查看 Pod 的状态信息，如图 14-6 所示。

```
kubectl get pod
```

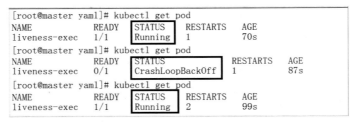

图 14-6

3. TCPSocketAction

对指定容器的 IP 地址和端口发起建立 TCP Socket 连接的请求，如果 Socket 连接被成功建立，则诊断的结果是成功的。

下面来演示如何使用 TCPSocketAction 方式的探针。

（1）创建 "tcpsocketaction.yaml" 文件，并在其中输入以下内容。

```
apiVersion: v1
kind: Pod
metadata:
  name: liveness-tcp
  labels:
    app: httpd
spec:
  containers:
  - name: liveness-tcp
    image: nginx
    ports:
    - containerPort: 80
```

```
      readinessProbe:
        tcpSocket:
          port: 8080
        initialDelaySeconds: 10
        periodSeconds: 5
      livenessProbe:
        tcpSocket:
          port: 8080
        initialDelaySeconds: 10
        periodSeconds: 5
```

> tcpsocketaction.yaml 文件基于 Nginx 的镜像创建 Pod。Nginx 提供的服务端口是 80。然后配置两个探针——readinessProbe 和 livenessProbe。这两个探针使用 TCPSocketAction 的方式连接 8080 端口。第 1 次监测时间为 Pod 容器启动后的 10 s，以后每隔 5 s 监测一次。由于探针无法连接容器的 8080 端口而导致检测失败，所以容器一直重启。

（2）使用"kubeclt apply"命令应用"tcpsocketaction.yaml"文件。

```
kubectl apply -f tcpsocketaction.yaml
```

（3）使用"kubectl describe"命令查看 Pod 的信息，如图 14-7 所示。

```
kubectl describe pod/liveness-tcp
```

```
Events:
  Type     Reason     Message
  ----     ------     -------
  Normal   Scheduled  Successfully assigned default/liveness-tcp to node2
  Normal   Pulled     Successfully pulled image "nginx"
  Normal   Created    Created container liveness-tcp
  Normal   Started    Started container liveness-tcp
  Normal   Pulling    Pulling image "nginx"
  Warning  Unhealthy  Liveness probe failed: dial tcp 10.244.2.22:8080: connect: connection refused
  Warning  Unhealthy  Readiness probe failed: dial tcp 10.244.2.22:8080: connect: connection refused
  Normal   Killing    Container liveness-tcp failed liveness probe, will be restarted
[root@master yaml]#
```

图 14-7

14.5　Pod 的调度策略

Pod 的调度是指，Kubernetes 在创建 Pod 时，将其创建到最合适的 node 节点上，然后由 node 节点上的 kubelet 来运行 Pod。在默认情况下，调度器 scheduler 会根据特定的算法和策略将 Pod 调度到 node 节点上，这样可以满足绝大多数的需求。例如，调度 Pod 到资源满足要求的 node 节点上运行；或者分散到不同 node 节点上，以达到资源的均衡使用。

在一些特殊的场景下，scheduler 的默认调度算法策略并不能满足实际的需求。Kubernetes 也允许用户调整调度约束字段来指定将 Pod 调度到哪些 node 节点上。

14.5.1 Pod 的创建过程

要理解 Kubernetes 是如何调度 Pod 的，首先需要了解 Pod 是如何被创建的。图 14-8 说明了 Pod 的创建过程。

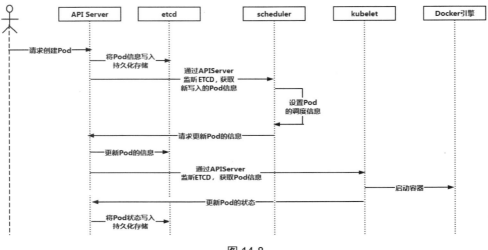

图 14-8

scheduler 在对 Pod 进行调度时，会根据 Pod 和 Pod 中每个容器对资源的需求，选择一个最优的 node 节点去运行这个 Pod。"满足 Pod 资源调度请求的 node 节点"被称为可调度节点。如果在集群中不存在这样的可调度节点，则 Pod 将被设置为未调度状态，直到集群中出现可调度节点。

14.5.2 【实战】自定义 Pod 调度的约束策略

scheduler 有默认支持的算法和策略用于对 Pod 进行调度。Kubernetes 也允许用户自定义 Pod 调度的约束策略，从而将 Pod 运行在指定的 node 节点上。

调度约束可以通过以下两个字段来进行定义。

- nodeName：将 Pod 调度到指定名称的 node 节点上。
- nodeSelector：将 Pod 调度到匹配 Label 标签的 node 节点上。

下面通过示例来演示它们的用法。

1. 指定 nodeName 的调度约束

（1）创建"nodename-demo.yaml"文件，并在其中输入以下内容：

```yaml
apiVersion: v1
kind: Pod
metadata:
  name: nodename-demo
  labels:
    app: nginx
spec:
  containers:
  - name: nginx
    image: nginx
    imagePullPolicy: IfNotPresent
  #通过 nodeName 指定 node1 节点，nodeName 是节点名
  nodeName: node1
```

（2）使用"kubectl apply"命令应用 YAML 文件。

```
kubectl apply -f nodename-demo.yaml
```

（3）查看 Pod 的详细信息，可以看出"nodename-demo"被调度器调度到 node1 节点上了，如图 14-9 所示。

```
kubectl get pod -o wide
```

```
NAME            READY   STATUS    RESTARTS   AGE   IP           NODE
nodename-demo   1/1     Running   0          50s   10.244.1.18  node1
[root@master yaml]#
```

图 14-9

2. 指定 nodeSelector 的调度约束

（1）给 node2 节点添加一个标签。

```
kubectl label node node2 nodeselector-key=nodeselector-value
```

（2）创建"nodeselector-demo.yaml"文件，并在其中输入以下内容：

```yaml
apiVersion: v1
kind: Pod
metadata:
  name: nodeselector-demo
  labels:
    app: nginx
spec:
  containers:
  - name: nginx
    image: nginx
    imagePullPolicy: IfNotPresent
  #根据 nodeSelector 指定的标签将 Pod 调度到对应的节点上
```

```
    nodeSelector:
      nodeselector-key: nodeselector-value
```

（3）使用"kubectl apply"命令应用 YAML 文件。

```
kubectl apply -f nodeselector-demo.yaml
```

（4）查看 Pod 的详细信息，可以看出"nodeselector-demo"被调度器调度到 node2 节点上了，如图 14-10 所示。

```
kubectl get pod -o wide
```

```
NAME               READY  STATUS   RESTARTS  AGE  IP           NODE
nodeselector-demo  1/1    Running  0         24s  10.244.2.24  node2
[root@master yaml]#
```

图 14-10

14.6 Pod 资源的使用限制

当 Kubernetes 调度并创建 Pod 后，Pod 是否有足够的资源来运行容器是非常重要的。资源分为两种类型——容器请求的资源和容器被限制的资源。

> 请求和限制是 Kubernetes 控制集群的 CPU 和内存等资源的重要方式，它们是两种不同的机制。

- 容器请求的资源是指，容器向 Kubernetes 集群请求的资源。
- 容器被限制的资源是指，Kubernetes 集群限制容器运行时使用的最多资源。

通过在 YAML 文件中使用"spec.containers[ ].resources"字段，可以控制 Pod 中容器对资源的使用，见表 14-5。

表 14-5

资源的参数名称	描述
limits.cpu	Pod 中所有非终止状态容器可使用 CPU 总量的阈值
limits.memory	Pod 中所有非终止状态容器可使用内存总量的阈值
requests.cpu	Pod 中所有非终止状态容器可使用 CPU 总量的阈值
requests.memory	Pod 中所有非终止状态容器可使用内存总量的阈值

下面来演示 Kubernetes 是如何实现 Pod 资源的使用限制的。

（1）创建"resourcelimit.yaml"文件，并在其中输入以下内容。

```yaml
apiVersion: v1
kind: Pod
metadata:
  name: resourcelimit-demo
spec:
  containers:
  - name: db
    image: mysql
    imagePullPolicy: IfNotPresent
    env:
    - name: MYSQL_ROOT_PASSWORD
      value: "password"
    resources:
      requests:
        # 设置请求的内存为 64MB
        memory: "64Mi"
        # 设置请求的 CPU 资源为 0.25 核
        # 数字 1 表示 1 核 CPU，1.5 表示 1.5 核 CPU
        cpu: "250m"
      limits:
        # 设置请求的内存为 128MB
        memory: "128Mi"
        # 设置请求的 CPU 资源为 0.5 核
        cpu: "500m"
  - name: wp
    image: wordpress
    imagePullPolicy: IfNotPresent
    resources:
      requests:
        memory: "64Mi"
        cpu: "250m"
      limits:
        memory: "128Mi"
        cpu: "500m"
```

（2）使用"kubeclt apply"命令应用"resourcelimit.yaml"文件。

```
kubectl apply -f resourcelimit.yaml
```

（3）查看 Pod 的信息，如图 14-11 所示。

```
kubectl get pods -o wide
```

```
NAME                READY   STATUS      RESTARTS   AGE   NODE
resourcelimit-demo  1/2     OOMKilled   2          39s   node1
[root@master yaml]#
```

图 14-11

由于资源的使用超过了限制，所以容器将被 Kubernetes 自动终止，并且 resourcelimit-demo 被调度到 node1 节点上了。

（4）查看 node1 的资源使用情况，如图 14-12 所示。

```
kubectl describe nodes node1
```

```
Name                          CPU Requests   CPU Limits   Memory Requests   Memory Limits
----                          ------------   ----------   ---------------   -------------
resourcelimit-demo            500m (25%)     1 (50%)      128Mi (6%)        256Mi (13%)
kube-flannel-ds-amd64-p7ndm   100m (5%)      100m (5%)    50Mi (2%)         50Mi (2%)
kube-proxy-g7584              0 (0%)         0 (0%)       0 (0%)            0 (0%)
```

图 14-12

从图 14-12 中可以看到 node1 节点上 resourcelimit-demo 资源的请求和限制的信息。例如，内存被设定最多只能使用 256MB。

（5）查看 Pod 的详细信息，如图 14-13 所示。

```
kubectl describe pod resourcelimit-demo
```

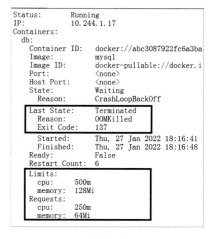

图 14-13

从图 14-13 中可以看出,"db"容器最新的状态信息为"Terminated",造成这种情况的原因是"OOMKilled"(即因内存溢出而被操作系统停止);同时也可以看到"db"容器资源的请求和限制信息。

14.7 Pod 的镜像拉取策略

Kubernetes 在创建 Pod 时,会使用相应的镜像来创建容器。表 14-6 中列出了 Kubernetes 支持的 3 种镜像拉取策略。

表 14-6

镜像拉取策略	说 明
IfNotPresent	只有当宿主机上不存在镜像时才拉取
Always	每次创建 Pod 时都会重新拉取一次镜像。如果镜像有了更新,则拉取最新的镜像
Never	Pod 不会主动拉取镜像,直接使用本地的镜像。如果本地不存在该镜像,则启动 Pod 失败

下面以 13.3 节中的 javawebdemo 为例,来查看它的镜像拉取策略。

(1)查看 Kubernetes 启动的 Pod 信息。

```
kubectl get pod
```

输出的信息如下:

```
NAME                            READY   STATUS    RESTARTS   AGE
javawebdemo-64758f5f66-mj6qp    1/1     Running   0          7m8s
javawebdemo-64758f5f66-sqscv    1/1     Running   0          7m8s
javawebdemo-64758f5f66-xz9sr    1/1     Running   0          7m8s
```

(2)执行以下命令查看 Pod 的镜像拉取策略。

```
kubectl get pod/javawebdemo-64758f5f66-mj6qp -o yaml | \
grep imagePullPolicy
```

输出的信息如下:

```
imagePullPolicy: Always
```

可以看到,在通过 YAML 文件启动 Pod 时,默认采用的是"Always"镜像拉取策略。但是如果通过"kubectl run"命令来部署应用,则采用"IfNotPresent"作为默认的镜像拉取策略。例如以下示例:

(1)基于"nginx:1.14"镜像部署一个应用。

```
kubectl run nginx --replicas=3 --image=nginx:1.14 --port=80
```

(2)查看 Kubernetes 启动的 Pod 信息。

```
kubectl get pod
```

输出的信息如下：

```
NAME                      READY   STATUS    RESTARTS   AGE
nginx-65fc77987d-6wqtr    1/1     Running   0          76s
nginx-65fc77987d-dfpfc    1/1     Running   0          76s
nginx-65fc77987d-ds4hx    1/1     Running   0          76s
```

(3)执行以下命令查看 Pod 的镜像拉取策略。

```
kubectl get pod/nginx-65fc77987d-6wqtr -o yaml | \
grep imagePullPolicy
```

输出的信息如下：

```
imagePullPolicy: IfNotPresent
```

14.8 Pod 的配置管理

当镜像制作完成后被用来创建 Pod 时，如果需要修改镜像中的一些参数值则比较麻烦——需要重新制作镜像。能否让镜像根据实际的需要，自动读取相应的配置信息呢？这时就需要使用到 Kubernetes 的配置管理。

14.8.1 为什么需要配置管理

在实际的应用开发过程会遇到这样的情况：在开发 Web 应用时，在开发环境中需要连接 MySQL 数据库；而在生产环境中需要连接 Redis 数据库，这是两套相互独立的环境。这就需要为应用指定不同的参数来满足实际的要求。如果不能很好地管理这些配置信息，则运维工作将变得无比烦琐。

为了解决这样的问题，Kubernetes 提供了自己的解决方案，它将配置信息作为一种独立的资源存入配置中心，并将其以注入的方式提供给 Pod 使用。如果更新了配置中心中的配置信息，则 Pod 会自动加载更新后的配置信息。Kubernetes 主要通过 ConfigMap 和 Secret 两种方式来实现配置信息的管理。

14.8.2 【实战】使用 ConfigMap 管理 Pod 的配置信息

ConfigMap 是用来存储配置信息的 Kubernetes 资源对象。ConfigMap 采用明文的方式将所有的配置信息都存储在 ETCD 中。在 ConfigMap 创建成功后，就可以在 Pod 中使用 ConfigMap 了。

1. 创建 ConfigMap

创建 ConfigMap 有 4 种方式：

- 在命令行中通过指定 ConfigMap 的参数进行创建。

（1）执行"kubectl create configmap"命令，并指定"--from-literal"参数。

```
kubectl create configmap demo-configmap1 \
--from-literal=db.host=192.168.1.2 \
--from-literal=db.port=3306 \
--from-literal=user=admin \
--from-literal=password=123456
```

（2）查看创建的 ConfigMap。

```
kubectl get configmap
```

输出的信息如下：

```
NAME              DATA    AGE
demo-configmap1   4       22s
```

"kubectl get configmap"命令可以简写成"kubectl get cm"。

（3）查看"demo-configmap1"的详细信息。

```
kubectl describe configmap demo-configmap1
```

输出的信息如下：

```
Name:         demo-configmap1
Namespace:    default
Labels:       <none>
Annotations:  <none>

Data
====
db.host:
----
192.168.1.2
db.port:
----
3306
password:
----
123456
user:
----
admin
```

```
Events:  <none>
```

- 通过指定的配置文件创建 ConfigMap。

（1）创建一个存放配置文件的目录。

```
mkdir configmap
```

（2）编辑"configmap/redis.properties"文件，在其中输入以下内容：

```
redis.host=127.0.0.1
redis.port=6379
redis.password=123456
```

（3）编辑"configmap/mysql.properties"文件，在其中输入以下内容：

```
mysql.host=127.0.0.1
mysql.port=3306
mysql.password=123456
```

这里创建了两个配置文件——redis.properties 和 mysql.properties，可以把这两个配置文件中的配置信息保存在一个 ConfigMap 中。

（4）执行"kubectl create configmap"命令，通过指定"--from-file"参数创建 ConfigMap。

```
kubectl create configmap demo-configmap2 \
--from-file=./configmap/
```

（5）查看"demo-configmap2"的详细信息。

```
kubectl describe configmap demo-configmap2
```

输出的信息如下：

```
Name:         demo-configmap2
Namespace:    default
Labels:       <none>
Annotations:  <none>

Data
====
mysql.properties:
----
mysql.host=127.0.0.1
mysql.port=3306
mysql.password=123456

redis.properties:
```

```
----
redis.host=127.0.0.1
redis.port=6379
redis.password=123456

Events:  <none>
```

- 通过一个文件内的多个键值对创建 ConfigMap。

(1)执行以下命令创建 env-config.txt 文件。

```
cat << EOF > env-config.txt
db.host=192.168.0.1
db.port=3306
user=admin
password=123456
EOF
```

(2)执行 "kubectl create configmap" 命令，通过 "--from-env-file" 参数创建 ConfigMap。

```
kubectl create configmap demo-configmap3 --from-env-file=env-config.txt
```

- 通过 YAML 文件创建 ConfigMap。

(1)创建 "config.yaml" 文件。

```
apiVersion: v1
kind: ConfigMap
metadata:
  name: demo-configmap4
data:
  db.host: 192.168.0.200
  db.port: "3306"
  user: "admin"
  password: "123456"
```

(2)执行 "kubectl create -f" 命令创建 ConfigMap。

```
kubectl create -f config.yaml
```

(3)查看创建的 ConfigMap。

```
kubectl get cm
```

输出的信息如下：

```
NAME              DATA    AGE
demo-configmap1   4       46m
demo-configmap2   2       39m
demo-configmap3   4       5m20s
demo-configmap4   4       28s
```

2. 使用 ConfigMap

成功创建 ConfigMap 后，就可以在 Pod 中使用它来实现配置信息管理。具体可以通过两种方式：环境变量和数据卷 volume。

下面来演示如何使用它们。

- 通过环境变量直接将 ConfigMap 的配置信息传递给 Pod。

（1）创建"configmap-usage01.yaml"文件，使用环境变量将"demo-configmap4"中的"db.host"和"db.port"分别映射到 Pod 容器中的"HOST"和"PORT"这两个环境变量中。

```yaml
apiVersion: v1
kind: Pod
metadata:
  name: configmap-usage01
spec:
  containers:
    - name: busybox
      image: busybox
      imagePullPolicy: IfNotPresent
      command: [ "/bin/sh", "-c", "echo $(HOST) $(PORT)" ]
      env:
        - name: HOST
          valueFrom:
            configMapKeyRef:
              name: demo-configmap4
              key: db.host
        - name: PORT
          valueFrom:
            configMapKeyRef:
              name: demo-configmap4
              key: db.port
  restartPolicy: Never
```

（2）使用"kubectl apply -f"命令创建 Pod。

```
kubectl apply -f configmap-usage01.yaml
```

（3）查看 Pod 的标准输出日志。

```
kubectl logs configmap-usage01
```

输出的信息如下：

```
192.168.0.200 3306
```

- 通过数据卷 volume 将 ConfigMap 的配置信息挂载到 Pod 内。

具体的做法是：将 ConfigMap 的配置信息"demo-configmap2"通过数据卷挂载到容器中的"/etc/config"目录下，根据 key 在挂载目录下创建 redis.properties 和 mysql.properties 文件，文件内容是对应的 valus 值。

（1）创建"configmap-usage02.yaml"文件，并在其中输入以下内容。

```yaml
apiVersion: v1
kind: Pod
metadata:
  name: configmap-usage02
spec:
  containers:
    - name: busybox
      image: busybox
      imagePullPolicy: IfNotPresent
      command: [ "/bin/sh","-c","cat /etc/config/redis.properties" ]
      volumeMounts:
      - name: config-volume
        mountPath: /etc/config
  volumes:
    - name: config-volume
      configMap:
        name: demo-configmap2
  restartPolicy: Never
```

（2）使用"kubectl apply -f"命令创建 Pod。

```
kubectl apply -f configmap-usage02.yaml
```

（3）查看 Pod 的标准输出日志。

```
kubectl logs configmap-usage02
```

输出的信息如下：

```
redis.host=127.0.0.1
redis.port=6379
redis.password=123456
```

3. ConfigMap 的动态更新

在创建 ConfigMap 成功后，Kubernetes 支持动态更新 ConfigMap。ConfigMap 可以通过环境变量和数据卷这两种方式来使用。

- 如果通过环境变量使用 ConfigMap，则在配置信息更新后，环境变量不会同步更新。

- 如果通过数据卷使用 ConfigMap，则在配置信息更新后，数据卷中的配置信息会同步更新，但有一定的延迟。

下面来演示 ConfigMap 的动态更新。

（1）这里以"demo-configmap2"为例。将 ConfigMap 导出为 YAML 文件。

```
kubectl get cm demo-configmap2 -o yaml > demo-configmap2.yaml
```

（2）将"demo-configmap2.yaml"文件中的"redis.host"的值从"127.0.0.1"修改为"localhost"。

```
apiVersion: v1
data:
  mysql.properties: |
    mysql.host=127.0.0.1
    mysql.port=3306
    mysql.password=123456
  redis.properties: |
    redis.host=localhost
    redis.port=6379
    redis.password=123456
kind: ConfigMap
metadata:
  creationTimestamp: "2022-01-28T04:00:31Z"
  name: demo-configmap2
  namespace: default
  resourceVersion: "64797"
  selfLink: /api/v1/namespaces/default/configmaps/demo-configmap2
  uid: c8382408-1b2c-4da3-b468-1ca5f88d3f02
```

（3）执行"kubectl apply -f"命令生效"demo-configmap2.yaml"文件。

```
kubectl apply -f demo-configmap2.yaml
```

（4）查看修改后的"demo-configmap2"。

```
kubectl describe cm demo-configmap2
```

输出的信息如下：

```
Name:         demo-configmap2
Namespace:    default
Labels:       <none>
Annotations:  kubectl.kubernetes.io/last-applied-configuration:
                {"apiVersion":"v1","data":{"mysql.properties":"mysql.host=
127.0.0.1\nmysql.port=3306\nmysql.password=123456\n","redis.properties":"red
is.h...
```

```
Data
====
mysql.properties:
----
mysql.host=127.0.0.1
mysql.port=3306
mysql.password=123456

redis.properties:
----
redis.host=localhost
redis.port=6379
redis.password=123456

Events:  <none>
```

 如果"demo-configmap2"文件以数据卷方式被 Pod 使用，则 Pod 中挂载的配置信息在经过一段时间的延迟后会自动更新。

14.8.3 【实战】使用 Secret 管理 Pod 的配置信息

Secret 也是 Kubernetes 配置管理的一种方式，它采用 Base 64 编码机制保存配置信息。与 ConfigMap 不同的是，Secret 中包含敏感的信息，例如用户的登录密码、Token 等。这些敏感信息使用 Secret 来保存，可以更好地控制它们的用途，并降低意外暴露的风险。在成功创建 Secret 后，可以通过环境变量或数据卷的方式在 Pod 中使用它。

1. 创建 Secret

可以通过使用账号密码文件和 YAML 文件这两种方式创建 Secret。

- 使用账号密码文件命令创建 Secret。

（1）执行以下命令生成账号文件和密码文件。

```
echo 'collenzhao' > username.txt
echo 'password' > password.txt
```

（2）使用"kubectl create secret"命令创建 Secret。

```
kubectl create secret generic demo-secret1 \
--from-file=username.txt --from-file=password.txt
```

（3）获取创建的 Secret。

```
kubectl get secret
```

输出的信息如下：

```
NAME                   TYPE                                    DATA   AGE
default-token-gp47w    kubernetes.io/service-account-token     3      3d13h
demo-secret1           Opaque                                  2      4s
```

Kubernetes 允许用户创建自己的 Secret，同时系统也创建了一些 Secret（例如 default-token-gp47w）。

（4）查看"demo-secret1"的详细信息。

```
kubectl describe secret demo-secret1
```

输出的信息如下：

```
Name:         demo-secret1
Namespace:    default
Labels:       <none>
Annotations:  <none>

Type:  Opaque

Data
====
password.txt:  9 bytes
username.txt:  11 bytes
```

- 使用 YAML 文件创建 Secret。

由于 Secret 是采用 Base 64 编码机制来保存配置信息的，因此在创建 YAML 文件前，需要先得到配置信息的 Base 64 编码。

在以下例子中，先将账号和密码进行 Base 64 编码，然后通过 YAML 文件将编码结果保存到了 Secret 中。

（1）得到账号"admin"和密码"hello123"的 Base 64 编码。

```
echo 'admin' | base64
YWRtaW4K

echo 'hello123' | base64
aGVsbG8xMjMK
```

（2）创建"demo-secret2.yaml"文件，并在其中输入以下内容。

```
apiVersion: v1
kind: Secret
```

```
metadata:
  name: demo-secret2
type: Opaque
data:
  username: YWRtaW4K
  password: aGVsbG8xMjMK
```

（3）执行"kubectl create"命令创建 Secret。

```
kubectl create -f demo-secret2.yaml
```

（4）获取创建的 Secret。

```
kubectl get secret
```

输出的信息如下：

```
NAME                 TYPE                                  DATA   AGE
default-token-gp47w  kubernetes.io/service-account-token   3      3d13h
demo-secret1         Opaque                                2      9m37s
demo-secret2         Opaque                                2      18s
```

2. 使用 Secret

在成功创建 Secret 后，就可以在 Pod 中使用它来实现配置信息的管理了。Secret 的使用也有两种方式：环境变量和数据卷 volume。

- 通过环境变量直接将 Secret 的配置信息传递给 Pod。

（1）创建"secret-usage01.yaml"文件，并在其中输入以下内容。

```
apiVersion: v1
kind: Pod
metadata:
  name: secret-usage01
spec:
  containers:
  - name: secret-usage01
    image: nginx
    imagePullPolicy: IfNotPresent
    env:
      - name: SECRET_USERNAME
        valueFrom:
          secretKeyRef:
            name: demo-secret2
            key: username
      - name: SECRET_PASSWORD
        valueFrom:
          secretKeyRef:
```

```
            name: demo-secret2
            key: password
  restartPolicy: Never
```

(2)使用 "kubectl apply" 命令创建 Pod。

```
kubectl apply -f secret-usage01.yaml
```

(3)使用 "kubectl exec" 命令进入容器。

```
kubectl exec -it secret-usage01 bash
```

(4)查看 Secret 的配置信息,如图 14-14 所示。

```
echo $SECRET_USERNAME
echo $SECRET_PASSWORD
```

```
root@secret-usage01:/# echo $SECRET_USERNAME
admin
root@secret-usage01:/# echo $SECRET_PASSWORD
hello123
root@secret-usage01:/#
```

图 14-14

- 通过数据卷 volume 将 Secret 的配置信息挂载到 Pod 内。

(1)创建 "secret-usage02.yaml" 文件将创建的 Secret "demo-secret2" 挂载到容器指定的目录下。目录下的文件以 key 作为文件名,文件内容为 key 的值。

```
apiVersion: v1
kind: Pod
metadata:
  name: secret-usage02
spec:
  containers:
  - name: secret-usage02
    image: redis
    imagePullPolicy: IfNotPresent
    volumeMounts:
    - name: foo
      mountPath: "/etc/foo"
      readOnly: true
  volumes:
  - name: foo
    secret:
      secretName: demo-secret2
```

(2)执行 "kubectl apply" 命令创建 Pod。

```
kubectl apply -f secret-usage02.yaml
```

（3）使用"kubectl exec"命令进入容器。

```
kubectl exec -it secret-usage02 bash
```

（4）切换到容器内的"/etc/foo/"目录下，查看挂载的 Secret 配置信息，如图 14-15 所示。

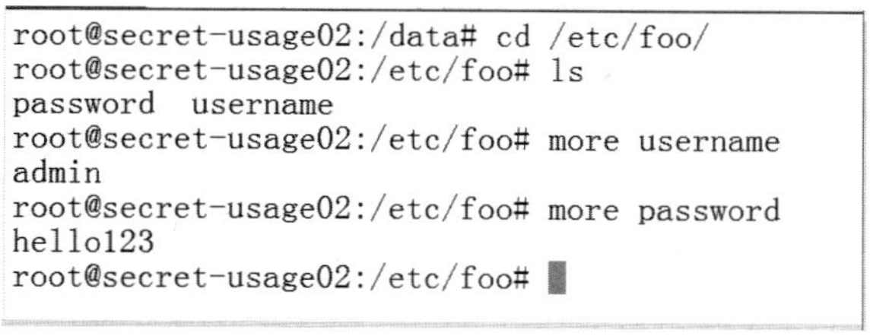

```
root@secret-usage02:/data# cd /etc/foo/
root@secret-usage02:/etc/foo# ls
password   username
root@secret-usage02:/etc/foo# more username
admin
root@secret-usage02:/etc/foo# more password
hello123
root@secret-usage02:/etc/foo#
```

图 14-15

第 15 章
使用控制器管理 Pod

Kubernetes 通过创建控制器来管理 Pod 的生命周期。为了满足不同需求的场景，Kubernetes 提供了不同的控制器，如 Deployment、DaemonSet、Job、CronJob、StatefuleSet 等。

15.1 为什么需要控制器

Pod 在 Kubernetes 中是存在生命周期的，因此需要有一种方式去操作和管理其状态和生命周期。这就需要用到 Kubernetes 提供的控制器了。

试想一下以下两种场景：

- 在双十一期间，用户访问量暴增，服务器正承受着巨大的压力；
- node 节点突然宕机了，运行在其上的 Pod 不能正常提供服务了。

如何解决这两种问题？

开发人员当然可以手动增加 node 节点，以启动更多的 Pod 来应对暴增的访问量；或者通过手动重启 node 节点来达到重新启动 Pod 的目的。但是，对于一个大型且复杂的系统来说，采用人工的方式去解决这样的问题太不现实了。

利用 Kubernetes 的控制器，可以非常方便地解决问题。

- 当 Pod 数量不够时，控制器会自动增加 Pod 的副本，以应对客户端的请求；
- 当 Pod 出现故障时，控制器会自动在其他合适的 node 节点启动新的 Pod。

表 15-1 列举了 Kubernetes 提供的几种控制器，其中最常用的是 Deployment。

表 15-1

控制器	作用
Deployment	定义无状态管理的 Pod。该控制器可以定义 Pod 的数量、更新方式、使用的镜像，资源限制等
DaemonSet	将 Pod 定义为节点的守护进程。该控制器适合运行那些在后台运行的应用
Job	将 Pod 定义为一次性任务。该控制器可以保证批处理任务的 Pod 能成功运行
CronJob	将 Pod 定义为周期性任务计划。该控制器可以保证批处理任务的 Pod 能成功运行
StatefuleSet	定义有状态管理的 Pod。该控制器可以保证 Pod 具有固定的网络标记、持久化存储、顺序部署和扩展、顺序滚动更新等

15.2 Deployment 控制器

Deployment 控制器将 Pod 部署成无状态的应用，它只关心 Pod 的数量、更新方式、使用的镜像和资源限制等。由于是无状态的管理方式，因此在 Deployment 控制器中没有"角色"和"顺序"的概念，即在 Deployment 控制器中没有状态。

通过使用 Deployment 控制器，开发人员可以部署 Pod、设置 Pod 的副本、实现 Pod 的升级与回滚。在 YAML 文件中描述清楚了 Deployment 控制器的目标是什么，Deployment 控制器会自动完成对 Pod 和 ReplicaSet 的管理。Kubernetes 可以直接运行一个新的 Deployment 控制器，也可以用一个新的 Deployment 控制器替换旧的 Deployment 控制器。

> ReplicaSet 是下一代复本控制器，可以独立使用。但在 Kubernetes 中，它主要被 Deployment 用来进行 Pod 的创建、更新和删除。在使用 Deployment 控制器时，它会自动创建 ReplicaSets，并对其进行管理。

Deployment 控制器、ReplicaSet 和 Pod 之间的关系如图 15-1 所示。

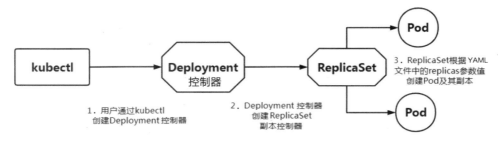

图 15-1

下面介绍创建和更新 Deployment 的过程。

（1）创建 Deployment 的过程。

在用户创建 Deployment 时，Kubernetes 会创建一个 ReplicaSet。ReplicaSet 在后台根据指定的副本数创建 Pod，并检查 Pod 的状态，以确定 Pod 是启动成功还是失败。

（2）更新 Deployment 的过程。

当用户更新 Deployment 时，Kubernetes 会创建一个新的 ReplicaSet。Deployment 会将 Pod 从旧的 ReplicaSet 中迁移到新的 ReplicaSet 中。

- 如果迁移失败或 Pod 不稳定，则 Deployment 会自动回滚到上一个版本。
- 如果迁移成功，则 Deployment 将清除旧的、不必要的 ReplicaSet。

15.2.1 【实战】创建和使用 Deployment 控制器

下面来演示如何使用 Deployment 控制器管理 Pod。

（1）创建 "deployment-demo.yaml" 文件，并在其中输入以下内容：

```yaml
apiVersion: apps/v1
kind: Deployment
metadata:
  name: deployment-demo
  labels:
    app: nginx
spec:
  replicas: 3
  selector:
    matchLabels:
      app: nginx
  template:
    metadata:
      labels:
        app: nginx
    spec:
      containers:
      - name: nginx
        image: nginx:1.7.9
        imagePullPolicy: IfNotPresent
        ports:
        - containerPort: 80
```

其中的参数含义见表 15-2。

表 15-2

参数	含义
.metadata.name	创建名为"deployment-demo"的 Deployment
.spec.replicas	指示 Deployment 创建具有 3 个副本的 Pod
.spec.selector	定义 Deployment 如何查找要管理的 Pod。在这种情况下，只需在 Pod 模板中定义对应的标签，则 Kubernetes 会进行自动匹配
.spec.template	定义 Pod 的相关属性
.spec.template.metadata.labels	给 Pod 添加标签 Label
.spec.template.spec	使用 Nginx 1.15.4 版本的镜像创建一个名为"nginx"的容器，并暴露容器的 80 端口

（2）运行以下命令创建 Deployment。

```
kubectl apply -f deployment-demo.yaml
```

（3）查看创建的 Deployment 和 ReplicaSet。

```
kubectl get deployments,replicaset,pod
```

 这条命令还可以简写为以下形式：

kubectl get deploy,rs,pod

输出的信息如图 15-2 所示。

图 15-2

（4）查看每个 Pod 自动生成的标签，如图 15-3 所示。

```
kubectl get pods --show-labels
```

图 15-3

（5）修改 YAML 文件来实现配置信息的更新。以下 YAML 文件将 Nginx 的版本升级到 1.9.1，并将 Pod 的副本数改成了 4。

```yaml
apiVersion: apps/v1
kind: Deployment
metadata:
  name: deployment-demo
  labels:
    app: nginx
spec:
  replicas: 4
  selector:
    matchLabels:
      app: nginx
  template:
    metadata:
      labels:
        app: nginx
    spec:
      containers:
      - name: nginx
        image: nginx:1.9.1
        imagePullPolicy: IfNotPresent
        ports:
        - containerPort: 80
```

> 也可以使用命令行的方式来实现更新，例如：
> kubectl set image deploy deployment-demo nginx=nginx:1.9.1

（6）运行以下命令重新应用 Deployment。

```
kubectl apply -f deployment-demo.yaml
```

（7）获取 Deployment 的信息。

```
kubectl describe deployments deployment-demo
```

输出的部分信息如下：

```
Pod Template:
  Labels:  app=nginx
  Containers:
   nginx:
    Image:      nginx:1.9.1
    Port:       80/TCP
```

```
    Host Port:      0/TCP
    Environment:    <none>
    Mounts:         <none>
   Volumes:         <none>
```

(8)执行以下命令将 Nginx 的版本更新为 1.7.9。

```
kubectl --record deployment.apps/deployment-demo \
set image deployment.apps/deployment-demo nginx=nginx:1.7.9
```

> 为了能够追溯修改历史记录，这里使用了 --record 参数。

(9)获取 Deployment 的信息。

```
kubectl describe deployments deployment-demo
```

输出的部分信息如下。可以看出，Nginx 的版本又回到了 1.7.9。

```
Pod Template:
  Labels:  app=nginx
  Containers:
   nginx:
    Image:          nginx:1.7.9
    Port:           80/TCP
    Host Port:      0/TCP
    Environment:    <none>
    Mounts:         <none>
```

(10)执行以下命令追溯修改的历史。

```
kubectl rollout history deployment deployment-demo
```

输出的信息如下：

```
deployment.extensions/deployment-demo
REVISION    CHANGE-CAUSE
2           <none>
3           kubectl deployment.apps/deployment-demo \
            set image deployment.apps/deployment-demo \
            nginx=nginx:1.7.9 --record=true
```

(11)执行以下命令将 Nginx 回滚到 2 版本。

```
kubectl rollout undo deployment deployment-demo --to-revision=2
```

(12)重新获取 Deployment 的信息。

```
kubectl describe deployments deployment-demo
```

输出的部分信息如下。可以看到，Nginx 又回滚到 1.9.1 版本。

```
Pod Template:
  Labels:     app=nginx
  Containers:
   nginx:
    Image:          nginx:1.9.1
    Port:           80/TCP
    Host Port:      0/TCP
    Environment:    <none>
    Mounts:         <none>
  Volumes:          <none>
```

（13）以下命令将对 Deployment 进行缩放。

```
kubectl scale deployment deployment-demo --replicas=7
```

（14）以下命令将暂停和恢复 Deployment。

```
kubectl rollout pause deployment deployment-demo
kubectl rollout resume deployment deployment-demo
```

可以在更新 Deployment 前暂停 Deployment，等到更新完成后再恢复它。这允许在"从暂停到恢复"这段时间内修改应用，而不会触发不必要的 Deployment。

15.2.2 【实战】验证 Deployment 控制器的不同状态

Deployment 控制器在整个生命周期中存在 3 种状态：已完成（Complete）、进行中（Progressing）和失败（Failed）。

通过观察 Deployment 的当前特征，可以判断 Deployment 的状态。表 15-3 展示了不同状态时 Deployment 的特征。

表 15-3

Deployment 的状态	Deployment 的特征
已完成（Complete）	• Deployment 管理的所有 Pod 副本都已更新到指定的版本了 • Deployment 管理的所有 Pod 副本都可用 • 所有旧的 Pod 副本都已停止
进行中（Progressing）	• Deployment 正在通过 ReplicaSet 创建 Pod • Deployment 正在通过 ReplicaSet 进行扩容/缩容 • Deployment 管理的 Pod 已经就绪或者可用

续表

Deployment 的状态	Deployment 的特征
失败（Failed）	• 就绪探针检测失败 • 资源的配额不足 • 镜像下载错误 • 权限与应用配置错误

下面通过一个例子来验证 Deployment 在生命周期内的不同状态。

（1）修改"deployment-demo.yaml"文件，将其副本数设置为 1。完整的 YAML 文件如下：

```
apiVersion: apps/v1
kind: Deployment
metadata:
  name: deployment-demo
  labels:
    app: nginx
spec:
  replicas: 1
  selector:
    matchLabels:
      app: nginx
  template:
    metadata:
      labels:
        app: nginx
    spec:
      containers:
      - name: nginx
        image: nginx:1.7.9
        imagePullPolicy: IfNotPresent
        ports:
        - containerPort: 80
```

（2）运行以下命令创建 Deployment。

```
kubectl apply -f deployment-demo.yaml
```

（3）运行以下命令获取 Deployment 的详细信息。

```
kubectl describe deploy deployment-demo
```

输出的信息如下：

```
Conditions:
  Type          Status  Reason
  ----          ------  ------
```

```
  Available       True    MinimumReplicasAvailable
  Progressing     True    NewReplicaSetAvailable
```

从输出的信息可以看出，Deployment 的"Progressing"和"Available"都是 True。这说明 Deployment 已经满足"已完成"（Complete）状态的特征，因此此时 Deployment 已经进入"已完成"（Complete）状态。

（4）执行 Deployment 的扩容操作，将其副本数设置为 5。

```
kubectl scale deployment deployment-demo --replicas=5
```

（5）使用"kubectl rollout status"命令监视 Deployment 扩容的进度。

```
kubectl rollout status deployment deployment-demo
```

输出的信息如下：

```
Waiting for deployment "deployment-demo" rollout to finish: 1 of 5 updated replicas are available...
Waiting for deployment "deployment-demo" rollout to finish: 2 of 5 updated replicas are available...
Waiting for deployment "deployment-demo" rollout to finish: 3 of 5 updated replicas are available...
Waiting for deployment "deployment-demo" rollout to finish: 4 of 5 updated replicas are available...
```

（6）在 Deployment 扩容过程中，运行以下命令获取 Deployment 的详细信息。

```
kubectl describe deploy deployment-demo
```

输出的信息如下：

```
Conditions:
  Type          Status  Reason
  ----          ------  ------
  Progressing   True    NewReplicaSetAvailable
  Available     False   MinimumReplicasUnavailable
```

从参数"Progressing"和"Available"的值（Progressing=True，Available=False）可以看出，Deployment 正处于"进行中"（Progressing）状态。

（7）更新 Nginx 的版本为 nginx:1.123。

```
kubectl set image deployment deployment-demo nginx=nginx:1.123
```

nginx:1.123 是一个不存在的版本,所以会导致 Deployment 更新失败。

(8)为了尽快看到 Deployment 的出错信息,可以将".spec.progressDeadlineSeconds"字段设置得小一些,例如 120 s。

```
kubectl patch deployment deployment-demo \
-p '{"spec":{"progressDeadlineSeconds":120}}'
```

".spec.progressDeadlineSeconds"字段表示在 Deployment 完成前需要等待的最长时间,默认值是 600 s。

(9)等待 120 s 后,运行以下命令获取 Deployment 的详细信息。

```
kubectl describe deploy deployment-demo
```

输出的信息如下:

```
Conditions:
  Type          Status  Reason
  ----          ------  ------
  Available     True    MinimumReplicasAvailable
  Progressing   False   ProgressDeadlineExceeded
```

由于 Deployment 无法下载镜像信息,因此进入"失败"(Failed)状态。这时参数 Processing 的值是 False。

(10)使用"kubectl rollout status"命令监视 Deployment 的更新进度。

```
kubectl rollout status deployment deployment-demo
```

输出的信息如下:

```
Waiting for deployment "deployment-demo" rollout to finish: 3 out of 5 new
replicas have been updated...
error: deployment "deployment-demo" exceeded its progress deadline
```

15.2.3 【实战】Deployment 控制器的清理策略

在 Deployment 中配置".spec.revisionHistoryLimit"字段,可以指定其清理策略。该字段用于指定 Deployment 保留旧 ReplicaSet 的个数,即更新 Pod 前的版本个数。该字段的默认值是 10。

下面通过一个示例来说明 ".spec.revisionHistoryLimit" 字段的作用。

（1）创建 "revisionhistory-demo.yaml" 文件，并在其中输入以下内容。

```yaml
apiVersion: apps/v1
kind: Deployment
metadata:
  name: revisionhistory-demo
  labels:
    app: nginx
spec:
  revisionHistoryLimit: 1
  replicas: 1
  selector:
    matchLabels:
      app: nginx
  template:
    metadata:
      labels:
        app: nginx
    spec:
      containers:
      - name: nginx
        image: nginx:1.14
        imagePullPolicy: IfNotPresent
        ports:
        - containerPort: 80
```

这里将 ".spec.revisionHistoryLimit" 字段设置成 1，即只保留 1 个旧版本。

（2）运行以下命令创建 Deployment。

```
kubectl apply -f revisionhistory-demo.yaml
```

（3）将镜像的版本从 nginx:1.14 升级到 nginx:1.7.9。

```
kubectl --record deployment.apps/revisionhistory-demo \
set image deployment.apps/revisionhistory-demo nginx=nginx:1.7.9
```

（4）执行以下命令追溯修改的历史。

```
kubectl rollout history deployment revisionhistory-demo
```

输出的信息如下：

```
REVISION    CHANGE-CAUSE
1           <none>
2           kubectl deployment.apps/revisionhistory-demo \
            set image deployment.apps/revisionhistory-demo \
            nginx=nginx:1.7.9 --record=true
```

（5）将镜像的版本从 nginx:1.7.9 升级到 nginx:1.9.1。

```
kubectl --record deployment.apps/revisionhistory-demo \
set image deployment.apps/revisionhistory-demo nginx=nginx:1.9.1
```

（6）执行以下命令追溯修改的历史。

```
kubectl rollout history deployment revisionhistory-demo
```

输出的信息如下：

```
REVISION    CHANGE-CAUSE
2           kubectl deployment.apps/revisionhistory-demo \
            set image deployment.apps/revisionhistory-demo \
            nginx=nginx:1.7.9 --record=true
3           kubectl deployment.apps/revisionhistory-demo \
            set image deployment.apps/revisionhistory-demo \
            nginx=nginx:1.9.1 --record=true
```

> 由于 ".spec.revisionHistoryLimit" 字段被设置成了 1，所以 Deployment 控制器只会保留 1 个旧版本，"revision 1"版本已经被自动清除了。

（7）将 Deployment 回滚到"revision 1"版本。

```
kubectl rollout undo deployment revisionhistory-demo --to-revision=1
```

这时将出现以下错误信息：

```
error: unable to find specified revision 1 in history
```

（8）将 Deployment 回滚到"revision 2"版本。

```
kubectl rollout undo deployment revisionhistory-demo --to-revision=2
```

（9）重新查看 Deployment 的修改历史。

```
kubectl rollout history deployment revisionhistory-demo
```

输出的信息如下：

```
REVISION    CHANGE-CAUSE
3           kubectl deployment.apps/revisionhistory-demo \
```

4	set image deployment.apps/revisionhistory-demo \\ nginx=nginx:1.9.1 --record=true kubectl deployment.apps/revisionhistory-demo \\ set image deployment.apps/revisionhistory-demo \\ nginx=nginx:1.7.9 --record=true

可以看到，即使执行了回滚操作，也会在修改历史记录中增加一个新的版本。

15.2.4 应用的部署

在项目迭代开发过程中，经常需要对应用进行上线部署。上线部署策略主要有 3 种：金丝雀部署、蓝绿部署和滚动部署。

1. 金丝雀部署

金丝雀部署也被叫作灰度部署。金丝雀部署过程如图 15-4 所示：先让一部分用户继续使用旧版本，而另一部分用户开始使用新版本；如果新版本没有发生问题，则逐步扩大新版本的使用范围，直到使用旧版本的用户都使用新版本。

通过使用金丝雀部署，可以最大限度地保证系统的整体稳定性，并能够在部署的早期就发现和解决问题。

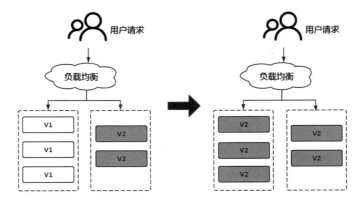

图 15-4

下面通过示例来演示金丝雀部署升级。

（1）创建"canary-demo-v1.yaml"文件，并在其中输入以下内容。

```
apiVersion: v1
kind: Service
```

```yaml
metadata:
  name: canary-demo
  labels:
    app: canary-demo
spec:
  type: NodePort
  ports:
  - name: http
    port: 80
    targetPort: http
  selector:
    app: canary-demo
---
apiVersion: apps/v1
kind: Deployment
metadata:
  name: canary-demo-v1
  labels:
    app: canary-demo
spec:
  replicas: 10
  selector:
    matchLabels:
      app: canary-demo
      version: v1.0.0
  template:
    metadata:
      labels:
        app: canary-demo
        version: v1.0.0
    spec:
      containers:
      - name: canary-demo
        image: collenzhao/k8s-deployment-strategies
        ports:
        - name: http
          containerPort: 8080
        env:
        - name: VERSION
          value: v1.0.0
```

这里使用 Service（服务）来实现 Deployment 的负载均衡。这方面的内容会在第 16 章中介绍。

（2）使用以下命令应用"canary-demo-v1.yaml"文件。

```
kubectl apply -f canary-demo-v1.yaml
```

（3）获取 Service（服务）的信息。

```
kubectl get service canary-demo
```

输出的信息如下：

```
NAME          TYPE       CLUSTER-IP      EXTERNAL-IP   PORT(S)        AGE
canary-demo   NodePort   10.1.119.250    <none>        80:30952/TCP   4s
```

（4）通过 Service（服务）访问 Deployment。

```
curl 10.1.119.250:80
```

输出的信息如下：

```
Host: canary-demo-v1-78b6cd78db-skjng, Version: v1.0.0
```

可以看出，目前应用的版本是 v1.0.0。

（5）使用金丝雀部署来升级应用，创建"canary-demo-v2.yaml"文件并在其中输入以下内容。

```
apiVersion: apps/v1
kind: Deployment
metadata:
  name: canary-demo-v2
  labels:
    app: canary-demo
spec:
  replicas: 1
  selector:
    matchLabels:
      app: canary-demo
      version: v2.0.0
  template:
    metadata:
      labels:
        app: canary-demo
        version: v2.0.0
    spec:
      containers:
      - name: canary-demo
        image: collenzhao/k8s-deployment-strategies
        ports:
        - name: http
          containerPort: 8080
```

```
    env:
    - name: VERSION
      value: v2.0.0
```

（6）开启两个命令行窗口，使用"watch"命令来分别监控 Deployment 和 Pod 的变化。

```
kubectl get --watch deployment
kubectl get --watch pod
```

（7）执行应用的升级。

```
kubectl apply -f canary-demo-v2.yaml
```

（8）观察 Deployment 和 Pod 的变化，如图 15-5 和图 15-6 所示。

```
[root@master ~]# kubectl get --watch deployment
NAME            READY    UP-TO-DATE   AVAILABLE    AGE
canary-demo-v1  10/10    10           10           2m32s
canary-demo-v2  1/1      1            1            29s
```

图 15-5

```
[root@master ~]# kubectl get --watch pod
NAME                                   READY   STATUS    RESTARTS   AGE
canary-demo-v1-78b6cd78db-67cg5        1/1     Running   0          2m36s
canary-demo-v1-78b6cd78db-7zjwf        1/1     Running   0          2m36s
canary-demo-v1-78b6cd78db-9dphd        1/1     Running   0          2m36s
canary-demo-v1-78b6cd78db-dskpc        1/1     Running   0          2m36s
canary-demo-v1-78b6cd78db-fdrwp        1/1     Running   0          2m36s
canary-demo-v1-78b6cd78db-gd9kf        1/1     Running   0          2m36s
canary-demo-v1-78b6cd78db-hr9x8        1/1     Running   0          2m36s
canary-demo-v1-78b6cd78db-lpjft        1/1     Running   0          2m36s
canary-demo-v1-78b6cd78db-nbbjx        1/1     Running   0          2m36s
canary-demo-v1-78b6cd78db-nmn2g        1/1     Running   0          2m36s
canary-demo-v2-7c4c5f5444-g69jr        1/1     Running   0          33s
```

图 15-6

> 从图 15-5 和图 15-6 可以看到，v1.0.0 版本共有 10 个实例，而 v2.0.0 版本只有 1 个实例。

（9）执行以下脚本请求应用。

```
for a in {1..11}
  do
    sleep 1;
    curl "10.1.119.250:80";
  done
```

输出的信息如下：

```
Host: canary-demo-v1-78b6cd78db-nbbjx, Version: v1.0.0
Host: canary-demo-v1-78b6cd78db-nbbjx, Version: v1.0.0
Host: canary-demo-v1-78b6cd78db-67cg5, Version: v1.0.0
Host: canary-demo-v1-78b6cd78db-gd9kf, Version: v1.0.0
Host: canary-demo-v1-78b6cd78db-7zjwf, Version: v1.0.0
Host: canary-demo-v1-78b6cd78db-gd9kf, Version: v1.0.0
Host: canary-demo-v1-78b6cd78db-dskpc, Version: v1.0.0
Host: canary-demo-v1-78b6cd78db-gd9kf, Version: v1.0.0
Host: canary-demo-v1-78b6cd78db-67cg5, Version: v1.0.0
Host: canary-demo-v1-78b6cd78db-fdrwp, Version: v1.0.0
Host: canary-demo-v2-7c4c5f5444-g69jr, Version: v2.0.0
```

 for 循环一共循环了 11 次，其中，10 次访问的是 v1.0.0 版本；只有 1 次访问的是 v2.0.0 版本。

（10）将 v2.0.0 版本的实例扩到 5 个，将 v1.0.0 版本的实例缩到 5 个。

```
kubectl scale --replicas=5 deploy canary-demo-v2
kubectl scale --replicas=5 deploy canary-demo-v1
```

（11）观察 Deployment 的变化，如图 15-7 所示。

```
kubectl get --watch deployment
```

```
[root@master ~]# kubectl get --watch deployment
NAME             READY   UP-TO-DATE   AVAILABLE   AGE
canary-demo-v1   5/5     5            5           19m
canary-demo-v2   5/5     5            5           17m
```

图 15-7

（12）重新执行以下脚本。

```
for a in {1..10}
  do
    sleep 1;
    curl "10.1.119.250:80";
  done
```

输出的信息如下：

```
Host: canary-demo-v1-78b6cd78db-67cg5, Version: v1.0.0
Host: canary-demo-v1-78b6cd78db-9dphd, Version: v1.0.0
Host: canary-demo-v2-7c4c5f5444-1cbhw, Version: v2.0.0
```

```
Host: canary-demo-v1-78b6cd78db-hr9x8, Version: v1.0.0
Host: canary-demo-v1-78b6cd78db-7zjwf, Version: v1.0.0
Host: canary-demo-v2-7c4c5f5444-lcbhw, Version: v2.0.0
Host: canary-demo-v1-78b6cd78db-fdrwp, Version: v1.0.0
Host: canary-demo-v2-7c4c5f5444-9hbwr, Version: v2.0.0
Host: canary-demo-v2-7c4c5f5444-9hbwr, Version: v2.0.0
Host: canary-demo-v1-78b6cd78db-hr9x8, Version: v1.0.0
```

（13）停止 v1.0.0 版本，并把 v2.0.0 版本的实例扩到 10 个。

```
kubectl delete deployment.apps/canary-demo-v1
kubectl scale --replicas=5 deploy canary-demo-v2
```

这时应用将全部升级到 v2.0.0 版本。如果再执行 for 循环，则所有请求的返回信息如下。

```
Host: canary-demo-v2-7c4c5f5444-g69jr, Version: v2.0.0
Host: canary-demo-v2-7c4c5f5444-lcbhw, Version: v2.0.0
Host: canary-demo-v2-7c4c5f5444-hs4k2, Version: v2.0.0
Host: canary-demo-v2-7c4c5f5444-lcbhw, Version: v2.0.0
Host: canary-demo-v2-7c4c5f5444-hs4k2, Version: v2.0.0
Host: canary-demo-v2-7c4c5f5444-g69jr, Version: v2.0.0
Host: canary-demo-v2-7c4c5f5444-9hbwr, Version: v2.0.0
Host: canary-demo-v2-7c4c5f5444-5h5sm, Version: v2.0.0
Host: canary-demo-v2-7c4c5f5444-g69jr, Version: v2.0.0
Host: canary-demo-v2-7c4c5f5444-5h5sm, Version: v2.0.0
```

（14）清理测试的数据。

```
kubectl delete all -l app=canary-demo
```

2. 蓝绿部署

蓝绿部署如图 15-8 所示：同时部署旧版本和新版本（旧版本被叫作蓝色版本，新版本被叫作绿色版本）；在绿色版本中对应用进行测试，待测试完全通过后，将应用从蓝色版本路由到绿色版本，然后将绿色版本变成新的生产环境。

蓝绿部署可以在线进行，具有较小的风险，但是会对用户的体验有影响。另外，蓝绿部署由于同时部署了新旧两套版本，所以需要双倍的资源。

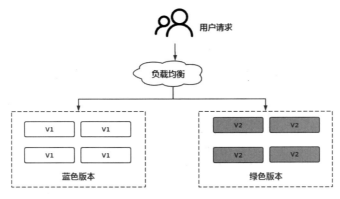

图 15-8

下面来演示如何实现应用的蓝绿部署升级。

（1）创建"blue-green-demo-v1.yaml"文件，并在其中输入以下内容。

```yaml
apiVersion: v1
kind: Service
metadata:
  name: blue-green-demo
  labels:
    app: blue-green-demo
spec:
  type: NodePort
  ports:
  - name: http
    port: 80
    targetPort: http
  selector:
    app: blue-green-demo
    version: v1.0.0
---
apiVersion: apps/v1
kind: Deployment
metadata:
  name: blue-green-demo-v1
  labels:
    app: blue-green-demo
spec:
  replicas: 3
  selector:
    matchLabels:
      app: blue-green-demo
```

```yaml
      version: v1.0.0
  template:
    metadata:
      labels:
        app: blue-green-demo
        version: v1.0.0
    spec:
      containers:
      - name: blue-green-demo
        image: collenzhao/k8s-deployment-strategies
        ports:
        - name: http
          containerPort: 8080
        env:
        - name: VERSION
          value: v1.0.0
```

（2）部署应用的 v1.0.0 版本。

```
kubectl apply -f blue-green-demo-v1.yaml
```

（3）查看创建的 Service（服务）。

```
kubectl get service
```

输出的信息如下：

```
NAME              TYPE       CLUSTER-IP   EXTERNAL-IP   PORT(S)        AGE
blue-green-demo   NodePort   10.1.88.3    <none>        80:30347/TCP   74s
```

（4）执行以下脚本请求应用。

```
for a in {1..3}
  do
    sleep 1;
    curl "10.1.88.3:80";
  done
```

这时所有的请求将访问 v1.0.0 版本，输出的信息如下：

```
Host: blue-green-demo-v1-6ddd7f6b74-gwbds, Version: v1.0.0
Host: blue-green-demo-v1-6ddd7f6b74-jsmdb, Version: v1.0.0
Host: blue-green-demo-v1-6ddd7f6b74-jsmdb, Version: v1.0.0
```

（5）创建"blue-green-demo-v2.yaml"文件，并在其中输入以下内容。

```yaml
apiVersion: apps/v1
kind: Deployment
metadata:
  name: blue-green-demo-v2
```

```
      labels:
        app: blue-green-demo
spec:
  replicas: 3
  selector:
    matchLabels:
      app: blue-green-demo
      version: v2.0.0
  template:
    metadata:
      labels:
        app: blue-green-demo
        version: v2.0.0
    spec:
      containers:
      - name: blue-green-demo
        image: collenzhao/k8s-deployment-strategies
        ports:
        - name: http
          containerPort: 8080
        env:
        - name: VERSION
          value: v2.0.0
```

（6）将应用升级到 v2.0.0 版本。

```
kubectl apply -f blue-green-demo-v2.yaml
```

（7）查看 Pod 的信息。

```
kubectl get pod
```

输出的信息如下：

```
NAME                                      READY   STATUS    RESTARTS   AGE
blue-green-demo-v1-6ddd7f6b74-92rjh       1/1     Running   0          7m18s
blue-green-demo-v1-6ddd7f6b74-gwbds       1/1     Running   0          7m18s
blue-green-demo-v1-6ddd7f6b74-jsmdb       1/1     Running   0          7m18s
blue-green-demo-v2-569957d476-4f96f       1/1     Running   0          53s
blue-green-demo-v2-569957d476-spbf2       1/1     Running   0          53s
blue-green-demo-v2-569957d476-xhf5z       1/1     Running   0          53s
```

这里可以看到前 3 个 Pod 是 v1.0.0 版本，而后 3 个 Pod 是 v2.0.0 版本。

（8）将 Service（服务）全部切换到 v2.0.0 版本。

```
kubectl patch service blue-green-demo \
-p '{"spec":{"selector":{"version":"v2.0.0"}}}'
```

（9）重新执行以下脚本请求应用。

```
for a in {1..3}
  do
    sleep 1;
    curl "10.1.88.3:80";
  done
```

这时所有的请求将访问 v2.0.0 版本，输出的信息如下：

```
Host: blue-green-demo-v2-569957d476-4f96f, Version: v2.0.0
Host: blue-green-demo-v2-569957d476-spbf2, Version: v2.0.0
Host: blue-green-demo-v2-569957d476-xhf5z, Version: v2.0.0
```

（10）删除应用的 v1.0.0 版本。

```
kubectl delete deploy blue-green-demo-v1
```

3. 滚动部署

滚动部署如图 15-9 所示：先将新版本部署到集群中的一台或几台服务器上，在更新版本后将这些服务器重新投入使用；循环这个过程，直到集群中的所有服务器都更新成新版本。

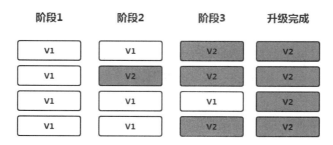

图 15-9

相对于蓝绿部署，这种部署方式更加节约资源。

下面来演示如何实现应用的滚动部署升级。

（1）创建 "rolling-update-demo-v1.yaml" 文件，并在其中输入以下内容。

```
apiVersion: v1
kind: Service
metadata:
  name: rolling-update-demo
```

```yaml
    labels:
      app: rolling-update-demo
spec:
  type: NodePort
  ports:
  - name: http
    port: 80
    targetPort: http
  selector:
    app: rolling-update-demo
---
apiVersion: apps/v1
kind: Deployment
metadata:
  name: rolling-update-demo
  labels:
    app: rolling-update-demo
spec:
  replicas: 3
  selector:
    matchLabels:
      app: rolling-update-demo
  template:
    metadata:
      labels:
        app: rolling-update-demo
        version: v1.0.0
    spec:
      containers:
      - name: rolling-update-demo
        image: collenzhao/k8s-deployment-strategies
        ports:
        - name: http
          containerPort: 8080
        env:
        - name: VERSION
          value: v1.0.0
```

（2）部署应用的 v1.0.0 版本。

```
kubectl apply -f rolling-update-demo-v1.yaml
```

（3）使用"watch"命令监控 Pod 的变化，如图 15-10 所示。

```
kubectl get --watch pod
```

```
NAME                                    READY    STATUS    RESTARTS    AGE
rolling-update-demo-787dfc98bd-bdq2v    1/1      Running   0           47s
rolling-update-demo-787dfc98bd-m5l5p    1/1      Running   0           47s
rolling-update-demo-787dfc98bd-vcnkd    1/1      Running   0           47s
```

图 15-10

（4）创建 "rolling-update-demo-v2.yaml" 文件，并在其中输入以下内容。

```yaml
apiVersion: apps/v1
kind: Deployment
metadata:
  name: rolling-update-demo
  labels:
    app: rolling-update-demo
spec:
  replicas: 3
  selector:
    matchLabels:
      app: rolling-update-demo
  template:
    metadata:
      labels:
        app: rolling-update-demo
        version: v2.0.0
    spec:
      containers:
      - name: rolling-update-demo
        image: collenzhao/k8s-deployment-strategies
        ports:
        - name: http
          containerPort: 8080
        env:
        - name: VERSION
          value: v2.0.0
```

（5）将应用升级到 v2.0.0 版本。

```
kubectl apply -f rolling-update-demo-v2.yaml
```

（6）这时会发现 v1.0.0 版本的 Pod 正在逐步被替换成 v2.0.0 版本的 Pod，如图 15-11 所示。

```
NAME                                       READY  STATUS             RESTARTS  AGE
rolling-update-demo-787dfc98bd-bdq2v       1/1    Running            0         47s
rolling-update-demo-787dfc98bd-m515p       1/1    Running            0         47s
rolling-update-demo-787dfc98bd-vcnkd       1/1    Running            0         47s
rolling-update-demo-59956d484b-rsh56       0/1    Pending            0         0s
rolling-update-demo-59956d484b-rsh56       0/1    Pending            0         0s
rolling-update-demo-59956d484b-rsh56       0/1    ContainerCreating  0         0s
rolling-update-demo-59956d484b-rsh56       1/1    Running            0         20s
rolling-update-demo-787dfc98bd-vcnkd       1/1    Terminating        0         2m21s
rolling-update-demo-59956d484b-mgp86       0/1    Pending            0         0s
rolling-update-demo-59956d484b-mgp86       0/1    Pending            0         0s
rolling-update-demo-59956d484b-mgp86       0/1    ContainerCreating  0         0s
rolling-update-demo-787dfc98bd-vcnkd       0/1    Terminating        0         2m22s
rolling-update-demo-787dfc98bd-vcnkd       0/1    Terminating        0         2m26s
rolling-update-demo-787dfc98bd-vcnkd       0/1    Terminating        0         2m26s
rolling-update-demo-59956d484b-mgp86       1/1    Running            0         20s
rolling-update-demo-787dfc98bd-m515p       1/1    Terminating        0         2m41s
rolling-update-demo-59956d484b-m6bpg       0/1    Pending            0         0s
rolling-update-demo-59956d484b-m6bpg       0/1    Pending            0         0s
rolling-update-demo-59956d484b-m6bpg       0/1    ContainerCreating  0         0s
rolling-update-demo-787dfc98bd-m515p       0/1    Terminating        0         2m42s
rolling-update-demo-787dfc98bd-m515p       0/1    Terminating        0         2m43s
rolling-update-demo-787dfc98bd-m515p       0/1    Terminating        0         2m43s
rolling-update-demo-59956d484b-m6bpg       1/1    Running            0         21s
rolling-update-demo-787dfc98bd-bdq2v       1/1    Terminating        0         3m2s
rolling-update-demo-787dfc98bd-bdq2v       0/1    Terminating        0         3m3s
rolling-update-demo-787dfc98bd-bdq2v       0/1    Terminating        0         3m6s
rolling-update-demo-787dfc98bd-bdq2v       0/1    Terminating        0         3m6s
```

图 15-11

15.2.5 编写 Deployment 控制器的规则

从 "deployment-demo.yaml" 文件和 "revisionhistory-demo.yaml" 文件可以看出，与创建 Kubernetes 的其他资源一样，创建 Deployment 也需要指定 apiVersion、kind 和 metadata 字段，另外还需要指定 .spec 字段。

表 15-4 中列出了在创建 Deployment 时各个字段的作用。

表 15-4

字段名称	字段含义
apiVersion	指定创建 Deployment 对象所使用的 Kubernetes API 版本
kind	指定创建的对象的类别，这里是 "Deployment"
metadata	指定 Deployment 对象的名称和标签信息等
.spec	指定 Pod 的匹配信息和 Pod 的模板信息。其中只有 ".spec.selector" 和 ".spec.template" 是必需的字段。 ● ".spec.selector" 字段 该字段通过标签的匹配，将 Deployment 对象与对应的 Pod 进行关联。因此，在定义标签时应保证其唯一性。 ● ".spec.template" 字段 该字段用于创建 Pod 的模板，语法规则与创建 Pod 的语法规则完全相同

除以上字段外，Deployment 还可以定义 Pod 的重启策略。但值得注意的是，".spec.template.spec.restartPolicy"字段的值只能是"Always"。

15.3 DaemonSet 控制器

使用 DaemonSet 控制器相当于在节点上启动了一个守护进程。通过 DaemonSet 控制器可以确保在每个 node 节点上运行 Pod 的一个副本。如果有新的 node 节点加入集群，则 DaemonSet 控制器会自动给新加入的节点增加一个 Pod 的副本；反之，当有 node 节点被从集群中移除时，DaemonSet 控制器也会自动回收这些 Pod 的副本。在删除 DaemonSet 控制器时，会删除 DaemonSet 控制器所创建的所有 Pod 的副本。

下面列举了 DaemonSet 控制器的一些典型应用场景：

- 使用 DaemonSet 控制器在节点上运行监控程序。
- 使用 DaemonSet 控制器在节点上运行日志收集程序，如日志收集程序 fluentd 和 logstash。
- 使用 DaemonSet 控制器运行 Kubernetes 的存储守护进程。

利用 Kubernetes 可以在节点上针对不同类型的守护进程单独启动一个 DaemonSet；也可以针对不同的硬件指标（如 CPU 和内存）部署多个 DaemonSet。

15.3.1 DaemonSet 控制器的创建

下面来演示如何创建 DaemonSet 控制器。

（1）创建"daemonset-demo.yaml"文件，并在其中输入以下内容：

```
apiVersion: apps/v1
kind: DaemonSet
metadata:
   name: daemonset-demo
   namespace: default
spec:
   minReadySeconds: 5
   selector:
      matchLabels:
         app: daemonset
   template:
      metadata:
```

```
        name: daemonset-demo
        namespace: default
        labels:
            app: daemonset
    spec:
        containers:
            - name: daemonset-demo
              image: nginx
              imagePullPolicy: IfNotPresent
              ports:
                - name: httpd
                  containerPort: 80
```

(2)执行以下命令创建 DaemonSet 控制器。

```
kubectl apply -f daemonset-demo.yaml
```

(3)查看 DaemonSet 控制器的信息。

```
kubectl get daemonset
```

> 该命令可以简写成以下形式：
>
> kubectl get ds

输出的信息如图 15-12 所示。可以看出，由于在 Kubernetes 集群中有 2 个 node 节点，因此 DaemonSet 控制器会在每个 node 节点上启动 1 个 Pod。

```
NAME              DESIRED   CURRENT   READY   UP-TO-DATE
daemonset-demo    2         2         2       2
[root@master yaml]#
```

图 15-12

(4)查看 Pod 的信息。

```
kubectl get pods -o wide
```

输出的信息如图 15-13 所示。

```
NAME                     READY   STATUS    AGE     IP             NODE
daemonset-demo-8v2qq     1/1     Running   3m51s   10.244.1.29    node1
daemonset-demo-rnxcn     1/1     Running   3m51s   10.244.2.39    node2
[root@master yaml]#
```

图 15-13

（5）删除 DaemonSet 控制器。

```
kubectl delete daemonset daemonset-demo
```

15.3.2 DaemonSet 控制器的调度

Kubernetes 的 scheduler 调度器能够确保所有符合条件的 node 节点都可以运行一个由 DaemonSet 控制器管理的 Pod。这也是 DaemonSet 控制器中 Pod 默认的调度方式。

但是，Kubernetes 也允许用户使用 DaemonSet 控制器自己的调度器来创建和调度 Pod。使用 DaemonSet 控制器自己的调度器来进行 Pod 的调度，可能会造成各个 node 节点上 Pod 行为的不一致。因为在正常情况下，Pod 在被创建后应该处于 Pending（等待被调度）状态；而由 DaemonSet 控制器创建的 Pod 不会处于 Pending 状态。

15.4 Job 控制器

Job 控制器是一次性任务的控制器。它控制 Pod 中的容器在执行完成任务后不会重启，并将容器的状态设置为 Completed。

- 如果 Pod 中的容器异常终止了，则 Job 控制器会根据设置的重启策略进行 Pod 的重启。
- 如果因为 node 节点故障导致 Pod 无法正常运行，则 Job 控制器会通过调度器将 Pod 调度到其他的节点上运行。

Job 控制器的运行方式分为：单工作队列的 Job 串行方式和多工作队列的 Job 并行方式。

15.4.1 【实战】单工作队列的 Job 串行方式

下面来演示如何使用单工作队列的 Job 串行方式。

（1）创建"job-demo1.yaml"文件，并在其中输入以下内容：

```
apiVersion: batch/v1
kind: Job
metadata:
  name: job-demo1
spec:
  template:
    spec:
      containers:
      - name: busybox
        image: busybox:latest
        imagePullPolicy: IfNotPresent
```

```
        command: [ "/bin/sh", "-c", "sleep 120s" ]
#重启策略：在发生错误时不进行重启
restartPolicy: Never
```

（2）执行以下命令创建 Job。

```
kubectl apply -f job-demo1.yaml
```

（3）查看 Job 信息和 Pod 信息，如图 15-14 所示。

```
kubectl get job,pod
```

```
NAME                   COMPLETIONS   DURATION   AGE
job.batch/job-demo1    0/1           14s        14s

NAME                    READY   STATUS    RESTARTS   AGE
pod/job-demo1-nhclj     1/1     Running   0          14s
[root@master yaml]#
```

图 15-14

（4）等待 120 s 后再次查看 Job 信息和 Pod 信息，如图 15-15 所示。

```
kubectl get job,pod
```

```
[root@master yaml]# kubectl get job,pod
NAME                   COMPLETIONS   DURATION   AGE
job.batch/job-demo1    1/1           2m2s       2m6s

NAME                    READY   STATUS      RESTARTS   AGE
pod/job-demo1-nhclj     0/1     Completed   0          2m6s
[root@master yaml]#
```

图 15-15

（5）对比图 15-14 和图 15-15 可以发现：由于 Job 控制器执行的是一次性任务，所以，当 Pod 中的容器运行 120 s 后，Pod 将被设置成 Completed 状态，并且不再重启。

15.4.2　【实战】多工作队列的 Job 并行方式

在创建多工作队列的 Job 并行方式时，需要指定以下两个参数。

- .spec.parallelism：Job 并行执行的数量，即队列的数量。
- .spec.completions：需要完成的 Job 总数量。

在以下例子中，需要完成的 Job 总数量是 5 个，并且创建了 3 个队列，即 Job 的并行度是 3。所有 Job 执行完需要 2 分钟。

(1)创建"job-demo2.yaml"文件,并在其中输入以下内容:

```yaml
apiVersion: batch/v1
kind: Job
metadata:
   name: job-demo2
spec:
   completions: 5
   parallelism: 3
   template:
     spec:
       containers:
       - name: job-demo2
         image: nginx
         imagePullPolicy: IfNotPresent
         # 使用 sleep 命令模拟 Job 需要执行 60 s
         command: ["/bin/bash","-c","sleep 60"]
       #重启策略:不进行重启
       restartPolicy: Never
```

(2)执行以下命令创建 Job。

```
kubectl apply -f job-demo2.yaml
```

(3)查看 Job 信息和 Pod 信息,如图 15-16 所示。

```
kubectl get job,pod
```

```
NAME                    COMPLETIONS   DURATION   AGE
job.batch/job-demo2     0/5           5s         5s

NAME                      READY   STATUS    RESTARTS   AGE
pod/job-demo2-4rrzn       1/1     Running   0          5s
pod/job-demo2-rp59n       1/1     Running   0          5s
pod/job-demo2-zqncm       1/1     Running   0          5s
[root@master yaml]#
```

图 15-16

> 由于需要完成的 Job 总数量是 5 个,但只有 3 个队列在并行执行。因此在图 15-6 中看到 Job 控制器启动了 3 个 Pod。

(4)等待 60 s 后再次查看 Job 信息和 Pod 信息,如图 15-17 所示。

```
kubectl get job,pod
```

```
NAME                    COMPLETIONS   DURATION   AGE
job.batch/job-demo2     3/5           67s        67s

NAME                    READY   STATUS       RESTARTS   AGE
pod/job-demo2-4rrzn     0/1     Completed    0          67s
pod/job-demo2-fb7kk     1/1     Running      0          6s
pod/job-demo2-gxnr7     1/1     Running      0          6s
pod/job-demo2-rp59n     0/1     Completed    0          67s
pod/job-demo2-zqncm     0/1     Completed    0          67s
[root@master yaml]#
```

图 15-17

经过 60 s 后，第一批执行的 3 个 Job 已经完成，Job 控制器会使用空闲的 2 个队列来执行剩下的 2 个 Job。

（5）等待 60 s 后再次查看 Job 信息和 Pod 信息，如图 15-18 所示。

```
kubectl get job,pod
```

```
NAME                    COMPLETIONS   DURATION   AGE
job.batch/job-demo2     5/5           2m3s       2m16s

NAME                    READY   STATUS       RESTARTS   AGE
pod/job-demo2-4rrzn     0/1     Completed    0          2m16s
pod/job-demo2-fb7kk     0/1     Completed    0          75s
pod/job-demo2-gxnr7     0/1     Completed    0          75s
pod/job-demo2-rp59n     0/1     Completed    0          2m16s
pod/job-demo2-zqncm     0/1     Completed    0          2m16s
[root@master yaml]#
```

图 15-18

经过 120 s 后，5 个 Job 都成功执行完成了。

15.4.3　Job 的终止与清理

在 Job 执行完成后，不会再创建新的 Pod，也不会删除已经创建的 Pod。Job 控制器会保留这些已经创建的 Pod，以方便用户查看日志或进行问题的诊断。

同样，Kubernetes 也不会删除已经执行完成的 Job 对象。Kubernetes 把删除和清理 Job 的操作留给用户来完成，例如：

```
kubectl delete job job-demo1
```

在删除 Job 对象时，由 Job 对象创建的 Pod 也会被同时删除。

另一种终止 Job 方式是，通过设置 ".spec.activeDeadlineSeconds" 字段来终止。该字段的单位是秒，代表了 Job 的活跃期限。当 Job 的运行时间达到 ".spec.activeDeadlineSeconds" 字段所设定的期限后，Job 的状态将自动更新为 "Failed"，与之相关联的 Pod 都会被终止。

下面通过示例来说明 ".spec.activeDeadlineSeconds" 字段的作用。

（1）创建 "job-demo3.yaml" 文件，并在其中输入以下内容：

```yaml
apiVersion: batch/v1
kind: Job
metadata:
  name: job-demo3
spec:
  activeDeadlineSeconds: 20
  template:
    spec:
      containers:
      - name: busybox
        image: busybox:latest
        imagePullPolicy: IfNotPresent
        command: [ "/bin/sh", "-c", "sleep 120s" ]
      #重启策略：在发生错误时不重启
      restartPolicy: Never
```

上面的 Job 将基于 busybox 的镜像创建一个 Pod，并让 Pod 睡眠 120 s。但由于 ".spec.activeDeadlineSeconds" 字段只被设置为 20 s，所以当 Pod 还在执行中时，Job 因为超过了生命周期的活跃期而变成了 "Failed" 状态，从而 Pod 被自动终止。

（2）执行以下命令创建 Job。

```
kubectl apply -f job-demo3.yaml
```

（3）查看 Job 和 Pod 的信息。

```
kubectl get job,pod
```

输出的信息如下：

```
NAME                        COMPLETIONS   DURATION   AGE
job.batch/job-demo3         0/1           8s         8s

NAME                        READY   STATUS    RESTARTS   AGE
pod/job-demo3-pw2t4         1/1     Running   0          8s
```

（4）等待 20 s 后再次查看 Job 和 Pod 的信息。

```
kubectl get job,pod
```

输出的信息如下：

```
NAME                        COMPLETIONS   DURATION   AGE
job.batch/job-demo3         0/1           33s        33s

NAME                        READY   STATUS        RESTARTS   AGE
pod/job-demo3-pw2t4         1/1     Terminating   0          33s
```

 由于 Job 的存活时间超过了 ".spec.activeDeadlineSeconds" 字段中设定的期限，所以 Job 中的 Pod 被自动终止了。

（5）查看 Job 的详细信息。

```
kubectl describe job.batch/job-demo3
```

输出的信息如下：

```
Events:
Type     Reason             Message
----     ------             -------
Normal   SuccessfulCreate   Created pod: job-demo3-pw2t4
Normal   SuccessfulDelete   Deleted pod: job-demo3-pw2t4
Warning  DeadlineExceeded   Job was active longer than specified deadline
```

15.5　CronJob 控制器

Job 控制器管理的 Job 在控制器资源被创建后会立即执行。而 CronJob 控制器用于管理和调度 Job 的运行时间，从而实现定时作业和周期作业的目的。这种控制器管理和调度 Job 的方式类似 Linux 中的 crontab 命令。CronJob 控制器管理和调度 Job 有以下方式：

- 在未来某个时间运行 Job 一次。
- 在指定的时间点重复运行 Job。

15.5.1 【实战】运行第一个 CronJob 控制器

下面通过示例来演示如何使用 CronJob 控制器。该示例会在每分钟打印出当前时间和问候消息。

（1）创建 "cronjob-demo.yaml" 文件，并在其中输入以下内容。

```yaml
apiVersion: batch/v1beta1
kind: CronJob
metadata:
  name: cronjob-demo
spec:
  #CronJob 控制器将 1 分钟调度 1 次
  schedule: "*/1 * * * *"
  jobTemplate:
    spec:
      template:
        spec:
          containers:
          - name: cronjob-demo
            image: busybox
            imagePullPolicy: IfNotPresent
            command:
            - /bin/sh
            - -c
            - date; echo Hello from the CronJob
          restartPolicy: OnFailure
```

（2）执行以下命令创建 CronJob。

```
kubectl apply -f cronjob-demo.yaml
```

（3）查看 CronJob 和 Pod 的信息，如图 15-19 所示。

```
kubectl get cronjob,pod
```

```
NAME                          SCHEDULE      SUSPEND   ACTIVE   LAST SCHEDULE
cronjob.batch/cronjob-demo    */1 * * * *   False     1        14s

NAME                                      READY   STATUS      RESTARTS   AGE
pod/cronjob-demo-1643438580-9v556         0/1     Completed   0          10s
[root@master yaml]#
```

图 15-19

> 这时 CronJob 控制器将创建第 1 个 Pod。

（4）查看 Pod 的标准输出信息。

```
kubectl logs pod/cronjob-demo-1643438580-9v556
```

输出的信息如下：

```
Sat Jan 29 06:43:05 UTC 2022
Hello from the CronJob
```

（5）等待 1 分钟后再次查看 CronJob 和 Pod 的信息，如图 15-20 所示。

```
kubectl get cronjob,pod
```

```
NAME                          SCHEDULE      SUSPEND   ACTIVE   LAST SCHEDULE
cronjob.batch/cronjob-demo    */1 * * * *   False     1        10s

NAME                                  READY   STATUS      RESTARTS   AGE
pod/cronjob-demo-1643438580-9v556     0/1     Completed   0          66s
pod/cronjob-demo-1643438640-6kczd     0/1     Completed   0          6s
[root@master yaml]#
```

图 15-20

这时 CronJob 控制器将创建第 2 个 Pod。

（6）等待 1 分钟后，再次查看 CronJob 和 Pod 的信息，如图 15-21 所示。

```
kubectl get cronjob,pod
```

```
NAME                          SCHEDULE      SUSPEND   ACTIVE   LAST SCHEDULE   AGE
cronjob.batch/cronjob-demo    */1 * * * *   False     1        5s              2m41s

NAME                                  READY   STATUS              RESTARTS   AGE
pod/cronjob-demo-1643438580-9v556     0/1     Completed           0          2m1s
pod/cronjob-demo-1643438640-6kczd     0/1     Completed           0          61s
pod/cronjob-demo-1643438700-ttg54     0/1     ContainerCreating   0          0s
[root@master yaml]#
```

图 15-21

这时 CronJob 控制器将创建第 3 个 Pod，以后每隔 1 分钟创建 1 个新的 Pod。

15.5.2　CronJob 控制器中的时间表示

CronJob 控制器中的时间表示方式，与 Linux 中 crontab 命令的时间表示方式类似。

下面以 "cronjob-demo.yaml" 文件为例，通过 ".spec.schedule" 字段设置 CronJob 控制器 1 分钟调度 1 次。具体的参数含义如图 15-22 所示。

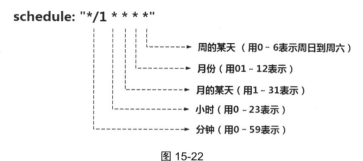

图 15-22

按照 CronJob 的时间表示方式，以下的 CronJob 控制器将在每月 21 号的 0 点，以及每个星期六的 0 点开始任务。

```
schedule: "0 0 21 * 6"
```

表 15-5 列举了 CronJob 常见的时间表示方式。

表 15-5

".spec.schedule" 字段的值	描述
0 0 1 1 *	每年 1 月 1 日的 0 点运行一次
0 0 1 * *	每月第一天的 0 点运行一次
0 0 * * 0	每周的周日 0 点运行一次
0 0 * * *	每天 0 点运行一次
0 * * * *	每小时运行一次

15.5.3　CronJob 控制器的限制

CronJob 控制器根据预先设定的时间，定时创建一个 Job 并执行该 Job。但 CronJob 并不严格保证 Job 一定会被创建，某些情况下可能不创建 Job，某些情况下可能会创建两个 Job。当 "startingDeadlineSeconds" 字段为默认值或者很大值，且 "concurrencyPolicy" 字段为 "Allow" 时，CronJob 控制器将保证 Job 至少运行一次。

由于 CronJob 控制器每隔 10 s 执行一次检查，因此不能将 "startingDeadlineSeconds" 字段的值设置得过小。如果该字段的值低于 10 s，则 CronJob 可能无法被调度。

15.6 StatefulSets 控制器

StatefulSets 控制器可以将 Pod 部署成有状态的服务。使用 StatefulSets 控制器,可以为 Pod 提供持久存储和持久的唯一性标识符。

StatefulSets 控制器与 Deployment 控制器不同的是:StatefulSets 控制器为管理的 Pod 维护了一个有黏性的标识符;无论这些 Pod 如何被调度,每个 Pod 的标识符都是永久不变的。这个特点可以满足一些特殊场景的需要,例如:在使用存储卷为 Kubernetes 集群提供持久型存储时,可以使用 StatefulSets 控制器作为解决方案的一种。

表 15-6 对比了无状态的控制器 Deployment 与有状态的控制器 StatefulSets 的主要异同。

表 15-6 无状态的控制器 Deployment 与有状态控的制器 StatefulSets 的主要异同

无状态的控制器 Deployment	有状态的控制器 StatefulSets
所有 Pod 都是一样的	Pod 之间存在一定的关系或者依赖,数据和配置信息可能不一致
所有 Pod 没有启动和关闭的顺序要求	启动和关闭可能需要按照顺序进行
所有 Pod 不用考虑在哪个 node 节点上运行	Pod 具有唯一的网络标识符,需要考虑在哪个 node 节点上运行
所有 Pod 可以执行随意扩容/缩容	有序、优雅地部署、扩展、删除、终止和更新

下面列举了 StatefulSets 控制器的一些典型应用场景。

- 需要唯一的、稳定的网络标识符,即在 Pod 被重新调度后其 Pod 名称和主机名不变。
- 需要持久的、稳定的持久化存储,即在 Pod 被重新调度后还能访问相同的持久化数据。
- 需要优雅的、有序地部署应用和扩容/缩容,即 Pod 的部署和启动是有顺序要求的,在部署或者扩容/缩容时要依据定义的顺序依次进行。
- 需要自动、有序地更新/回滚应用。

15.6.1 【实战】创建 StatefulSets 控制器

下面通过一个示例来说明如何创建 StatefulSets 控制器。

(1)创建"statefulsets-demo.yaml"文件,在其中输入以下内容:

```
apiVersion: apps/v1
kind: StatefulSet
metadata:
  #StatefulSets 控制器的名称
  name: statefulset-demo
spec:
  selector:
    matchLabels:
```

```
      #通过标签与 Pod 关联
      app: nginx
  #指定 Service 的名称
  serviceName: "nginx-service"
  replicas: 4
  template:
    metadata:
      labels:
        #定义 Pod 的标签
        app: nginx
    spec:
      terminationGracePeriodSeconds: 10
      containers:
      #定义 Pod 中容器的名称
      - name: nginx
        image: nginx:1.7.9
        ports:
        - containerPort: 80
          name: web
```

参数".spec.terminationGracePeriodSeconds"表示在关闭或删除 Pod 前需要等待的时间。对于 StatefulSets 控制器来说，不应将该参数设置为 0。这种做法是不安全的，非常不建议这么设置。

（2）执行以下命令创建 StatefulSets 控制器。

```
kubectl apply -f statefulsets-demo.yaml
```

（3）查看 StatefulSets 控制器的信息，如图 15-23 所示。

```
kubectl get statefulset,pod -o wide
```

```
NAME                                    READY    AGE    CONTAINERS    IMAGES
statefulset.apps/statefulset-demo       3/4      5s     nginx         nginx:1.7.9

NAME                        READY    STATUS              RESTARTS    NODE
pod/statefulset-demo-0      1/1      Running             0           node2
pod/statefulset-demo-1      1/1      Running             0           node1
pod/statefulset-demo-2      1/1      Running             0           node2
pod/statefulset-demo-3      0/1      ContainerCreating   0           node1
[root@master yaml]#
```

图 15-23

15.6.2　StatefulSets 控制器的扩容/缩容

由于 StatefulSets 控制器中的所有 Pod 都具有唯一的网络标识符，因此在对 StatefulSets 控制器进行扩容/缩容时，Kubernetes 将严格按照以下顺序进行：

- 在创建具有 N 个副本的 StatefulSets 控制器时，其中的每一个 Pod 都按照 0→(N-1)的顺序依次进行。
- 在删除具有 N 个副本的 StatefulSets 控制器时，其中的每一个 Pod 都按照(N-1)→0 的逆序依次进行。
- 当 StatefulSets 控制器的扩容操作被应用到某一个 Pod 时，Kubernetes 将保证前面所有 Pod 的状态必须是 Running 或 Ready 状态。
- 当 StatefulSets 控制器的缩容操作被应用到某一个 Pod 时，Kubernetes 将保证前面所有 Pod 必须是完全关闭的状态。

1. StatefulSets 控制器的扩容过程

以 15.6.1 节中的 statefulsets-demo.yaml 文件为例，在 statefulset-demo 的控制器被创建后，StatefulSets 会按照 pod/statefulset-demo-0 → pod/statefulset-demo-1 → pod/statefulset-demo-2 → pod/statefulset-demo-3 的顺序部署 4 个 Pod。

在启动 pod/statefulset-demo-1 时，pod/statefulset-demo-0 的状态一定是 Running 或者 Ready 状态。同理，在 pod/statefulset-demo-1 进入 Running 或者 Ready 状态前，不会部署 pod/statefulset-demo-2。

如果 pod/statefulset-demo-1 已经处于 Running 或者 Ready 状态而 pod/statefulset-demo-2 尚未部署，若此时发生 pod/statefulset-demo-0 运行失败，则 pod/statefulset-demo-2 将不会被部署，要等到 pod/statefulset-demo-0 部署完成且进入 Running 或者 Ready 状态后才会部署 pod/statefulset-demo-2。

2. StatefulSets 控制器的缩容过程

在用户进行缩容操作时，例如将"replicas"参数设置为 1，则首先被终止的是 pod/statefulset-demo-3。在 pod/statefulset-demo-3 没有被完全停止和删除前，pod/statefulset-demo-2 不会被终止。当 pod/statefulset-demo-3 已被终止和删除，但 pod/statefulset-demo-2 尚未被终止时，若发生 pod/statefulset-demo-1 运行失败，则不会终止 pod/statefulset-demo-2，必须等到 pod/statefulset-demo-0 进入 Running 或者 Ready 状态后才会终止 pod/statefulset-demo-2。

15.6.3　StatefulSets 控制器的更新/回滚

StatefulSet 的更新/滚动是通过设置".spec.updateStrategy"字段来实现的。该参数有以下两个值：

（1）RollingUpdate。

这是默认的更新策略，该策略将对 StatefulSets 控制器管理的所有 Pod 执行自动的滚动更新。

（2）OnDelete。

该策略不会自动更新 StatefulSets 控制器管理的所有 Pod。要实现 Pod 的更新，则必须手动删除旧的 Pod，以便让 StatefulSets 控制器创建新的 Pod。该策略将按照与"Pod 终止顺序"相同的顺序进行更新，每次更新一个 Pod。

第 16 章
通过 Service 访问 Pod

Service（服务）是 Kubernetes 中非常重要的概念。使用 Service（服务）能够为应用提供一个统一的访问地址（入口地址），并且，Service（服务）提供了负载均衡功能，从而将客户端的请求分发到后端的各个容器中。

16.1 Service 的概念与使用

在 Kubernetes 中部署的应用可能对应一个或者多个 Pod，而每个 Pod 又具有独立的 IP 地址。Service（服务）能够为一组功能相同的 Pod 提供统一不变的访问地址，使得集群具有稳定的 IP 地址（即 Cluster IP 地址），从而使在集群内部能够通过该 Cluster IP 地址将客户端请求路由到集群中的一个 Pod 上，从而实现客户端与 Pod 的通信。

Service 是 Kubernetes 中非常重要的组成部分，它主要为集群提供请求的负载均衡和 Pod 的自动发现功能，在 15.2.4 节中便使用 Service 实现了请求的负载均衡。图 16-1 说明了 Service 与 Pod 的关系。

Service 主要有以下两个作用：

- 通过标签 Label 与 Pod 关联，实现与 Pod 的通信。
- 提供不同的访问策略，以实现访问 Pod 请求的负载均衡。

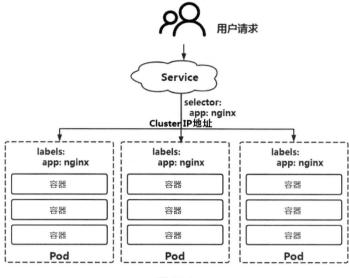

图 16-1

16.1.1 【实战】通过 Service 向外部暴露 Pod

创建 Service 与创建应用的 Pod，可以被定义在同一个 YAML 文件中，也可以将它们在不同 YAML 文件中分开定义。下面通过使用两个 YAML 文件来分别定义 Service 和 Pod。

（1）使用 Deployment 控制器来部署应用，创建文件"service-demo1.yaml"，并在其中输入以下内容。

```yaml
apiVersion: apps/v1
kind: Deployment
metadata:
  labels:
    app: nginx
  name: nginx
spec:
  replicas: 3
  selector:
    matchLabels:
      app: nginx
  template:
    metadata:
      labels:
        app: nginx
    spec:
      containers:
```

```
        - image: nginx
          name: nginx
          imagePullPolicy: IfNotPresent
```

(2) 使用 "kubectl apply" 命令部署应用。

```
kubectl apply -f service-demo1.yaml
```

(3) 查看 Pod 的信息。

```
kubectl get pod -o wide
```

输出的信息如下：

```
NAME                       READY   STATUS    IP             NODE
nginx-55fc968d9-c5csg      1/1     Running   10.244.2.186   node2
nginx-55fc968d9-kr7bb      1/1     Running   10.244.1.135   node1
nginx-55fc968d9-vwxw4      1/1     Running   10.244.2.187   node2
```

(4) 创建 Service：编辑文件 "service-demo2.yaml"，在其中输入以下内容。

```
apiVersion: v1
kind: Service
metadata:
  name: service-demo2
  namespace: default
spec:
  #这里使用 NodePort 类型的 Service 将应用暴露给外部
  type: NodePort
  ports:
  - name: http
    port: 80
    protocol: TCP
    targetPort: 80
  #定义标签选择器，将 Service 与匹配便签的一组 Pod 关联起来
  selector:
    app: nginx
```

(5) 使用 "kubectl apply" 命令创建 Service。

```
kubectl apply -f service-demo2.yaml
```

(6) 查看 Pod、Service 和 Endpoint 的信息。

```
kubectl get pods,service,endpoints -o wide
```

这条命令可以简写成以下形式：
kubectl get pods,svc,ep -o wide

输出的信息如下：

```
NAME                            READY   STATUS    IP              NODE
pod/nginx-55fc968d9-c5csg       1/1     Running   10.244.2.186    node2
pod/nginx-55fc968d9-kr7bb       1/1     Running   10.244.1.135    node1
pod/nginx-55fc968d9-vwxw4       1/1     Running   10.244.2.187    node2

NAME                      TYPE        CLUSTER-IP    PORT(S)        SELECTOR
service/kubernetes        ClusterIP   10.1.0.1      443/TCP        <none>
service/service-demo2     NodePort    10.1.48.136   80:30430/TCP   app=nginx

NAME                        ENDPOINTS
endpoints/kubernetes        192.168.79.11:6443
endpoints/service-demo2     10.244.1.135:80,
                            10.244.2.186:80,
                            10.244.2.187:80
```

 从输出的信息可以看出，在创建 Service 时会自动创建应用的接入点 Endpoint，并将每个 Pod 的 IP 地址自动加入 Endpoint 中。Endpoint 也会自动感知后端 Pod 的 IP 地址，从而实现动态的负载均衡。

（7）访问任意节点的 30430 端口都可以访问应用，如图 16-2 所示。

图 16-2

16.1.2　Service 的多端口设置

在 service-demo2.yaml 文件中定义的 Service 只暴露了一个端口，但在很多情况下需要 Service 暴露多个端口，例如同时暴露 HTTP 端口和 HTTPS 端口。Kubernetes 允许在定义 Service 时指定多个端口，但每个端口必须指定一个唯一的名称，以避免产生歧义。

下面是 Kubernetes 官方提供的一个示例。

```
apiVersion: v1
```

```
kind: Service
metadata:
  name: my-service
spec:
  selector:
    app: MyApp
  ports:
    - name: http
      protocol: TCP
      port: 80
      targetPort: 9376
    - name: https
      protocol: TCP
      port: 443
      targetPort: 9377
```

16.1.3 集群内部的 DNS 服务

Service 在实现请求代理和负载均衡时，默认采用的是 Cluster IP 地址。但是 Cluster IP 地址不是永远不变的，因此建议在应用中不要使用 Cluster IP 地址，而使用 Service 的名称。Kubernetes 集群提供的 DNS 服务可以将 Service 的名称解析为 Cluster IP 地址。

下面对 Kubernetes 集群内部的 DNS 服务进行一个简单的验证。

（1）查看系统命名空间中的 Pod。

```
kubectl get pods -n kube-system | grep dns
```

输出的信息如下：

```
coredns-bccdc95cf-68dgw            1/1      Running    3      320d
coredns-bccdc95cf-7dp9n            1/1      Running    3      320d
```

 在 kube-system 的命名空间中自动启动了 Pod 来运行 DNS 服务。

（2）创建 Service。

```
kubectl apply -f service-demo1.yaml
```

输出的信息如下：

```
service/service-demo2 created
```

（3）使用 busybox 的 1.28.4 版本创建一个 Pod，并进入 Pod 的内部。

```
kubectl run -it --image=busybox:1.28.4 --rm --restart=Never sh
```

（4）在 busybox 的 Pod 中查找"service-demo2"服务，集群内部的 DNS 服务将返回 Service 的域名信息，如果 16-3 所示。

```
nslookup service-demo2
```

```
If you don't see a command prompt, try pressing enter.
/ # nslookup service-demo2
Server:    10.1.0.10
Address 1: 10.1.0.10 kube-dns.kube-system.svc.cluster.local

Name:      service-demo2
Address 1: 10.1.16.199 service-demo2.default.svc.cluster.local
/ #
```

图 16-3

16.1.4 【实战】无头 Service

每一个 Service 都会有一个 Service 名称，并最终由 DNS 解析成 Cluster IP 地址。连接到 Service 的客户端最终通过 Cluster IP 地址被转发到后端一个随机选择的 Pod 上。因此，这时客户端并不清楚后端 Pod 的 IP 地址。但是对于一些有状态的客户端来说，需要清楚地知道后端每个 Pod 的 IP 地址才能与其直接进行通信。这时就需要使用无头 Service（Headless Service）。

图 16-4 说明了无头 Service 的运行机制，其核心是：去掉了 DNS 解析 Cluster IP 地址这个过程，直接返回后端 Pod 的 IP 地址。

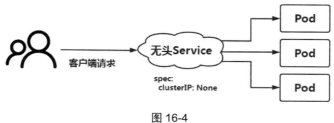

图 16-4

下面来演示如何使用无头 Service。

（1）创建部署描述文件 headless-service.yaml，并在其中输入以下内容。

```
apiVersion: v1
kind: Service
metadata:
  name: headless-service
spec:
  selector:
    name: busybox
  # 设置为无头 Service
```

```yaml
  clusterIP: None
  ports:
    - name: demo
      port: 1234
      targetPort: 1234
---
apiVersion: v1
kind: Pod
metadata:
  name: headless-service-pod-1
  labels:
    name: busybox
spec:
  hostname: headless-service-pod-1
  containers:
    - image: busybox
      command:
        - sleep
        - "3600"
      name: busybox
---
apiVersion: v1
kind: Pod
metadata:
  name: headless-service-pod-2
  labels:
    name: busybox
spec:
  hostname: headless-service-pod-2
  containers:
    - image: busybox
      command:
        - sleep
        - "3600"
      name: busybox
```

（2）执行"kubectl apply -f"命令。

```
kubectl apply -f headless-service.yaml
```

（3）查看 Service 和 Pod 的信息。

```
kubectl get svc,pod -o wide
```

输出的信息如下：

NAME	TYPE	CLUSTER-IP	PORT(S)	SELECTOR
service/headless-service	ClusterIP	None	1234/TCP	name=busybox

```
service/kubernetes              ClusterIP   10.96.0.1    443/TCP    <none>

NAME                           READY   STATUS    IP           NODE
pod/headless-service-pod-1     1/1     Running   10.244.1.6   node1
pod/headless-service-pod-2     1/1     Running   10.244.1.7   node1
```

 可以看到,"service/headless-service"的 Cluster IP 的值是 None(即这是一个无头 Service),在它的后面有两个 Pod,以及每个 Pod 实际的 IP 地址。

(4)使用 "kubectl exec" 命令进入其中一个 Pod 内部。

```
kubectl exec pod/headless-service-pod-1 -it /bin/sh
```

(5)在 Pod 内部查看所使用的 DNS 服务器信息。

```
more /etc/resolv.conf
```

输出的信息如下:

```
nameserver 10.96.0.10
search default.svc.cluster.local svc.cluster.local cluster.local
options ndots:5
```

在这里可以看到,使用的 DNS 服务器的 IP 地址是 10.96.0.10。

(6)在宿主机上使用 dig 命令访问无头 Service。

```
dig @10.96.0.10 headless-service.default.svc.cluster.local
```

输出的信息如下:

```
; <<>> DiG 9.11.3-1ubuntu1.13-Ubuntu <<>> @10.96.0.10 headless-service.default.svc.cluster.local
; (1 server found)
;; global options: +cmd
;; Got answer:
;; WARNING: .local is reserved for Multicast DNS
;; You are currently testing what happens when an mDNS query is leaked to DNS
;; ->>HEADER<<- opcode: QUERY, status: NOERROR, id: 1327
;; flags: qr aa rd; QUERY: 1, ANSWER: 2, AUTHORITY: 0, ADDITIONAL: 1
;; WARNING: recursion requested but not available

;; OPT PSEUDOSECTION:
; EDNS: version: 0, flags:; udp: 4096
; COOKIE: 2f6ded90a4739c23 (echoed)
;; QUESTION SECTION:
```

```
;headless-service.default.svc.cluster.local. IN A

;; ANSWER SECTION:
headless-service.default.svc.cluster.local. 30 IN A 10.244.1.6
headless-service.default.svc.cluster.local. 30 IN A 10.244.1.7

;; Query time: 0 msec
;; SERVER: 10.96.0.10#53(10.96.0.10)
;; WHEN: Thu Feb 03 04:36:26 UTC 2022
;; MSG SIZE  rcvd: 199
```

可以看到，在"ANSWER SECTION"中直接返回了后端两个 Pod 的 IP 地址。

16.2 Service 的发布类型

在 16.1.1 节中，Service 的发布使用的是 NodePort 类型。除此之外，Service 的发布还支持 ClusterIP、LoadBalancer 和 ExternalName 这 3 种类型。

16.2.1 NodePort

在把 Service 的 type 字段设置为 NodePort 时，Kubernetes 将在每个节点上随机分配一个端口作为外部用户访问的入口（该端口号的默认范围是 30000 ～ 32767）。该端口允许外部用户访问集群内部的 Pod 应用。如果用户想指定 30000 ～ 32767 内某个具体的端口，则需要增加一个 nodePort 字段。图 16-5 说明了 NodePort 的工作机制。

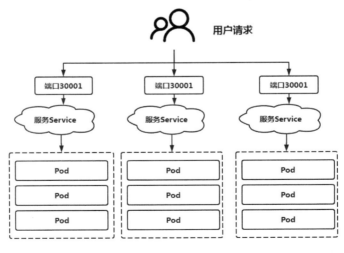

图 16-5

在以下 YAML 文件中，使用 nodePort 字段指定了一个自定义的端口号 31234。

```yaml
apiVersion: v1
kind: Service
metadata:
  name: service-demo2
  namespace: default
spec:
  #这里使用 NodePort 类型的 Service 将应用暴露给外部
  type: NodePort
  ports:
  - name: http
    #为方便起见，一般将 targetPort 字段的值设置为与 port 字段相同的值
    port: 80
    targetPort: 80
    #可选字段，自定义端口号
    nodePort: 31234
    protocol: TCP
  #定义标签选择器，将 Service 与匹配便签的一组 Pod 关联起来
  selector:
    app: nginx
```

16.2.2 ClusterIP

这是 Service 默认的发布类型。它将在集群内部分配一个可以访问的虚拟 IP 地址，通过该地址暴露服务。因此，这种类型的 Service 只能够实现同一个集群内部应用之间的相互访问。图 16-6 说明了 ClusterIP 的工作机制。

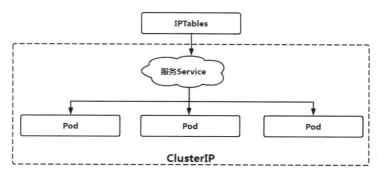

图 16-6

下面来演示 ClusterIP 的使用。

（1）通过 Deployment 创建应用的部署描述文件 "service-clusterip1.yaml"，并在其中输入以下内容。

```yaml
apiVersion: apps/v1
kind: Deployment
metadata:
  labels:
    app: nginx
  name: service-clusterip
spec:
  replicas: 3
  selector:
    matchLabels:
      app: nginx
  template:
    metadata:
      labels:
        app: nginx
    spec:
      containers:
      - image: nginx
        name: nginx
```

（2）使用"kubectl apply"命令创建 Deployment。

```
kubectl apply -f service-clusterip1.yaml
```

（3）创建 Service 的描述文件"service-clusterip2.yaml"，并在其中输入以下内容：

```yaml
apiVersion: v1
kind: Service
metadata:
  name: service-clusterip
  namespace: default
spec:
  ports:
  - name: http
    port: 1234
    protocol: TCP
    targetPort: 80
  selector:
    app: nginx
```

> 这里没有指定 Service 的发布类型，默认就是 ClusterIP。同时这里将容器中的 80 端口暴露成了 1234 端口。
>
> 也可以使用以下命令暴露应用：
> kubectl expose deployment nginx-test --port=1234\
> --target-port=80 --name=my-service-2

（3）使用"kubectl apply"命令创建 Service。

```
kubectl apply -f service-clusterip2.yaml
```

（4）查看 Pod、Service 和 Endpoint 的详细信息。

```
kubectl get pod,svc,ep -o wide
```

输出的信息如下：

```
NAME                                     READY  STATUS   IP            NODE
pod/service-clusterip-554bf9-bdjmr       1/1    Running  10.244.2.189  node2
pod/service-clusterip-554b7f9-nb2qq      1/1    Running  10.244.2.188  node2
pod/service-clusterip-554bf9-qx6lc       1/1    Running  10.244.1.136  node1

NAME                          TYPE       CLUSTER-IP     PORT(S)
service/kubernetes            ClusterIP  10.1.0.1       443/TCP
service/service-clusterip     ClusterIP  10.1.146.190   1234/TCP

NAME                          ENDPOINTS
endpoints/kubernetes          192.168.79.11:6443
endpoints/service-clusterip   10.244.1.136:80,
                              10.244.2.188:80,
                              10.244.2.189:80
```

（5）通过 Cluster IP 地址访问应用，将返回 Nginx 的首页。

```
curl 10.1.146.190:1234
```

16.2.3 LoadBalance

如果要使用外部的负载均衡器来访问应用（如 Google Cloud、AWS 和 OpenStack 等），则可以通过使用 LoadBalancer 类型的 Service 将 Kubernetes 集群中的 IP 地址和端口号自动加入公有云的 LoadBalancer 中，从而异步地实现负载均衡。图 16-7 说明了 LoadBalancer 的工作机制。

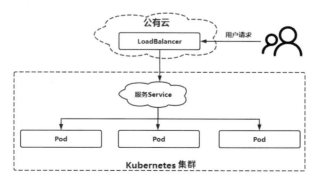

图 16-7

在以下 YAML 示例文件中，使用阿里云作为 Kubernetes 集群的外部负载均衡器。

```yaml
apiVersion: v1
kind: Service
metadata:
  name: service-loadbalancer
  labels:
    app: service-loadbalancer
spec:
  ports:
  - port: 80
    targetPort: 80
    protocol: TCP
    name: main-port
  selector:
    app: service-loadbalancer
  type: LoadBalancer
---
apiVersion: extensions/v1beta1
kind: Deployment
metadata:
  name: alicloud-controller-manager
spec:
  replicas: 1
  template:
    metadata:
      labels:
        app: alicloud-controller-manager
    spec:
      containers:
      - image: collenzhao/alicloud-controller-manager
        name: alicloud-controller-manager
        env:
          - name: ACCESS_KEY_ID
            value: 你的阿里云 ACCESS_KEY_ID
          - name: ACCESS_KEY_SECRET
            value: 你的阿里云 ACCESS_KEY_SECRET
```

16.2.4　ExternalName

ExternalName 类型的 Service 可以将一个已经存在的 Service 映射到外部的 DNS 服务，从而达到通过使用外部 DNS 服务解析服务应用的目的。

下面是 Kubernetes 官方提供的一个示例。

```yaml
apiVersion: v1
```

```
kind: Service
metadata:
  name: my-service
  namespace: prod
spec:
  type: ExternalName
  externalName: my.database.example.com
```

在查找主机"my-service.prod.svc.cluster.local"时，Kubernetes 集群的 DNS 服务将返回"my.database.example.com"。

16.3 虚拟 IP 与 Service 的代理模式

由于 Service 的默认发布类型是 ClusterIP，因此也可以把 Cluster IP 地址叫作虚拟 IP 地址。在 Kubernetes 创建 Service 时，每个节点上运行的 kube-proxy 会自动为 Service 分配一个虚拟 IP 地址，即通过转发代理 kube-proxy 来实现路由转发功能。kube-proxy 在具体实现流量代理转发与负载均衡时，有 3 种模式：

- userspace 代理模式。
- iptables 代理模式。
- IPVS 代理模式。

Cluster IP 地址是一个虚拟的 IP 地址，它是 Kubernetes 集群拥有的独立网络空间。它具有以下 3 个特征：
- Cluster IP 地址仅作用于 Kubernetes 的 Service 对象，并由 Kubernetes 进行管理和分配。
- Cluster IP 地址无法被直接访问，也没有实体的网络元素与其对应。
- 不同 Service 中的 Pod 在集群内部可以通过 Cluster IP 地址进行相互访问。

16.3.1 userspace 代理模式

在 userspace 代理模式下，访问 Service 的请求首先访问 node 节点的 iptable 表，再回到 Kubernetes 的命名空间中被 kube-proxy 转发到后端的 Pod 中。在默认情况下，userspace 代理模式下的 kube-proxy 通过轮询算法选择后端的 Pod。图 16-8 说明了 userspace 代理模式的工作机制。

图 16-8

userspace 代理模式最大的问题是：请求会存在一次状态转换过程（即从 node 节点的命名空间到 Kubernetes 命名空间的转换），从而有性能上的损耗。

16.3.2　iptables 代理模式

iptables 代理模式是目前 Service 实现代理的默认方式。iptables 代理模式是通过添加或移除 iptable 表中的路由规则，从而实现路由转发功能。在这种模式下，kube-proxy 会通过 API Server 监听集群中 Service 对象和 Endpoint 对象的创建和删除，从而创建 iptable 表中的规则，以实现将访问 Service 的请求重定向到后端的一组 Pod 中。

iptables 代理模式的工作机制如图 16-9 所示。iptables 代理模式的默认重定向策略会使用随机选择一个后端的 Pod。Kubernetes 也允许用户将 ".service.spec.sessionAffinity" 字段的值设置为 "ClientIP"，以实现基于客户端会话的 IP 地址亲和性的重定向策略。

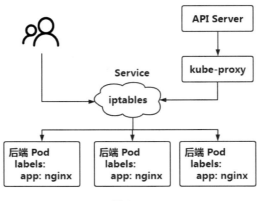

图 16-9

由于在 iptables 中记录了路由的规则，因此 iptables 代理模式会占用较少的系统资源。当 kube-proxy 运行在 iptables 代理模式下且与后端的 Pod 连接失败时，它会自动对后端的其他 Pod 进行连接重试。

下面以 16.1.2 节中的示例来验证 iptables 代理模式。

（1）执行以下语句创建 Deployment 和 Service。

```
kubectl apply -f service-demo1.yaml
kubectl apply -f service-demo2.yaml
```

（2）查看 Service 的信息。

```
kubectl get svc
```

输出的信息如下：

```
NAME             TYPE        CLUSTER-IP    EXTERNAL-IP   PORT(S)        AGE
kubernetes       ClusterIP   10.1.0.1      <none>        443/TCP        8d
service-demo2    NodePort    10.1.79.37    <none>        80:31234/TCP   28m
```

（3）查看节点的 iptable 表中的路由规则。

```
iptables-save |grep 10.1.79.37
```

输出的 iptable 表中的路由规则信息如下：

```
-A KUBE-SERVICES ! -s 10.244.0.0/16 -d 10.1.79.37/32 \
        -p tcp -m comment \
        --comment "default/service-demo2:http cluster IP" \
        -m tcp --dport 80 -j KUBE-MARK-MASQ
-A KUBE-SERVICES -d 10.1.79.37/32 -p tcp -m comment \
        --comment "default/service-demo2:http cluster IP" \
        -m tcp --dport 80 -j KUBE-SVC-PK3TNCFZ7PNC24KY
```

（4）删除 Service。

```
kubectl delete -f service-demo2.yaml
```

（5）重新查看节点的 iptable 表中的路由规则。

```
iptables-save |grep 10.1.79.37
```

这时在 iptable 表中没有任何路由信息了。

iptables 代理模式尽管使用简单,占用的资源也较少。但它却有以下的不足:
- 当存在大量的 Service 时,需要在 iptable 表中创建大量的路由规则,从而难以维护。
- 在 iptable 表中进行路由规则匹配时,可能会造成转发的延时。

这时可以将 kube-proxy 运行在 IPVS 代理模式下。下面来介绍 IPVS 代理模式。

16.3.3 IPVS 代理模式

与 iptables 代理模式一样,IPVS 代理模式通过 API Server 监听集群中 Service 对象和 Endpoint 对象的创建和删除,从而创建 IPVS 路由转发规则,并定期与 Service 对象和 Endpoint 对象同步 IPVS 路由转发规则,以达到路由转发的目的。请求在访问 Service 时,会被重定向到后端的一个 Pod 中。

IPVS 代理模式的工作机制如图 16-10 所示。

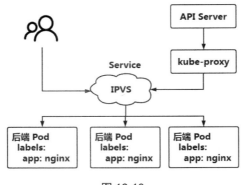

图 16-10

IPVS 代理模式创建的是 IPVS 路由转发规则,而不是 iptables 路由转发规则。
kube-proxy 会定期与 Service 对象和 Endpoints 对象同步 IPVS 路由转发规则。

以下步骤将在 Kubernetes 集群中使用 Service 的 IPVS 代理模式。

(1)在所有节点上安装 IPVS 模块。

```
yum -y install ipvsadm ipvsset
```

(2)让所有节点启用 IPVS 模块。

```
cat > /etc/sysconfig/modules/ipvs.modules <<EOF
modprobe -- ip_vs
modprobe -- ip_vs_rr
```

```
modprobe -- ip_vs_wrr
modprobe -- ip_vs_sh
modprobe -- nf_conntrack_ipv4
EOF
```

(3)给"/etc/sysconfig/modules/ipvs.modules"文件授权。

```
chmod 755 /etc/sysconfig/modules/ipvs.modules && \
bash /etc/sysconfig/modules/ipvs.modules && lsmod | \
grep -e ip_vs -e nf_conntrack_ipv4
```

(4)修改 kube-proxy 的配置。

```
kubectl edit configmap -n kube-system
```

将"mode"字段改为以下形式:

```
 mode: "ipvs"
```

(5)重启 kube-proxy。

```
kubectl get pod -n kube-system | \
grep kube-proxy | \
awk '{system("kubectl delete pod "$1" -n kube-system")}'
```

(6)编辑"service-ipvs.yaml"文件创建 Service 和 Deployment。

```
apiVersion: apps/v1
kind: Deployment
metadata:
  labels:
    app: nginx
  name: nginx
spec:
  replicas: 3
  selector:
    matchLabels:
      app: nginx
  template:
    metadata:
      labels:
        app: nginx
    spec:
      containers:
      - image: nginx
        name: nginx
        imagePullPolicy: IfNotPresent
---
apiVersion: v1
kind: Service
```

```yaml
metadata:
  name: service-ipvs
  namespace: default
spec:
  ports:
  - name: http
    port: 80
    protocol: TCP
    targetPort: 80
  selector:
    app: nginx
```

（7）执行 "kubectl apply –f" 命令。

```
kubectl apply -f service-ipvs.yaml
```

（8）查看 Pod 和 Service 的信息。

```
kubectl get pod,svc -o wide
```

输出的信息如下：

```
NAME                            READY   STATUS    IP            NODE
pod/nginx-55fc968d9-6829h       1/1     Running   10.244.1.78   node1
pod/nginx-55fc968d9-j45kv       1/1     Running   10.244.2.77   node2
pod/nginx-55fc968d9-s5gr5       1/1     Running   10.244.1.79   node1

NAME                        TYPE        CLUSTER-IP
service/kubernetes          ClusterIP   10.1.0.1
service/service-ipvs        ClusterIP   10.1.245.223
```

（9）查看 IPVS 路由转发规则。

```
ipvsadm -Ln
```

输出的信息如下：

```
IP Virtual Server version 1.2.1 (size=4096)
Prot LocalAddress:Port Scheduler Flags
  -> RemoteAddress:Port           Forward Weight ActiveConn InActConn
TCP  10.1.0.1:443 rr
  -> 192.168.79.11:6443           Masq    1      0          0
TCP  10.1.0.10:53 rr
  -> 10.244.0.2:53                Masq    1      0          0
  -> 10.244.0.3:53                Masq    1      0          0
TCP  10.1.0.10:9153 rr
  -> 10.244.0.2:9153              Masq    1      0          0
  -> 10.244.0.3:9153              Masq    1      0          0
TCP  10.1.245.223:80 rr
  -> 10.244.1.78:80               Masq    1      0          0
```

```
              -> 10.244.1.79:80            Masq    1       0          0
              -> 10.244.2.77:80            Masq    1       0          0
UDP  10.1.0.10:53 rr
              -> 10.244.0.2:53             Masq    1       0          0
              -> 10.244.0.3:53             Masq    1       0          0
```

16.4 集群外部的请求访问集群内应用的最佳方式——Ingress

Kubernetes 通过 Service 为 Pod 提供了统一的入口地址，并使用 NodePort 和 LoadBalance 类型的 Service 让外部的请求可以访问集群内部的应用。

但是，使用 NodePort 和 LoadBalance 类型有以下缺点：

- NodePort 类型通过在每个节点上暴露一个端口作为外部访问的入口，因此当 Service 很多时，这种方式会占用集群的大量端口。
- LoadBalance 类型需要为每一个 Service 都定义一个负载均衡器，会浪费资源。

因此，Kubernetes 提供了 Ingress 域名访问应用的方式，这也是从集群外部访问集群内应用的最佳方式。

16.4.1 Ingress 是什么

Ingress 的本质是，定义了一组从域名（或 URL）到 Service 的路由转发规则。Ingress 不会公开端口信息和所使用的协议，而是通过配置 HTTP 或 HTTPS 的路由规则，来实现从集群外部访问集群内应用的 URL 并实现负载均衡功能。

要创建和配置 Ingress 路由转发规则，则需要使用 Ingress Controller。它可以由任何具有反向代理功能的应用（如 Nginx 和 Haproxy）来实现。

Ingress 的工作机制如图 16-11 所示。

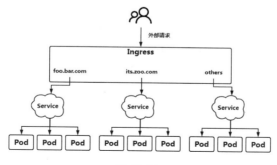

图 16-11

16.4.2 【实战】使用 Ingress Controller 创建 Ingress

下面来演示如何使用 Ingress Controller 创建 Ingress，从而实现从集群外部访问集群内部的应用。首先部署 Ingress Controller，再部署两个应用，最后创建 Ingress 的路由规则从集群外部访问集群内部的应用。

（1）在 GitHub 上搜索 "kubernetes/ingress-nginx"，下载 Ingress Controller 的部署文件 "mandatory.yaml"。如果无法正常下载，可以使用国内的镜像。

（2）按照以下行号及说明修改 "mandatory.yaml" 文件。

将 191 行的 kind 改为 DaemonSet。

```
kind: DaemonSet
```

将 199 行的 replicas 注释掉。

```
#replicas: 1
```

在 213 行后增加一行——使用宿主机网络。

```
hostNetwork: true
```

（3）创建 Ingress Controller。

```
kubectl apply -f mandatory.yaml
```

输出的信息如下：

```
namespace/ingress-nginx created
configmap/nginx-configuration created
configmap/tcp-services created
configmap/udp-services created
serviceaccount/nginx-ingress-serviceaccount created
clusterrole.rbac.authorization.k8s.io/nginx-ingress-clusterrole created
role.rbac.authorization.k8s.io/nginx-ingress-role created
rolebinding.rbac.authorization.k8s.io/nginx-ingress-role-nisa-binding created
clusterrolebinding.rbac.authorization.k8s.io/nginx-ingress-clusterrole-nisa-binding created
daemonset.apps/nginx-ingress-controller created
limitrange/ingress-nginx created
```

（4）查看创建的 Ingress Controller。

```
kubectl get pods -n ingress-nginx
```

输出的信息如下：

```
NAME                              READY   STATUS    IP               NODE
nginx-ingress-controller-4m2k5    1/1     Running   192.168.79.113   node2
```

```
nginx-ingress-controller-tmk8d    1/1    Running 192.168.79.112    node1
```
（5）创建第 1 个应用的 Dockerfile 文件。
```
FROM nginx
RUN echo '<h1>Application One</h1>' > /usr/share/nginx/html/index.html
```
（6）使用"docker build"命令进行编译。
```
docker build -t collenzhao/ingress-app1 .
```
（7）将生成的镜像上传到镜像仓库中。
```
docker push collenzhao/ingress-app1
```
（8）创建第 1 个应用的部署描述文件"ingress-app1.yaml"并应用。
```
apiVersion: apps/v1
kind: Deployment
metadata:
  labels:
    app: ingress-app1
  name: ingress-app1
spec:
  replicas: 3
  selector:
    matchLabels:
      app: ingress-app1
  template:
    metadata:
      labels:
        app: ingress-app1
    spec:
      containers:
      - image: collenzhao/ingress-app1
        name: ingress-app2
        imagePullPolicy: IfNotPresent
---
apiVersion: v1
kind: Service
metadata:
  name: service-app1
spec:
  ports:
  - name: http
    port: 80
    protocol: TCP
    targetPort: 80
  selector:
```

```
    app: ingress-app1
  type: NodePort
```

（9）创建第 2 个应用的 Dockerfile 文件和部署描述文件"ingress-app2.yaml"并应用。

```
# Dockerfile 2
FROM nginx
RUN echo '<h1>Application Two</h1>' > /usr/share/nginx/html/index.html
```

```
#ingress-app2.yaml
apiVersion: apps/v1
kind: Deployment
metadata:
  labels:
    app: ingress-app2
  name: ingress-app2
spec:
  replicas: 3
  selector:
    matchLabels:
      app: ingress-app2
  template:
    metadata:
      labels:
        app: ingress-app2
    spec:
      containers:
      - image: collenzhao/ingress-app2
        name: ingress-app2
        imagePullPolicy: IfNotPresent
---
apiVersion: v1
kind: Service
metadata:
  name: service-app2
spec:
  ports:
  - name: http
    port: 80
    protocol: TCP
    targetPort: 80
  selector:
    app: ingress-app2
  type: NodePort
```

（10）查看 Pod、Deployment、Service 和 Endpoint 的信息，如图 16-12 所示。

```
kubectl get pod,deploy,svc,ep
```

```
NAME                                     READY   STATUS    RESTARTS   AGE
pod/ingress-app1-6868cfb7fd-95hk9        1/1     Running   0          81s
pod/ingress-app1-6868cfb7fd-k47dj        1/1     Running   0          81s
pod/ingress-app1-6868cfb7fd-sg9v5        1/1     Running   0          81s
pod/ingress-app2-5c69fccbf6-gzp8m        1/1     Running   0          76s
pod/ingress-app2-5c69fccbf6-kjkhh        1/1     Running   0          76s
pod/ingress-app2-5c69fccbf6-p1x42        1/1     Running   0          76s

NAME                                     READY   UP-TO-DATE   AVAILABLE   AGE
deployment.extensions/ingress-app1       3/3     3            3           81s
deployment.extensions/ingress-app2       3/3     3            3           76s

NAME                     TYPE        CLUSTER-IP      EXTERNAL-IP   PORT(S)        AGE
service/kubernetes       ClusterIP   10.1.0.1        <none>        443/TCP        320d
service/service-app1     NodePort    10.1.34.198     <none>        80:30639/TCP   81s
service/service-app2     NodePort    10.1.254.133    <none>        80:31723/TCP   76s

NAME                         ENDPOINTS                                        AGE
endpoints/fuseim.pri-ifs     <none>                                           178d
endpoints/kubernetes         192.168.79.111:6443                              320d
endpoints/service-app1       10.244.1.85:80,10.244.1.86:80,10.244.2.83:80     81s
endpoints/service-app2       10.244.1.87:80,10.244.2.84:80,10.244.2.85:80     76s
[root@master ~]#
```

图 16-12

（11）创建 Ingress 路由规则描述文件 "ingress-app.yaml"。

```yaml
apiVersion: extensions/v1beta1
kind: Ingress
metadata:
  name: ingress-app
spec:
  rules:
  - host: develop.example.com
    http:
      paths:
      - path: /
        backend:
          serviceName: service-app1
          servicePort: 80
  - host: product.example.com
    http:
      paths:
      - path: /
        backend:
          serviceName: service-app2
          servicePort: 80
```

（12）创建 Ingress 路由规则。

```
kubectl apply -f ingress-app.yaml
```

（13）查看创建的 Ingress。

```
kubectl get ingress
```

输出的信息如下：

```
NAME         HOSTS                                      PORTS   AGE
ingress-app  develop.example.com,product.example.com    80      8s
```

（14）编辑 hosts 文件加入 IP 地址与域名的对应关系，如下：

```
192.168.79.11 develop.example.com
192.168.79.11 product.example.com
```

（15）通过 URL 分别访问两个应用，如图 16-13 所示。

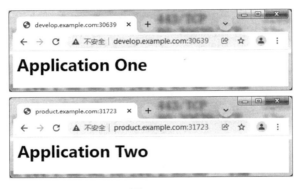

图 16-13

16.4.3 【实战】使用 Ingress 的注解

有时默认的配置并不一定能满足实际的需求，这时可以通过修改 Ingress 的注解来实现对特定参数的配置。在每次修改 Ingress 注解后，Ingress Controller 会立即加载新的配置。

1. Ingress 的常用注解

下面是官方提供的一个示例。该示例将完成文件上传功能。但默认情况下上传的请求大小是 1MB，修改注解可以改变请求的大小限制。例如，下面将请求的大小设置为 2MB。

```
apiVersion: extensions/v1beta1
kind: Ingress
metadata:
  name: openbayes-server-ing
  annotations:
    nginx.ingress.kubernetes.io/proxy-body-size: "2048"
    nginx.ingress.kubernetes.io/ssl-redirect: "false"
spec:
```

```
      rules:
      - http:
          paths:
          - path: /api
            backend:
              serviceName: openbayes-server-svc
              servicePort: 80
```

表 16-1 列举了一些常用的 Ingress 注解。

表 16-1

注解名称	作用
nginx.ingress.kubernetes.io/app-root	定义应用根目录
nginx.ingress.kubernetes.io/rewrite-target	执行 URL 重定向
nginx.ingress.kubernetes.io/use-regex	开启路径正则匹配
nginx.ingress.kubernetes.io/affinity	开启会话的粘连
nginx.ingress.kubernetes.io/ssl-redirect	开启 HTTP 到 HTTPS 的跳转
nginx.ingress.kubernetes.io/force-ssl-redirect	即使未启用 TLS 也会强制重定向到 HTTP

2. 使用 Ingress 注解实现 URL 的重定向

下面使用注解来实现 URL 的重定向。

（1）创建第 1 个应用的部署描述文件"ingress-rewrite-app1.yaml"，并在其中输入以下内容。

```
apiVersion: v1
kind: Service
metadata:
  name: ingress-rewrite-app1
spec:
  ports:
  - name: http
    port: 80
    targetPort: 80
  selector:
    app: ingress-rewrite-app1
  type: NodePort
---
apiVersion: extensions/v1beta1
kind: Deployment
metadata:
  name: ingress-rewrite-app1
spec:
  replicas: 1
```

```
  selector:
    matchLabels:
      app: ingress-rewrite-app1
  template:
    metadata:
      labels:
        app: ingress-rewrite-app1
    spec:
      containers:
      - image: nginx
        name: ingress-rewrite-app1
        imagePullPolicy: IfNotPresent
```

（2）创建第 2 个应用的部署描述文件 "ingress-rewrite-app2.yaml"，并在其中输入以下内容。

```
apiVersion: v1
kind: Service
metadata:
  name: ingress-rewrite-app2
spec:
  ports:
  - name: http
    port: 80
    targetPort: 80
  selector:
    app: ingress-rewrite-app2
  type: NodePort
---
apiVersion: extensions/v1beta1
kind: Deployment
metadata:
  name: ingress-rewrite-app2
spec:
  replicas: 1
  selector:
    matchLabels:
      app: ingress-rewrite-app2
  template:
    metadata:
      labels:
        app: ingress-rewrite-app2
    spec:
      containers:
      - image: nginx
        name: ingress-rewrite-app2
        imagePullPolicy: IfNotPresent
```

（3）使用"kubectl apply-f"命令应用"ingress-rewrite-app1.yaml"文件和"ingress-rewrite-app2.yaml"文件。

```
kubectl apply -f ingress-rewrite-app1.yaml
kubectl apply -f ingress-rewrite-app2.yaml
```

为了看到更好的效果，可以进入 Pod 中修改 Nginx 的 index.html 文件。例如，将 App1 的首页修改为"App1"，将 App2 的首页修改为"App2"。

（4）创建 Ingress 的描述文件"ingress-rewrite-app3.yaml"，实现 URL 的重定向。

```
apiVersion: extensions/v1beta1
kind: Ingress
metadata:
  name: ingress-rewrite-app3
  annotations:
    nginx.ingress.kubernetes.io/rewrite-target: /$1
    # /$1 这里重定向到 "/"
    # /abc/$1 这表示重定向到服务的 /abc/
spec:
  rules:
  - host: test.example.com
    http:
      paths:
        - path: /
          # 将所有的/** 重定向到 /abc/**
          backend:
            serviceName: ingress-rewrite-app1
            servicePort: 80
        - path: /app2
          backend:
            serviceName: ingress-rewrite-app2
            servicePort: 80
```

（5）应用文件"ingress-rewrite-app3.yaml"。

```
kubectl apply -f ingress-rewrite-app3.yaml
```

（6）在 hosts 文件中增加 IP 地址与域名的对应关系。

```
192.168.79.11 test.example.com
```

（7）访问这两个应用，当访问"test.example.com"时会返回"App1"，访问"test.example.com/app2"时会返回"App2"，如图 16-14 所示。

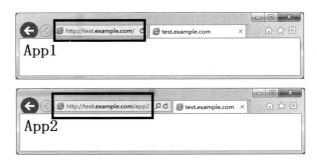

图 16-14

16.4.4 基于 Ingress 的高可用架构

Ingress 的路由规则是由 Ingress Controller 创建和维护的，因此，Ingress Controller 的高可用性就显得非常重要了。如果在 Kubernetes 集群中只存在一个 Ingress Controll，则必然会造成单点故障。为了解决这个问题，可以采用 Ingress Controller 的多副本部署，并将其作为 DaemonSet 的守护进程在每个节点上启动一个。这也是为什么在 16.4.2 节将"mandatory.yaml"文件中 191 行的 kind 改为 DaemonSet 的原因。

从 16.4.2 节第（4）步输出的信息可以看到，在 node1 和 node2 节点上各启动了一个 Ingress Controller。为了最大限度地利用 Ingress 进行路由，建议可以将 Ingress Controller 以独占方式部署，这样可以避免业务应用与 Ingress 服务争用资源。

图 16-15 展示了基于 Ingress 的高可用架构，其中引入了 KeepAlived 以实现从公网域名到机房的公网 IP 地址服务器的路由解析。

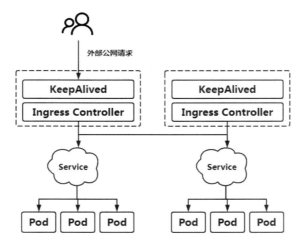

图 16-15

第 17 章

持久化存储

由于容器是一种无状态的服务，所以容器中的文件在宿主机上表现出来的都是临时存放（当容器崩溃或者重启时，容器中的文件会丢失）。另外，Kubernetes 也需要在 Pod 之间实现数据共享。为了解决这些问题，Kubernetes 与 Docker 一样，也通过使用数据卷的方式来实现数据持久化。

17.1 Kubernetes 持久化存储方式

在 Kubernetes 中，数据卷具有明确的生命周期，该生命周期与 Pod 的生命周期相同。即使 Pod 中的容器崩溃或者重启了，其挂载的数据集依然存在。因此，Kubernetes 中数据卷的生命周期比 Pod 中运行的任何容器的生命同期都要长。Pod 中的容器可以访问挂载的数据卷、读写数据卷中的文件。

Kubernetes 允许 Pod 使用任意数据的数据卷，并支持各种类型的数据卷驱动程序。下面列举了 Kubernetes 支持的数据卷驱动程序。

awsElasticBlockStore	emptyDir	iscsi	scaleIO
azureDisk	fc(fibre channel)	local	secret
azureFile	flexVolume	nfs	storageos
cephfs	flocker	persistentVolumeClaim	vsphereVolume
cinder	gcePersistentDisk	projected	scaleIO
configMap	gitRepo(deprecated)	portworxVolume	secret
csi	glusterfs	quobyte	
downwardAPI	hostPath	rbd	

要在 Kubernetes 中使用数据卷，则需要在 Pod 的描述文件中使用"spec.volumes"字段来

指定数据卷的类型和挂载的目录，还需要使用"spec.containers.volumeMounts"字段来指定数据卷映射到容器的位置。

根据数据卷的挂载方式的不同，Kubernetes 中的数据卷分为：节点数据卷（HostPath Volume）、网络数据卷（NFS Volume）和临时数据卷（EmptyDir Volume）。

17.1.1 【实战】使用节点数据卷

节点数据卷（HostPath Volume）是指，将 node 节点上的某个文件或者目录挂载到 Pod 中的一个数据卷。节点数据卷与 Docker 的数据卷类似——都把宿主机的目录挂载到容器下，因此要求在每个 node 节点上被挂载的目录必须存在。因为，Kubernetes 在创建 Pod 时，并不能确定将 Pod 分配到哪个 node 节点上。

下面来演示如何使用节点数据卷。

（1）创建 Pod 的描述文件"hostdir-demo.yaml"，并在其中输入以下内容。

```yaml
apiVersion: v1
kind: Pod
metadata:
  name: hostdir-demo
spec:
  containers:
  - name: container-demo
    image: nginx
    volumeMounts:
    - mountPath: /demo-pod
      name: volume-demo
  volumes:
  - name: volume-demo
    hostPath:
      path: /tmp
      type: Directory
```

> 这里在创建 Pod 的同时创建了一个节点数据卷，实现了将宿主机上的"/tmp"目录挂载到容器内部的"/demo-pod"目录下。

（2）使用"kubectl apply -f"命令创建 Pod。

```
kubectl apply -f hostdir-demo.yaml
```

（3）查看 Pod 的详细信息。

```
kubectl get pod -o wide
```

输出的信息如下。可以看到，Pod 被分配到 node2 节点上了。

```
NAME             READY    STATUS     RESTARTS    AGE    IP              NODE
hostdir-demo     1/1      Running    0           4s     10.244.2.214    node2
```

（4）进入 Pod 中的容器内。

```
kubectl exec -it hostdir-demo -c container-demo bash
```

（5）查看"/demo-pod"目录下的内容，如图 17-1 所示。

```
root@hostdir-demo:/# ls /demo-pod/
ks-script-qxI7j4
systemd-private-4357b5da3ed04dd4966c0923b1f6c300-vgauthd.service-6oeEhT
systemd-private-4357b5da3ed04dd4966c0923b1f6c300-vmtoolsd.service-1cy21L
tmp.1RrrxcWDz0
tmp.mdvg1dzX2m
yum.log
root@hostdir-demo:/#
```

图 17-1

（6）在 node2 节点上删除"/tmp"目录下的内容，再次查看容器内的"/demo-pod"目录。这时会发现容器内挂载的文件也随之被删除了。

17.1.2 【实战】使用网络数据卷

网络数据卷（NFS Volume）能将网络文件系统 NFS 直接挂载到 Pod 中。在删除 Pod 时，网络数据卷不会同时被删除，只是被从挂载的 Pod 上卸载了。通过使用网络数据卷，可以在 Pod 被创建之前预先填充数据，这些数据可以在创建 Pod 时被直接传递给 Pod。

下面来演示如何使用节点数据卷，将把 master 节点作为 NFS Server 来使用。

（1）执行以下命令在所有节点上安装并启动 NFS。

```
yum -y install nfs-utils
systemctl enable nfs
systemctl start nfs
```

（2）在 master 节点上编辑"/etc/exports"文件，输入以下配置信息。该节点将作为 NFS Server。

```
/nfs *(rw,sync,no_root_squash)
```

其中的参数说明如下。

- /nfs：NFS 共享的目录。
- *：可以访问所有的主机网段。
- rw：可读写权限。如果是只读权限，则是"ro"。

- sync：数据传输采用同步方式。采用同步方式可以保障数据的安全性，但传输速度较慢。如果采用异步方式，则是"async"。在异步方式下，数据传输效率高，但安全性差。
- no_root_squash：NFS 服务共享目录的属性。如果用户是 root，则它对这个目录就有 root 的权限了。

（3）在 master 节点上创建"/nfs"目录，并在该目录下生成一些测试文件。

```
mkdir /nfs
echo "<h1>Hello World and Hello NFS</h1>" > /nfs/index.html
```

（4）重启 master 节点上的 NFS 服务。

```
systemctl restart nfs
```

（5）创建 Pod 的描述文件"nfsdir-demo.yaml"，并在其中输入以下内容。

```
apiVersion: extensions/v1beta1
kind: Deployment
metadata:
  name: nfsdir-demo
spec:
  replicas: 1
  template:
    metadata:
      labels:
        app: nginx
    spec:
      containers:
      - name: nginx
        image: nginx
        imagePullPolicy: IfNotPresent
        volumeMounts:
        - name: webroot
          mountPath: /usr/share/nginx/html
        ports:
        - containerPort: 80
      volumes:
      - name: webroot
        nfs:
          server: 192.168.79.11
          path: /nfs
```

> 这里通过使用网络数据卷将 NFS Server 上的目录挂载到了容器内部的"/usr/share/nginx/html"目录下。

（6）使用"kubectl apply -f"命令创建 Pod。

```
kubectl apply -f nfsdir-demo.yaml
```

（7）查看 Pod 的 IP 地址。

```
kubectl get pod -o wide
```

输出的信息如下：

```
NAME                          READY   STATUS    IP
nfsdir-demo-5c579dffbc-fgnbm  1/1     Running   10.244.2.215
```

（8）使用 curl 命令访问 Pod IP 地址的 80 端口。

```
curl 10.244.2.215:80
```

将返回 Nginx 的首页，内容如下：

```
<h1>Hello World and Hello NFS</h1>
```

> 这里使用网络数据卷将事先准备好的 Nginx 首页 index.html 挂载到 Pod 容器的内部了。可以使用以下命令进入 Pod 容器的内部查看"/usr/share/nginx/html"目录下的 index.html 文件。
>
> kubectl exec -it nfsdir-demo-5c579dffbc-fgnbm -c nginx bash

17.1.3 【实战】使用临时数据卷

临时数据卷（EmptyDir Volume）是 Pod 生命周期中的一个临时目录。与其他数据卷所不同的是，当 Pod 的生命周期结束时（如 Pod 被删除），临时数据卷也会同时被删除。

利用临时数据卷，可以实现 Pod 内部多个容器之间的数据共享，也可以使用临时数据卷作为容器的临时目录来进行数据的缓存。

下面演示如何使用节点数据卷。

（1）创建 Pod 的描述文件"emptydir-demo.yaml"，并在其中输入以下内容。

```yaml
apiVersion: v1
kind: Pod
metadata:
  name: emptydir-demo
spec:
  containers:
  - name: tomcat
    image: tomcat
    imagePullPolicy: IfNotPresent
```

```
      ports:
      - containerPort: 8080
      volumeMounts:
      - name: app-logs
        mountPath: /usr/local/tomcat/logs
    - name: busybox
      image: busybox
      imagePullPolicy: IfNotPresent
      command: ["sh", "-c", "tail -f /logs/catalina*.log"]
      volumeMounts:
      - name: app-logs
        mountPath: /logs
  volumes:
  - name: app-logs
    emptyDir: {}
```

 这里的 Pod 创建了两个容器：tomcat 容器和 busybox 容器。通过定义一个临时数据卷，实现了 tomcat 容器的 "/usr/local/tomcat/logs" 目录与 busybox 容器的 "/logs" 目录的数据共享。这样就可以在 busybox 容器中使用 tail 命令来查看 tomcat 容器中的日志信息了。

（2）使用"kubectl apply –f"命令创建 Pod。

```
kubectl apply -f emptydir-demo.yaml
```

（3）进入 busybox 容器。

```
kubectl exec -it emptydir-demo -c busybox sh
```

（4）查看"/logs/catalina.2022-02-03.log"文件的内容，如图 17-2 所示。

```
tail /logs/catalina.2022-02-03.log
```

```
/ # tail /logs/catalina.2022-02-03.log
03-Feb-2022 08:23:11.181 INFO [main] org.apache.catalina.startup.Version
LoggerListener.log Command line argument: -Djava.io.tmpdir=/usr/local/to
mcat/temp
03-Feb-2022 08:23:11.204 INFO [main] org.apache.catalina.core.AprLifecyc
leListener.lifecycleEvent Loaded Apache Tomcat Native library [1.2.31] u
sing APR version [1.7.0].
03-Feb-2022 08:23:11.205 INFO [main] org.apache.catalina.core.AprLifecyc
leListener.lifecycleEvent APR capabilities: IPv6 [true], sendfile [true]
, accept filters [false], random [true], UDS [true].
03-Feb-2022 08:23:11.216 INFO [main] org.apache.catalina.core.AprLifecyc
leListener.initializeSSL OpenSSL successfully initialized [OpenSSL 1.1.1
k  25 Mar 2021]
03-Feb-2022 08:23:11.838 INFO [main] org.apache.coyote.AbstractProtocol.
```

图 17-2

17.2 持久卷

数据卷是在创建 Pod 时通过挂载目录来实现数据的共享和持久化的。但是在一个大型系统中，这种方式是非常不利于管理的，因为数据卷把数据的"持久存储"和"供应使用"封装在一起了。

那么，能否将数据的"持久存储"和"供应使用"分别进行管理和使用呢？为了解决这个问题，Kubernetes 提供了持久卷（Persistent Volume，PV）和持久化声明（Persistent Volume Claim，PVC）。

本节将重点介绍持久卷。在 17.3 节中将介绍持久卷声明。

17.2.1 持久卷是什么

持久卷（PersistentVolume，PV）是 Kubernetes 集群的一种存储方式。持久卷也是集群的一种资源，可以事先创建或者通过存储类（Storage Class）来动态提供。

持久卷通过卷插件的形式对外部的存储资源进行操作，并可以通过指定存储回收策略来控制持久卷回收对外部存储中的数据的影响。

持久卷对象一般由 Kubernetes 集群的管理员创建。它只代表集群为用户提供的一种存储资源。至于这种存储资源如何被使用，它不需要关心。

持久卷和数据卷一样也是使用卷插件来实现的，但二者的最大区别在于：

- 持久卷与 Pod 的生命周期相互独立，它不会因为 Pod 生命周期的结束而被销毁。
- 数据卷一般与 Pod 绑定，它的生命周期与 Pod 相同。

17.2.2 【实战】第一个持久卷示例

下面演示如何创建一个持久卷。

（1）创建持久卷的描述文件"pv-demo-1.yaml"，并在其中输入以下内容。

```yaml
apiVersion: v1
kind: PersistentVolume
metadata:
  name: pv-demo-1
spec:
  capacity:
    storage: 5Gi
  volumeMode: Filesystem
  accessModes:
```

```
    - ReadWriteMany
  persistentVolumeReclaimPolicy: Recycle
  storageClassName: slow
  nfs:
    server: 192.168.79.11
    path: /nfs
```

其中重要的字段说明如下。

- spec.capacity：用于指定持久卷的容量。上面代码中设置持久卷的容量是 5GB。
- spec.volumeMode：指定持久卷的模式。Kubernetes 支持两种类型的持久卷模式——文件系统（File System）和块存储设备（Raw Block Devices）。
  - 如果该参数没有被指定，则默认采用文件系统。
  - 如果将该字段设置为块存储设备，且该存储设备为空，则 Kubernetes 会在第 1 次使用持久卷时在该设备上创建文件系统。这也是在 Pod 中以最快的速度来访问存储的方式。
- spec.accessModes：指定持久卷的访问模式。通过设置持久卷的访问模式，可以让持久卷以不同的读写方式挂载到宿主系统上，从而让每个持久卷能够拥有对存储资源的读写能力。持久卷的访问模式会在 17.2.3 节详细介绍。
- spec.persistentVolumeReclaimPolicy：在定义持久卷时可以通过该字段指定存储资源的回收策略。该回收策略用来指定在删除持久卷或者持久卷声明时，Kubernetes 如何处理存储资源上的数据文件。持久卷的回收策略会在 17.2.4 节详细介绍。
- spec.storageClassName：用来设置存储类。设置了存储类的持久卷只能提供给请求该存储类的持久卷声明使用。如果持久卷没有设置该字段，则该持久卷只能提供给没有存储类的持久卷声明使用。持久卷的存储类会在 17.3.3 节详细介绍。
- spec.nfs：指定持久卷的挂载选项。当 Pod 要使用持久卷时，能够指定附加的挂载选项。Kubernetes 的持久卷支持具有多种挂载选项的卷插件，如 AWSElasticBlockStore、CephFS、Cinder、NFS、RBD 等。在 pv-demo-1.yaml 文件中就是用 NFS（网络文件系统）作为挂载选项。Kubernetes 支持的卷插件请参考表 17-2。

（2）使用 "kubectl apply -f" 命令创建持久卷。

```
kubectl apply -f pv-demo-1.yaml
```

（3）查看创建的持久卷信息。

```
kubectl get pv
```

输出的信息如下：

```
NAME          CAPACITY    ACCESS MODES    RECLAIM POLICY    STATUS
pv-demo-1     5Gi         RWX             Recycle           Available
```

这里的 STATUS 是"Available",表示该持久卷还没有被挂载使用。

(4)获取持久卷的详细信息。

```
kubectl describe pv pv-demo-1
```

输出的信息如下:

```
Name:              pv-demo-1
Labels:            <none>
Annotations:       kubectl.kubernetes.io/last-applied-configuration:
{"apiVersion":"v1","kind":"PersistentVolume","metadata":{"annotations":{},"n
ame":"pv-demo-1"},"spec":{"accessModes":["ReadWriteMany"],"cap...
Finalizers:        [kubernetes.io/pv-protection]
StorageClass:      slow
Status:            Available
Claim:
Reclaim Policy:    Recycle
Access Modes:      RWX
VolumeMode:        Filesystem
Capacity:          5Gi
Node Affinity:     <none>
Message:
Source:
    Type:          NFS (an NFS mount that lasts the lifetime of a pod)
    Server:        192.168.79.11
    Path:          /nfs
    ReadOnly:      false
Events:            <none>
```

17.2.3　持久卷的访问模式

持久卷可以使用不同的存储资源,以所支持的任何方式挂载到宿主系统上。每个持久卷都会有自身的访问模式以描述持久卷的读写能力。"spec.accessModes"字段用于指定持久卷的访问模式。

表 17-1 列出了 Kubernetes 所支持的持久卷访问模式。

表 17-1

访问模式	命令行的缩写形式	描述
ReadWriteOnce	RWO	持久卷可以被一个 node 节点以读写方式挂载，同时也允许运行在同一个 node 节点上的不同 Pod 访问该持久卷
ReadOnlyMany	ROX	持久卷可以同时被多个 node 节点以只读方式挂载
ReadWriteMany	RWX	持久卷可以同时被多个 node 节点以读写方式挂载
ReadWriteOncePod	RWOP	持久卷只能被单个 Pod 以读写方式挂载。通过这种方式，可以保证在整个集群中只有一个 Pod 可以读写该持久卷

持久卷支持多种访问模式，但在同一个时刻只能使用一种访问模式挂载。例如，一个 NFS 的持久卷对象在某一个时刻可以被 node 节点以 ReadWriteOnce 访问模式挂载，或者被多个 node 节点以 ReadOnlyMany 访问模式挂载，但不允许同时使用两种及以上的访问模式挂载。

表 17-2 是 Kubernetes 官方提供的持久卷的不同卷插件所支持的访问模式。

表 17-2 持久卷不同卷插件所支持的访问模式

卷插件	ReadWriteOnce	ReadOnlyMany	ReadWriteMany	ReadWriteOncePod
AWSElasticBlockStore	√			
AzureFile	√	√	√	
AzureDisk	√			
CephFS	√	√	√	
Cinder	√			
CSI	取决于驱动程序	取决于驱动程序	取决于驱动程序	取决于驱动程序
FC	√	√		
FlexVolume	√	√		
Flocker	√			
GCEPersistentDisk	√	√		
Glusterfs	√	√	√	
HostPath	√			
iSCSI	√	√		
Quobyte	√	√	√	
NFS	√	√	√	
RBD	√	√		
VsphereVolume	√		Pod 运行于同一节点上时可行	

续表

卷插件	ReadWriteOnce	ReadOnlyMany	ReadWriteMany	ReadWriteOncePod
PortworxVolume	√		√	
StorageOS	√			

17.2.4 【实战】持久卷的回收策略

在定义持久卷时，可以通过"persistentVolumeReclaimPolicy"字段来指定存储资源的回收策略。

1. 了解持久卷回收策略

目前 Kubernetes 的持久卷回收策略有以下 3 种。

（1）Retain。

该回收策略允许用户手动执行回收资源的操作。在删除 PVC 对象时，PV 对象不会被真的删除，只是 PV 对象的状态会变成 Released。因为用户数据仍然存在于 PV 对象的存储资源中，所以，该 PV 对象暂时还不能够提供给其他 PVC 对象使用。管理员必须在删除旧的 PV 对象并清理对应的存储资源后，才可以创建新的 PV 对象继续使用。

如果存储资源中有非常重要的数据，推荐使用这种回收策略。

（2）Delete。

该回收策略在删除 PersistentVolumeClaim 时，会自动从 Kubernetes 中删除对应的 PV 对象，以及外部存储的资源。该回收策略也是持久卷动态供给时 PV 对象的默认回收策略。

持久卷的动态供给将在 17.4 节中介绍。

（3）Recycle。

该回收策略在删除 PV 对象时，会对持久卷执行清除操作（即执行"rm -rf /thevolume/*"命令）。通过这样方式回收的 PV 对象，可以再次用来处理新的 PVC 对象的申请。

> Recycle 回收策略已废弃，推荐使用 PV 动态供给方式。

2. 实例

下面来演示不同回收策略的行为特征。这里以 17.2.2 节中的 "pv-demo-1.yaml" 与 17.3.2 节中的 "pvc-demo-1.yaml" 为例。

（1）由于持久卷 pv-demo-1 对象使用的是 NFS 存储方式，因此首先到 master 节点的 NFS Server 上创建一些测试文件。

```
cd /nfs/
echo hello world > data.txt
```

（2）创建 pv-demo-1 对象和 pvc-demo-1 对象。

```
kubectl apply -f pv-demo-1.yaml
kubectl apply -f pvc-demo-1.yaml
```

（3）查看 PV 和 PVC 的信息，如图 17-3 所示。

```
kubectl get pv,pvc
```

```
[root@master yaml]# kubectl get pv,pvc
NAME                           CAPACITY   ACCESS MODES   RECLAIM POLICY   STATUS   CLAIM
persistentvolume/pv-demo-1     5Gi        RWX            Recycle          Bound    default/pvc-demo-1

NAME                                  STATUS   VOLUME      CAPACITY   ACCESS MODES   STORAGECLASS   AGE
persistentvolumeclaim/pvc-demo-1      Bound    pv-demo-1   5Gi        RWX            slow           3m
[root@master yaml]#
```

图 17-3

> 从图 17-3 中可以看出，pv-demo-1 对象的回收策略是 Recycle。这意味着：在删除 pvc-demo-1 对象时，会对 pv-demo-1 对象执行清空操作，把持久卷存储的数据彻底删除，即执行 "rm -rf /nfs" 命令。这时并不会删除 pv-demo-1 对象，只是其状态将由 "Bound" 变成 "Available"。

（4）修改 pv-demo-1 对象的回收策略为 Retain。

```
kubectl patch pv pv-demo-1 \
-p '{"spec":{"persistentVolumeReclaimPolicy":"Retain"}}'
```

（5）删除 pvc-demo-1 对象。

```
kubectl delete persistentvolumeclaim/pvc-demo-1
```

（6）查看 PV 和 PVC 对象的信息，这时 pv-demo-1 对象的状态将由 Bound 变成"Available"。

```
kubectl get pv,pvc
```

输出的信息如下：

```
NAME                            ... ... RECLAIM POLICY   STATUS      ... ...
persistentvolume/pv-demo-1... ...      Retain           Released    ... ...
```

> Retain 回收策略不会对外部的存储资源执行清空操作，因此这时 master 节点的 NFS Server 上的数据依然存在。检查 master 节点的"/nfs"目录：
>
> [root@master ~]# ls /nfs/
> data.txt
> [root@master ~]# more /nfs/data.txt
> hello world

17.3 持久卷声明

持久卷声明（Persistent Volume Claim，PVC），是用户对 Kubernetes 存储资源的一种请求。

通过使用持久卷声明（PVC），用户可以将实际的存储需求告诉给 Kubernetes，然后由 Kubernetes 在已有的持久卷（PV）中进行查找。当寻找到合适的持久卷（PV）时，Kubernetes 会将它提供给持久卷声明（PVC）使用。

因此，持久卷声明（PVC）可以被看成资源消费者，它消费的是持久卷（PV）这种资源。

17.3.1 持久卷和持久卷声明的区别

持久卷和持久卷声明都是 Kubernetes 提供给用户使用的两种资源对象。通过使用持久卷（PV）和持久卷声明（PVC），可以封装集群数据持久化时的存储细节。

二者也有明显的区别，主要体现在以下两方面。

（1）二者的使用者不同。

- Kubernetes 集群的管理员应当重点关注持久卷（PV）如何被创建、应该通过持久卷（PV）提供什么样的存储功能，却不需要关注它如何被使用。

- Kubernetes 集群的用户，应重点关注持久卷声明（PVC）。用户只需要将存储的需求通过持久卷声明（PVC）挂载到 Pod 中即可，不需要关注需求如何被具体实现。

（2）二者所承担的任务不同。

持久卷（PV）可以被看成资源的生产者，而持久卷声明（PVC）可以被看成资源的消费者。

- 持久卷（PV）负责生成存储的资源，并提供给持久卷声明（PVC）使用。
- 持久卷声明（PVC）在消费持久卷（PV）时，可以向持久卷（PV）申请存储资源的大小及访问模式等。

17.3.2 【实战】在 Pod 中使用持久卷声明

在 17.2.2 节中已经创建了一个持久卷"pv-demo-1"，下面在 Pod 中创建一个持久卷声明来使用它。

（1）编辑"pvc-demo-1.yaml"文件，在其中输入以下内容。

```yaml
apiVersion: v1
kind: PersistentVolumeClaim
metadata:
  name: pvc-demo-1
spec:
  accessModes:
    - ReadWriteMany
  resources:
    requests:
      storage: 1Gi
  storageClassName: slow
```

（2）执行"kubectl apply -f"命令创建一个 PVC 资源。

```
kubectl apply -f pvc-demo-1.yaml
```

（3）查看创建的持久卷（PV）和持久卷声明（PVC），如图 17-4 所示。

```
kubectl get pv,pvc
```

```
NAME                              CAPACITY    STATUS   CLAIM                  STORAGECLASS
persistentvolume/pv-demo-1         5Gi         …… ……   Bound    default/pvc-demo-1     slow

NAME                                   STATUS   VOLUME       STORAGECLASS
persistentvolumeclaim/pvc-demo-1       Bound    pv-demo-1    …… ……   slow
[root@master yaml]#
```

图 17-4

从图 17-4 中可以看到,在创建了持久卷声明 "pvc-demo-1" 后,持久卷 "pv-demo-1" 的状态会自动地从 "Available" 变成 "Bound",并且自动匹配了 "pvc-demo-1"。这种自动匹配是通过存储类 storageClass 来实现的,因为在 "pv-demo-1" 和 "pvc-demo-1" 的描述文件中都指定了 "storageClassName: slow" 字段。

(4)编辑 "pvc-pod-demo1.yaml" 文件以创建 Pod 来使用持久卷声明对象,在文件中输入以下内容。

```
apiVersion: v1
kind: Pod
metadata:
  name: pvc-pod-demo1
spec:
  containers:
    - name: pvc-pod-container
      image: busybox
      imagePullPolicy: IfNotPresent
      args:
      - /bin/sh
      - -c
      - sleep 30000
      volumeMounts:
      - mountPath: "/mydata"
        name: mydata
  volumes:
    - name: mydata
      persistentVolumeClaim:
        claimName: pvc-demo-1
```

在 "pvc-pod-demo1.yaml" 描述文件中,通过 "volumes.persistentVolumeClaim" 字段指定了持久卷声明(PVC)的对象为 "pvc-demo-1",并将其挂载到了容器内部的 "/mydata" 目录下,而持久卷声明对象 "pvc-demo-1" 又自动匹配了持久卷(PV)对象 "pv-demo-1"。因此,最终的效果是:把网络数据卷 NFS 挂载到了容器的 "/mydata" 目录下了。

(5)执行 "kubectl apply –f" 命令。

```
kubectl apply -f pvc-pod-demo1.yaml
```

(6)执行 "kubectl exec –it" 命令进入容器内。

```
kubectl exec -it pvc-pod-demo1 -c pvc-pod-container sh
```

（7）查看容器内的"/mydata"目录，如图 17-5 所示。

```
/ # ls /mydata/
index.html
/ # more /mydata/index.html
<h1>Hello World and Hello NFS</h1>
/ #
```

图 17-5

"/mydata"目录下的内容就是在 17.1.2 节中创建在 NFS Server 上的 Nginx 的首页 index.html。

17.3.3 storageClass 详解

持久卷（PV）对象可以属于某一个存储类 storageClass。

- 如果属于特定存储类的持久卷声明（PVC）对象请求存储资源，则它只能请求与其对应存储类所标识的持久卷（PV）对象。
- 如果持久卷（PV）对象没有设置存储类，则它只能被没有指定存储类的持久卷声明（PVC）对象请求。

示例 1

下面是 Kubernetes 官方提供的一个创建存储类 storageClass 的示例。该存储类使用 NFS（网络文件系统）的制备器来提供存储资源。

```
apiVersion: storage.k8s.io/v1
kind: StorageClass
metadata:
  name: example-nfs
provisioner: example.com/external-nfs
parameters:
  server: nfs-server.example.com
  path: /share
  readOnly: false
```

从示例中可以看出，在存储类 storageClass 的定义中包含"provisioner""parameters"和"reclaimPolicy"字段。这些字段会在存储类 storageClass 需要动态分配持久卷（PV）时被使用。

在这 3 个字段中，只有"provisioner"（制备器）字段是必需的，它决定在创建 PV 资源时使用哪种类型的卷插件作为存储资源。例如，在该示例中使用"example.com/external-nfs"作为 PV 对象的制备器，该制备器通过 NFS 的方式提供存储资源。因此，在存储类 storageClass 的最后通过"parameters"参数指定了 NFS 的地址信息。

每一个被创建的持久卷（PV）对象会有一个存储资源的回收策略（通过"reclaimPolicy" 字段指定）。该字段的值可以是"Delete"或"Retain"，默认值是"Delete"。

示例 2

下面是 Kubernetes 官方提供的另一个存储类示例，该存储类使用 Glusterfs 插件来提供相应的存储资源。

```yaml
apiVersion: storage.k8s.io/v1
kind: StorageClass
metadata:
  name: slow
provisioner: kubernetes.io/glusterfs
parameters:
  resturl: "http://127.0.0.1:8081"
  clusterid: "630372ccdc720a92c681fb928f27b53f"
  restauthenabled: "true"
  restuser: "admin"
  secretNamespace: "default"
  secretName: "heketi-secret"
  gidMin: "40000"
  gidMax: "50000"
  volumetype: "replicate:3"
```

17.4 【实战】实现持久卷的动态供给

在实际的生产环境中，有大量的业务运行在 Kubernetes 集群中，会产生大量的持久卷声明（PVC）需求。如果靠人工的方式去完成持久卷（PV）对象和持久卷声明（PVC）对象的匹配，显然很不现实。这就要求 Kubernetes 集群能够提供一种动态的自动挂载方式来匹配 PV 对象和 PVC 对象。

在 17.3.2 节中提到，PV 对象和 PVC 对象可以通过存储类 storageClass 实现自动的匹配和绑定。因此，存储类 storageClass 可以实现 PV 资源的动态供给，它是 PV 资源分配的一种策略。

基于存储类 storageClasss 实现持久卷的动态供给的过程如图 17-6 所示。

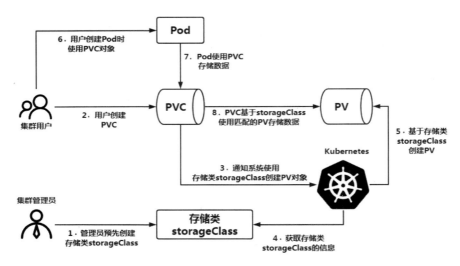

图 17-6

下面来演示如何实现持久卷的动态供给。

（1）编辑"nfs-client-provisioner.yaml"文件，创建一个制备器 Provisioner 提供给存储类 storageClass 使用。

```
kind: Deployment
apiVersion: extensions/v1beta1
metadata:
  name: nfs-client-provisioner
spec:
  replicas: 1
  strategy:
    type: Recreate
  template:
    metadata:
      labels:
        app: nfs-client-provisioner
    spec:
      containers:
        - name: nfs-client-provisioner
          image: vbouchaud/nfs-client-provisioner
          imagePullPolicy: IfNotPresent
          volumeMounts:
            - name: nfs-client-root
              mountPath: /persistentvolumes
          env:
            - name: PROVISIONER_NAME
```

```
                    value: fuseim.pri/ifs
                  - name: NFS_SERVER
                    value: 192.168.79.11
                  - name: NFS_PATH
                    value: /nfs
              volumes:
                - name: nfs-client-root
                  nfs:
                    server: 192.168.79.11
                    path: /nfs
```

存储类 storageClass 需要使用制备器字段（Provisioner）来决定使用哪个卷插件来创建 PV。制备器字段必须指定。"nfs-client-provisioner.yaml" 文件通过创建制备器来提供基于 NFS 的 PV 存储方式。

这里使用 "vbouchaud/nfs-client-provisioner" 镜像作为 NFS Server 的客户端程序来访问 NFS。

（2）编辑 "managed-nfs-storage.yaml" 文件创建存储类 storageClass。

```
apiVersion: storage.k8s.io/v1beta1
kind: StorageClass
metadata:
  name: managed-nfs-storage
provisioner: fuseim.pri/ifs
```

（3）执行 "kubectl apply -f" 命令。

```
kubectl apply -f nfs-client-provisioner.yaml
kubectl apply -f managed-nfs-storage.yaml
```

（4）查看创建的存储类 storageClass。

```
kubectl get storageclass
```

该命令可以简写成以下形式：
kubectl get sc

输出的信息如下：

```
NAME                   PROVISIONER       AGE
managed-nfs-storage    fuseim.pri/ifs    13s
```

（5）编辑 "test-pvc-claim.yaml" 文件创建测试的 PVC 对象，以检测存储类 storageClass

能否正常工作。

```
kind: PersistentVolumeClaim
apiVersion: v1
metadata:
  name: test-pvc-claim
spec:
  accessModes:
    - ReadWriteMany
  resources:
    requests:
      storage: 1Mi
  storageClassName: managed-nfs-storage
```

（6）执行"kubectl apply -f"命令。

```
kubectl apply -f test-pvc-claim.yaml
```

（7）查看 Deploy、Pod、SC 和 PVC 的信息，如图 17-7 所示。

```
kubectl get deploy,pod,sc,pvc
```

```
NAME                                            READY   UP-TO-DATE   AVAILABLE   AGE
deployment.extensions/nfs-client-provisioner    1/1     1            1           91s

NAME                                            READY   STATUS    RESTARTS   AGE
pod/nfs-client-provisioner-64554f884f-516z7     1/1     Running   0          91s

NAME                                                PROVISIONER      AGE
storageclass.storage.k8s.io/managed-nfs-storage     fuseim.pri/ifs   88s

NAME                                     STATUS         STORAGECLASS          AGE
persistentvolumeclaim/test-pvc-claim     Pending  ……    managed-nfs-storage   75s
[root@master yaml]#
```

图 17-7

 这里发现 PVC 一直处于"Pending"状态。

（8）查看 Pod 的日志信息。

```
kubectl log -f
```

输出的信息如下：

```
error retrieving resource lock default/fuseim.pri-ifs:
  endpoints "fuseim.pri-ifs" is forbidden: User
"system:serviceaccount:default:default"
  cannot get resource "endpoints" in API group "" in the namespace "default"
```

从日志信息中可以看出，default 命名空间中的服务账号（serviceaccount）没有权限访问 fuseim.pri-ifs 对象。

（9）编辑"nfs-provisioner-runner.yaml"文件创建 nfs-provisioner-runner 角色资源，并绑定角色与账户的关系。

```yaml
kind: ClusterRole
apiVersion: rbac.authorization.k8s.io/v1
metadata:
  name: nfs-provisioner-runner
rules:
  - apiGroups: [""]
    resources: ["persistentvolumes"]
    verbs: ["get", "list", "watch", "create", "delete"]
  - apiGroups: [""]
    resources: ["persistentvolumeclaims"]
    verbs: ["get", "list", "watch", "update"]
  - apiGroups: ["storage.k8s.io"]
    resources: ["storageclasses"]
    verbs: ["get", "list", "watch"]
  - apiGroups: [""]
    resources: ["events"]
    verbs: ["watch", "create", "update", "patch"]
  - apiGroups: [""]
    resources: ["services"]
    verbs: ["get"]
  - apiGroups: ["extensions"]
    resources: ["podsecuritypolicies"]
    resourceNames: ["nfs-provisioner"]
    verbs: ["use"]
  - apiGroups: [""]
    resources: ["endpoints"]
    verbs: ["get", "list", "watch", "create", "update", "patch"]
---
kind: ClusterRoleBinding
apiVersion: rbac.authorization.k8s.io/v1
metadata:
  name: run-nfs-provisioner
subjects:
  - kind: ServiceAccount
    name: default
    namespace: default
roleRef:
```

```
    kind: ClusterRole
    name: nfs-provisioner-runner
    apiGroup: rbac.authorization.k8s.io
```

（10）执行"kubectl apply -f"命令。

```
kubectl apply -f nfs-provisioner-runner.yaml
```

（11）重新查看 Deploy、Pod、SC 和 PVC 的信息，如图 17-8 所示。

```
kubectl get deploy,pod,sc,pvc
```

```
NAME                                         READY   UP-TO-DATE   AVAILABLE   AGE
deployment.extensions/nfs-client-provisioner  1/1     1            1           6m21s

NAME                                            READY   STATUS    RESTARTS   AGE
pod/nfs-client-provisioner-64554f884f-516z7     1/1     Running   0          6m21s

NAME                                            PROVISIONER      AGE
storageclass.storage.k8s.io/managed-nfs-storage fuseim.pri/ifs   6m18s

NAME                                 STATUS   VOLUME
CAPACITY   ACCESS MODES   STORAGECLASS        AGE
persistentvolumeclaim/test-pvc-claim  Bound   pvc-8f71a30f-dd42-4b42-8f76-31323644b161
1Mi        RWX            managed-nfs-storage  6m5s
[root@master yaml]#
```

图 17-8

 这时会发现 PVC 绑定成功。

（12）编辑"test-pvc-pod.yaml"文件创建 Pod 来使用 PVC 资源。

```
kind: Pod
apiVersion: v1
metadata:
  name: test-pod
spec:
  containers:
  - name: test-pod
    image: busybox
    args:
    - /bin/sh
    - -c
    - sleep 30000
    volumeMounts:
      - name: nfs-pvc
        mountPath: "/mnt"
  restartPolicy: "Never"
  volumes:
    - name: nfs-pvc
```

```
      persistentVolumeClaim:
        claimName: test-pvc-claim
```

（13）执行 "kubectl apply -f" 命令。

```
kubectl apply -f test-pvc-pod.yaml
```

（14）查看 Deploy、Pod、SC 和 PVC 的信息，观察新创建的 PVC 是否能够自动绑定存储类 storageClass，如图 17-9 所示。

```
kubectl get deploy,pod,sc,pvc
```

```
NAME                                            READY   UP-TO-DATE   AVAILABLE   AGE
deployment.extensions/nfs-client-provisioner    1/1     1            1           12m
NAME                                              READY   STATUS    RESTARTS   AGE
pod/nfs-client-provisioner-64554f884f-516z7       1/1     Running   0          12m
pod/test-pod                                      1/1     Running   0          71s
NAME                                            PROVISIONER       AGE
storageclass.storage.k8s.io/managed-nfs-storage  fuseim.pri/ifs    12m
NAME                                      STATUS   VOLUME
                     CAPACITY   ACCESS MODES   STORAGECLASS        AGE
persistentvolumeclaim/test-pvc-claim      Bound    pvc-8f71a30f-dd42-4b42-8f76-31323644b
161                  1Mi        RWX            managed-nfs-storage  11m
[root@master yaml]#
```

图 17-9

（15）在 master 节点上进入 NFS Server 的共享目录 "/nfs" 下，并创建数据文件 data.txt。这时 "/nfs" 目录如下所示。

```
[root@master ~]# tree /nfs
/nfs
└── default-test-pvc-claim-pvc-8f71a30f-dd42-4b42-8f76-31323644b161
    └── data.txt
```

由于 Kubernetes 实现了持久卷的动态供给，并将 NFS 作为存储的方式，因此会在 NFS Server 上自动为 PVC 创建存储的目录：

/nfs/default-test-pvc-claim-pvc-8f71a30f-dd42-4b42-8f76-31323644b161

（16）进入 Pod 中观察是否能够访问新创建的文件。

```
kubectl exec -it pod/test-pod sh
```

输出的信息如下：

```
[root@master ~]# kubectl exec -it pod/test-pod sh
/ # ls /mnt
data.txt
/ # more /mnt/data.txt
Hello World
```

第 18 章
Kubernetes 的安全认证

Kubernetes 作为一个分布式的虚拟化集群管理工具，保证其集群的安全性就显得非常重要。由于 API Server 是访问集群资源的唯一入口，因此 Kubernetes 的安全机制都是围绕保护 API Server 来设计的。

18.1 Kubernetes 的安全框架

Kubernetes 的安全框架主要由认证、鉴权和准入控制这 3 个阶段组成。这 3 个阶段的关系如图 18-1 所示。

1. 认证（Authentication）

当客户端与 Kubernetes 集群建立 HTTP 通信时，HTTP 请求会先进入认证阶段。由于 API Server 是操作集群资源的唯一入口，因此，可以在 API Server 上配置一个或者多个认证模块，API Server 将逐个验证每一个认证模块，直到其中一个认证成功。如果认证失败，则 API Server 返回 401 状态码给客户端，表示 Kubernetes 拒绝了客户端的连接请求。一般情况下，认证模块只会检查 HTTP 的头部信息（这里包含用户名、密码、客户端证书、令牌等信息），而不会检查整个 HTTP 请求。

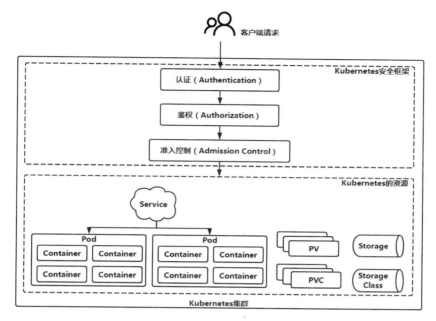

图 18-1

2. 鉴权（Authorization）

客户端请求在通过了认证阶段后，会进入鉴权阶段。这个阶段会检查请求者是否拥有相应的权限来执行操作。因此，在鉴权阶段需要提供请求者的用户名、请求的权限或行为，以及操作的资源对象。如果请求者无法提供，则 Kubernetes 将拒绝该请求。

下面是 Kubernetes 鉴权的一个例子。

```
{ "apiVersion": "abac.authorization.kubernetes.io/v1beta1",
  "kind": "Policy",
  "spec": {
      "user": "Jerry",
      "namespace": "project-dev",
      "resource": "pods",
      "readonly": true
  }
}
```

其中指定了 Jerry 能够在"project-dev"命名空间中读取 Pod。当 Jerry 执行以下操作后，就可以正常读取"project-dev"名称空间中的 Pod 对象了。

```
{ "apiVersion": "authorization.k8s.io/v1beta1",
  "kind": "SubjectAccessReview",
  "spec": {
```

```
  "resourceAttributes": {
    "namespace": "project-dev",
    "verb": "get",
    "group": "dev.example.org",
    "resource": "pods"
  }
 }
}
```

如果 Jerry 在"project-dev"命名空间中执行写操作（如 create 和 update），则会被鉴权拒绝。另外，Jerry 只对"project-dev"命名空间有读取 Pod 的权限，对于其他命名空间则没有任何权限。

3. 准入控制（Admission Control）

在客户端请求通过了认证阶段和鉴权阶段后，API Server 不会立即处理客户端请求。因为，这时客户端请求还要通过最后一个阶段——准入控制阶段。在该阶段中将修改客户端请求中的参数以完成一些特殊的任务。

另外，Kubernetes 为准入控制阶段维护了一个插件列表，发送给 API Server 的所有客户端请求都需要通过该列表中的所有插件的检查。如果某个插件拒绝了客户端请求，则该请求将立即被拒绝，而不会被后续的插件检查。

Kubernetes 允许用户自己开发每一个阶段的插件，并集成到相应的阶段中以实现用户的访问控制。所有插件都是由 API Server 来启用的。

18.2 Kubernetes 的用户认证

在 Kubernetes 集群中，用户主要分为两类：User Account 和 Service Account。本书前面的章节使用的都是 User Account。通过使用 User Account，能够让用户访问 Kubernetes 集群中的资源。简单来说，User Account 是为人而设计的；但 Service Account 却是为 Pod 而设计的。Service Account 的内容会在 18.4 节中详细介绍。

1. Kubernetes 的用户认证方式

通过指定 API Server 的启动参数，可以让 Kubernetes 使用不同的用户认证方式。表 18-1 列举了 Kubernetes 主要支持的 3 种用户认证方式。

表 18-1

认证方式	API Server 启动参数设置	说明
密码认证	--basic-auth-file=密码文件	最基本的认证方式,通过用户名和密码对用户进行认证。密码文件是一个 CSV 格式的文件,该文件中至少包含密码、用户名和用户 ID 这 3 列
证书认证	--client-ca-file=证书文件	通过证书文件进行用户的认证。证书中的 CN(common name)会作为用户名,而证书中的 O(organization)所指定的信息会成为用户的组
令牌认证	--token-auth-file=令牌文件	如果在 API Server 的启动参数中设置了"--token-auth-file"参数,则 Kubernetes 会从指定文件中读取用户的令牌信息,从而进行用户认证

通过以下命令可以看出,在默认情况下 Kubernetes 使用的是证书认证。

cat /etc/kubernetes/manifests/kube-apiserver.yaml |grep client-ca

输出的信息如下:

- --client-ca-file=/etc/kubernetes/pki/ca.crt
- --requestheader-client-ca-file=/etc/kubernetes/pki/front-proxy-ca.crt

2. 配置 Kubernetes 集群使用密码认证

(1)创建"/etc/kubernetes/pki/basic_auth_file.csv"文件用于保存用户名与密码,在文件中输入以下内容。

```
password123,myadmin,1
```

该 CSV 文件的格式为:密码,用户名,用户 ID

(2)按照以下内容修改"/etc/kubernetes/manifests/kube-apiserver.yaml"文件。

增加如下参数:

```
- --basic-auth-file=/etc/kubernetes/pki/basic_auth_file.csv
```

将下面这一行注释掉:

```
#- --client-ca-file=/etc/kubernetes/pki/ca.crt
```

这样的配置使得 Kubernetes 集群采用密码认证。如果使用的是用 kubeadmin 部署的 Kubernetes 集群,则 API Server 会自动重启以加载新的 kube-apiserver.yaml 文件。

（3）尝试访问 Kubernetes 集群。

```
kubectl get node
```

这时将出现以下错误：

```
error: You must be logged in to the server (Unauthorized)
```

（4）由于使用密码认证登录时需要集群的 CA 证书，所以先创建相关的目录。

```
mkdir -p /opt/kubernetes/bin/
cd /opt/kubernetes/bin/
```

（5）下载需要的软件，并授予可执行的权限。

```
wget https://pkg.cfssl.org/R1.2/cfssl_linux-amd64
wget https://pkg.cfssl.org/R1.2/cfssljson_linux-amd64
wget https://pkg.cfssl.org/R1.2/cfssl-certinfo_linux-amd64
chmod +x cfssl*
mv cfssl-certinfo_linux-amd64 /opt/kubernetes/bin/cfssl-certinfo
mv cfssljson_linux-amd64 /opt/kubernetes/bin/cfssljson
mv cfssl_linux-amd64 /opt/kubernetes/bin/cfssl
```

（6）编辑 "~/.bash_profile" 文件加入以下内容设置环境变量。

```
export PATH=/opt/kubernetes/bin/:$PATH
```

（7）生效环境变量。

```
source ~/.bash_profile
```

（8）初始化 cfssl，并创建临时证书目录。

```
cd ~
mkdir ssl && cd ssl
cfssl print-defaults config > config.json
cfssl print-defaults csr > csr.json
```

（9）创建用于生成 CA 证书文件的 JSON 配置信息。

```
cat > ca-config.json <<EOF
{
  "signing": {
    "default": {
      "expiry": "8760h"
    },
    "profiles": {
      "kubernetes": {
        "usages": [
          "signing",
          "key encipherment",
```

```
            "server auth",
            "client auth"
        ],
        "expiry": "8760h"
      }
    }
  }
}
EOF
```

（10）创建用于生成 CA 证书签名请求（CSR）的 JSON 配置信息。

```
cat > ca-csr.json <<EOF
{
  "CN": "kubernetes",
  "key": {
    "algo": "rsa",
    "size": 2048
  },
  "names": [
    {
      "C": "CN",
      "ST": "BeiJing",
      "L": "BeiJing",
      "O": "k8s",
      "OU": "System"
    }
  ]
}
EOF
```

（11）生成 CA 证书（ca.pem）和密钥（ca-key.pem）。

```
cfssl gencert -initca ca-csr.json | cfssljson -bare ca
```

（12）将生成的证书复制到 "/opt/kubernetes/ssl" 目录下。

```
mkdir -p /opt/kubernetes/ssl
cp ca.csr ca.pem ca-key.pem ca-config.json /opt/kubernetes/ssl
```

（13）创建 "myadmin" 用户的信息。

```
kubectl config set-credentials myadmin \
--username myadmin --password password123
```

（14）配置需要访问的集群。

```
kubectl config set-cluster k8s-cluster \
--server https://192.168.79.11:6443 \
```

```
--certificate-authority /opt/kubernetes/ssl/ca.pem
```

（15）设置"myadmin"用户访问的上下文信息。

```
kubectl config set-context k8s-cluster-ctx \
--cluster k8s-cluster --user myadmin
```

（16）切换到"myadmin"用户访问的上下文。

```
kubectl config use-context k8s-cluster-ctx
```

（17）再次访问 Kubernetes 集群。

```
kubectl get node
```

> 这时出现以下错误：
>
> ```
> Error from server (Forbidden): nodes is forbidden: User "myadmin" cannot list nodes at the cluster scope
> ```
>
> 这与第（3）步的错误不一样。这里的错误表示：用户已经通过了认证阶段，但不具备任何权限，因此也不能操作 Kubernetes 集群的资源。

18.3　Kubernetes 的鉴权管理

客户端请求通过认证阶段后，将进入鉴权阶段。这个阶段将包含两个内容：①审查客户端请求的属性；②确定请求的操作。

1. 审查客户端请求的属性

Kubernetes 审查客户端请求的属性主要包括以下几个方面。

- 用户与组：经过认证的用户名和所属组名的列表。
- API 和动作：对 Kubernetes 资源的操作，如 create、list、get 等请求动词。
- 请求路径和动作：指示各种非 Kubernetes 资源的路径（如"/api"），以及对该路径执行的 HTTP 操作（如 GET、POST、PUT 等）。
- 命名空间：正在访问的 Kubernetes 对象的命名空间。

2. 确定请求的操作

确定请求的操作是指，确定对 Kubernetes 资源对象的请求动词或 HTTP 操作，以及该动词或 HTTP 操作是针对单个资源还是一组资源。

表 18-2 列举常见的请求动词与 HTTP 操作，以及它们的对应关系。

表 18-2

请求动词	HTTP 操作
create	POST
get、list	GET、HEAD
update	PUT
patch	PATCH
delete、deletecollection	DELETE

根据 Kubernetes 鉴权时使用的模块，可以将 Kubernetes 的鉴权分为以下 4 种方式：

- 基于角色的访问控制（RBAC 鉴权）。
- 基于属性的访问控制（ABAC 鉴权）。
- 基于节点的访问控制（Node 鉴权）。
- 基于 Webhook 的访问控制。

> 基于角色的访问控制（RBAC 鉴权）是最重要的鉴权方式。

18.3.1 基于角色的访问控制（RBAC 鉴权）

基于角色的访问控制（Role-Based Access Control），通过为用户赋予不同的角色来控制其访问 Kubernetes 集群资源。它允许用户动态配置不同的角色策略。基于角色的访问控制需要使用 "rbac.authorization.k8s.io" API 组来执行。

1. 基于角色的访问控制中的概念

在基于角色的访问控制中涉及 3 个非常重要的概念：角色、角色绑定和主体。

（1）角色。

角色是一组权限的集合。Kubernetes 中的角色分为两种：Role 和 ClusterRole。

- Role 是某个命名空间中对象访问权限的集合。因此，在创建 Role 时，必须指定 Role 所属的命名空间。
- ClusterRole 是访问某个命名空间的权限的集合。

（2）角色绑定。

将包含各种权限的角色授予给一个主体，这个过程被叫作角色绑定。因为角色分为 Role 和 ClusterRole，所以角色绑定分为 RoleBinding 和 ClusterRoleBinding。

（3）主体。

使用角色的用户被叫作主体（Subject）。它可以是一个用户（User）、一个用户组（Group），也可以是一个服务账号（ServiceAccount）。

角色与角色绑定存在 3 种关系：RoleBind-Role、ClusterRoleBind-ClusterRole 和 RoleBind-ClusterRole，如图 18-2 所示。

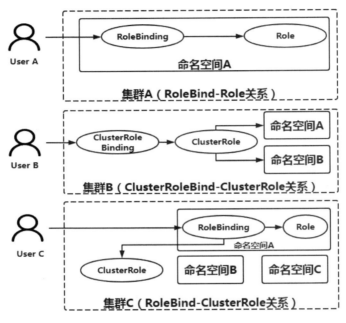

图 18-2

下面解释了这 3 种关系的区别：

- "User A"通过 RoleBinding 绑定到了 Role 上。因此，它就拥有了"命名空间 A"的操作权限。
- 在集群 B 上有两个命名空间——"命名空间 A"和"命名空间 B"。"User B"通过 ClusterRolebinding 绑定到 ClusterRole 上，因此它拥有了集群的操作权限（即访问"命名空间 A"和"命名空间 B"）。
- "User C"在使用 RoleBind 和 ClusterRole 进行绑定时，仅能获取当前名称空间的所有权限（即"User C"只能访问"命名空间 A"）。

角色、角色绑定、主体之间的约束关系如图 18-3 所示。

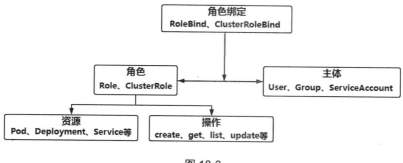

图 18-3

2. 实现基于角色的访问控制

下面来演示如何实现基于角色的访问控制。这里将实现"Jerry"用户只能对"mydemo"命名空间拥有读取 Pod 的权限。

（1）创建"mydemo"命名空间，并在该命名空间中创建 Pod。

```
kubectl create ns mydemo
kubectl run nginx --image=nginx --replicas=3 -n mydemo
```

（2）查看"mydemo"命名空间中的 Pod 信息。

```
kubectl get pods -n mydemo
```

输出的信息如下：

```
NAME                      READY   STATUS    RESTARTS   AGE
nginx-7bb7cd8db5-jwzlk    1/1     Running   0          23s
nginx-7bb7cd8db5-n7w44    1/1     Running   0          23s
nginx-7bb7cd8db5-tmh8t    1/1     Running   0          23s
```

（3）生成"只有读取 Pod 权限的角色"的描述信息。

```
kubectl create role mydemo-pod-reader-role \
--verb=get,list,watch --resource=pods --dry-run -o yaml
```

（4）在生成的描述信息中增加"namespace: mydemo"字段，并将描述信息保存为"mydemo-pod-reader-role.yaml"文件。

```
apiVersion: rbac.authorization.k8s.io/v1
kind: Role
metadata:
  creationTimestamp: null
  name: mydemo-pod-reader-role
  namespace: mydemo
rules:
- apiGroups:
```

```
- ""
  resources:
  - pods
  verbs:
  - get
  - list
  - watch
```

（5）创建"mydemo-pod-reader-role"角色。

```
kubectl apply -f mydemo-pod-reader-role.yaml
```

（6）查看"mydemo"命名空间中的角色信息。

```
kubectl get role -n mydemo
```

输出的信息如下：

```
NAME                        AGE
mydemo-pod-reader-role      4s
```

（7）编辑"mydemo-pod-reader-rolebinding.yaml"文件进行角色绑定，将主体与角色进行绑定。

```
apiVersion: rbac.authorization.k8s.io/v1
kind: RoleBinding
metadata:
  name: mydemo-pod-reader-rolebinding
  namespace: mydemo
subjects:
- kind: User
  #名字大小写敏感
  name: Jerry
  apiGroup: rbac.authorization.k8s.io
roleRef:
  kind: Role #this must be Role or ClusterRole
  # 名字必须与 Role 或 ClusterRole 的名字一致
  name: mydemo-pod-reader-role
  apiGroup: rbac.authorization.k8s.io
```

> 这里的主体是用户名称（即 Jerry），绑定的角色是"mydemo-pod-reader-role"。

（8）执行"kubectl apply -f"命令。

```
kubectl apply -f mydemo-pod-reader-rolebinding.yaml
```

(9) 查看 "mydemo" 命名空间中的角色信息与角色绑定信息。

```
kubectl get role,rolebinding -n mydemo
```

输出的信息如下:

```
NAME                                                         AGE
role.rbac.authorization.k8s.io/mydemo-pod-reader-role        3m46s

NAME
rolebinding.rbac.authorization.k8s.io/mydemo-pod-reader-rolebinding
```

(10) 创建 "Jerry" 用户和认证证书。

```
#生成用户的私钥
openssl genrsa -out Jerry.key 2048

#使用刚生成的私钥创建证书,并在-subj 中指定用户和组。证书文件格式:用户名.csr,
openssl req -new -key Jerry.key -out Jerry.csr -subj "/CN=Jerry/O=mydemo"

#在 "/etc/kubernetes/pki" 目录下,找到 Kubernetes 集群的证书 ca.crt 和 ca.key
#生成最终的证书 Jerry.crt,有效期为 30 天
openssl x509 -req -in Jerry.csr -CA /etc/kubernetes/pki/ca.crt \
-CAkey /etc/kubernetes/pki/ca.key -CAcreateserial \
-out Jerry.crt -days 30
```

(11) 把 "Jerry" 用户的凭证加入 kubeconfig。

```
kubectl config set-credentials Jerry \
--client-key=Jerry.key \
--client-certificate=Jerry.crt
```

(12) 查看 config 文件,看是否把密钥的内容写进去了,以便在命令行中切换用户。

```
tail ~/.kube/config
```

输出的信息如下:

```
users:
- name: Jerry
  user:
    client-certificate: /root/Jerry.crt
    client-key: /root/Jerry.key
```

(13) 创建用户上下文对象 "Jerry-contex"。

```
kubectl config set-context Jerry-context \
--cluster=kubernetes --namespace=mydemo --user=Jerry
```

(14) 切换到 "Jerry" 用户。

```
kubectl config use-context Jerry-context
```

输出的信息如下:

```
Switched to context "Jerry-context".
```

> 在 Kubernetes 中进行用户切换的常用命令有:
> #回到管理员
> kubectl config use-context kubernetes-admin@kubernetes
> #获取所有用户的 context 列表
> kubectl config get-contexts
> #获取当前用户的 context 信息
> kubectl config current-context

(15)测试"Jerry"用户能否读取"mydemo"命名空间中的 Pod 信息。

```
kubectl get pod -n mydemo
```

如果能够正常读取"mydemo"命名空间中的 Pod 信息,则输出如下信息:

```
NAME                       READY   STATUS    RESTARTS   AGE
nginx-7bb7cd8db5-jwzlk     1/1     Running   0          52m
nginx-7bb7cd8db5-n7w44     1/1     Running   0          52m
nginx-7bb7cd8db5-tmh8t     1/1     Running   0          52m
```

(16)执行以下命令测试"Jerry"用户能否在"mydemo"命名空间中创建 Pod:

```
kubectl run Jerry-nginx --image=nginx --replicas=3 -n mydemo
```

这时将出现以下错误信息:

```
Error from server (Forbidden):
deployments.apps is forbidden:
User "Jerry" cannot create resource "deployments" in API group "apps" in the namespace "mydemo"
```

(17)按照第(10)步的方法创建"Tom"用户和认证证书,并切换到"Tom"用户,测试"Tom"用户能否读取"mydemo"命名空间中的 Pod 信息。

18.3.2 基于属性的访问控制(ABAC 鉴权)

> ABAC 鉴权功能从 Kubernetes 1.6 版本被弃用。

基于属性的访问控制（Attribute-Based Access Control）通过将属性组合在一起来定义用户可以访问的范围。其策略文件是一个具有多行 JSON 格式的文件，该文件中的每一行都是一个策略（即一个 JSON 对象）。

表 18-3 列举了该 JSON 对象应具备的属性。

表 18-3

属　　性	属性中的字段
版本控制	• apiVersion 字符串类型。该字段表示匹配 Kubernetes API 的哪些版本。该字段允许的值为：abac.authorization.kubernetes.io/v1beta1 • kind 字符串类型，有效值为 Policy
spec	• user 字符串类型。该字段的值可以是验证通过的用户名，也可以是通过"--token-auth-file"指定的 Token 文件。 • group 字符串类型。如果指定了该字段，则它必须是经过身份验证的用户组。使用"system:authenticated"可以匹配所有经过身份验证的用户组；使用"system:unauthenticated"可以匹配所有没有经过身份验证的用户组
资源匹配	• apiGroup 字符串类型，表示匹配一个 Kubernetes API 资源组。使用"*"则匹配所有 API 资源组。也可以具体指定某一个资源组，例如"extensions"。 • namespace 字符串类型，表示匹配某一个命名空间。使用"*"则匹配所命名空间。也可以具体指定某一个命名空间，例如"kube-system" • resource 字符串类型，表示匹配的资源类型。使用"*"将匹配所资源。也可以具体指定某一个或者某几个资源，例如"pod, service"
非资源匹配	• nonResourcePath 字符串类型，表示请求的路径。例如："/version"或"/apis"。 该字段也可以使用通配符。例如，使用 * 则匹配所有非资源请求；使用"/dev/*"则匹配"/dev/"的所有子路径
readonly	布尔类型。当该字段被设置为 true 时，资源匹配策略仅适用于 GET、LIST 和 WATCH 操作，而非资源匹配策略仅适用于 GET 操作

下面是使用 ABAC 鉴权时的几个 JSON 策略。

- "Tom"用户可以对所有资源做任何事情。

```
{
"apiVersion": "abac.authorization.kubernetes.io/v1beta1",
"kind": "Policy",
```

```
"spec": {
    "user": "tom",
    "namespace": "*",
    "resource": "*",
    "apiGroup": "*"
}
}
```

- "Kubelet" 用户可以访问任何 Pod 和服务。

```
{
"apiVersion": "abac.authorization.kubernetes.io/v1beta1",
"kind": "Policy",
"spec": {
    "user": "kubelet",
    "namespace": "*",
    "resource": "pods, services",
    "readonly": true
}
}
```

- "Bob" 用户可以在命名空间 "dev-demo" 中访问 Pod。

```
{
"apiVersion": "abac.authorization.kubernetes.io/v1beta1",
"kind": "Policy",
"spec": {
    "user": "bob",
    "namespace": "dev-demo",
    "resource": "pods",
    "readonly": true
}
}
```

> 在实际情况中，通常会把一个 JSON 字符串放到一个 JSON 策略文件中，来作为访问控制的策略文件。在配置完 JSON 策略文件后，在重启 API Server 时需要指定该文件才能获取新的策略。

18.3.3 基于节点的访问控制（node 鉴权）

基于节点的访问控制（node 鉴权）是专门针对 kubectl 发出的 API 请求的一种鉴权。

> 要启用基于节点的访问控制，则需要在启动 API Server 时使用 "--authorization-mode=Node" 参数。

对于 kubectl 执行的 API 请求，Node 鉴权过程主要包括读取操作和写入操作。

1. 读取操作

当 kubectl 使用基于节点的方式操作 API 资源时，允许对以下资源进行读取操作：

- Pod、Service、Endpoint、Node。
- Secret、ConfigMap、PVC。
- 绑定到 kubelet 节点的与 Pod 相关的持久卷。

2. 写入操作

当 kubectl 使用基于节点的方式操作 API 资源时，允许对以下的资源进行写入操作：

- 节点和节点状态（启用 NodeRestriction 准入控制器，以限制 kubelet 只能修改本节点的信息）。
- Pod 和 Pod 状态（启用 NodeRestriction 准入控制器，以限制 kubelet 只能修改绑定到本节点的 Pod 信息）。
- Event（事件）。

> 由于写入操作需要启用 NodeRestriction（准入控制器），因此在启动 API Server 时，需要在 "--enable-admission-plugins" 中加入 "NodeRestriction"，例如：
> --enable-admission-plugins=...,NodeRestriction,...

18.3.4 基于 Webhook 的访问控制

基于 Webhook 的访问控制定义了 HTTP 的一个回调接口，从而实现在某些特定事件发生时，该接口的应用会向一个远端的授权服务器发送 HTTP POST 信息。因此，在启用基于 Webhook 的访问控制后，Kubernetes 会调用外部的服务来对用户访问的资源进行授权。

> 要启用基于 Webhook 的访问控制，则需要在启动 API Server 时使用 "--authorization-mode=webhook" 参数。

1. 基于 Webhook 的访问控制的架构

基于 Webhook 的访问控制的架构如图 18-4 所示。

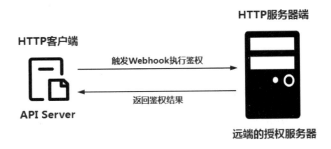

图 18-4

从架构来看，基于 Webhook 的访问控制其实是一种基于 HTTP 协议的客户端与服务器（Client-Server）架构。由于 Kubernetes 的访问控制是围绕 API Server 进行的，因此，这里的 HTTP 客户端就是 API Server，而 HTTP 服务器端就是远端的授权服务器。在配置 HTTP 客户端 API Server 时，需要使用一个配置文件来指定远端的授权服务器的信息。

下面是 Kubernetes 官方提供的一个 HTTP 客户端配置示例。

```
# Kubernetes API 版本
apiVersion: v1
# API 对象种类
kind: Config
# clusters 代表远程服务
clusters:
  - name: name-of-remote-authz-service
    cluster:
      # 对远端服务进行身份认证的 CA
      certificate-authority: /path/to/ca.pem
      # 远端服务的查询 URL。必须使用 HTTPS
      server: https://authz.example.com/authorize

# users 代表 API 服务器的 Webhook 配置
users:
  - name: name-of-api-server
    user:
      client-certificate: /path/to/cert.pem # webhook plugin 使用 cert
      client-key: /path/to/key.pem          # cert 所对应的 Key

# kubeconfig 文件必须有 context
current-context: webhook
contexts:
```

```
- context:
    cluster: name-of-remote-authz-service
    user: name-of-api-server
  name: webhook
```

其中的"clusters"字段代表的是远端的授权服务器。

有了客户端的配置文件后，在启动 API Server 时，除要指定使用 Webhook 的鉴权方式（即设置"--authorization-mode=webhook"参数）外，还需要用"--authorization-webhook-config-file"参数来指定 HTTP 客户端 API Server 的配置文件。

2. 基于 Webhook 的访问控制的运行机制

在 Kubernetes 配置好基于 Webhook 的访问控制后，在进行鉴权时，API Server 会自动向远端的授权服务器发送一个 HTTP POST 的报文，在该请求中包含一个 JSON 格式的"SubjectAccessReview"对象用来描述当前执行的动作请求。

以下是对 Kubernetes 的资源对象请求进行鉴权的 HTTP POST 报文格式。该报文希望"jane"用户可以获取 Pod 的列表。

```
{
  "apiVersion": "authorization.k8s.io/v1beta1",
  "kind": "SubjectAccessReview",
  "spec": {
    "resourceAttributes": {
      "namespace": "kittensandponies",
      "verb": "get",
      "group": "unicorn.example.org",
      "resource": "pods"
    },
    "user": "jane",
    "group": [
      "group1",
      "group2"
    ]
  }
}
```

可以看出，"SubjectAccessReview"对象中除包含被访问资源和请求的动作信息外，还包含用户的信息。

如果要对 Kubernetes 的非资源对象的请求进行鉴权，则报文格式如下：

```
{
  "apiVersion": "authorization.k8s.io/v1beta1",
  "kind": "SubjectAccessReview",
  "spec": {
    "nonResourceAttributes": {
      "path": "/logs",
      "verb": "get"
    },
    "user": "jane",
    "group": [
      "group1",
      "group2"
    ]
  }
}
```

远端的授权服务器在收到 HTTP POST 报文后，如果鉴权成功，则返回一个"SubjectAccessReview"对象，并在对象中填充"status"字段来允许访问。例如：

```
{
  "apiVersion": "authorization.k8s.io/v1beta1",
  "kind": "SubjectAccessReview",
  "status": {
    "allowed": true
  }
}
```

在远端的授权服务器拒绝请求的 HTTP POST 报文时，也是通过填充"status"字段来拒绝。例如：

```
{
  "apiVersion": "authorization.k8s.io/v1beta1",
  "kind": "SubjectAccessReview",
  "status": {
    "allowed": false,
    "reason": "user does not have read access to the namespace"
  }
}
```

如果远端的授权服务器拒绝了 API Server 的请求,同时也想拒绝其他授权者再次对该请求进行鉴权，则可以在"status"字段中增加一个"denied"参数。例如：

```
{
  "apiVersion": "authorization.k8s.io/v1beta1",
  "kind": "SubjectAccessReview",
  "status": {
    "allowed": false,
    "denied": true,
    "reason": "user does not have read access to the namespace"
  }
}
```

18.4　管理服务账号（Service Account）

服务账号（Service account）是为了方便 Pod 中的进程调用 Kubernetes API 资源或访问其他外部服务而设计的。

18.4.1　服务账号与用户账号

服务账号（Service Account）和用户账号（User Account）有着以下的不同：

- 服务账号是为 Pod 而设计的，目的是让 Pod 中的进程能够访问 Kubernetes API 的资源；用户账号是为用户而设计的，目的是让用户能够访问 Kubernetes 集群。
- 服务账号仅作用在当前的命名空间中；用户账号是全局性的，可以跨越不同的命名空间。
- 服务账号的创建应遵循权限最小化的原则，Kubernetes 允许为实现某个具体任务而创建服务账号；而创建用户账号通常需要使用外部的数据库，且可能需要组合不同的权限。
- 服务账号属于轻量级配置，只作用在当前命名空间中，因此，在一个复杂系统中可能包含各种服务账号的定义；用户账号的创建规则比较复杂，且可能涉及复杂的业务流程。

由于服务账号和用户账号的使用对象不同，因此对二者的监控也不同。

18.4.2　【实战】创建和使用服务账号

下面来演示如何创建和使用服务账号。

（1）创建一个新的命名空间"sa-demo"。

```
kubectl create namespace sa-demo
```

（2）查看新命名空间中的 Service Account。

```
kubectl get sa -n sa-demo
```

输出的信息如下：

```
NAME      SECRETS   AGE
default   1         5s
```

对于每一个命名空间会自动创建一个默认的 Service Account。

(3) 查看新命名空间中的 Secret。

```
kubectl get secret -n sa-demo
```

输出的信息如下：

```
NAME                   TYPE                                  DATA   AGE
default-token-cbxf4    kubernetes.io/service-account-token   3      2m39s
```

在创建完命名空间后，Service Account 会自动创建一个 Secret。它将用在"将当前的 Service Account 连接至 API Server 时所使用的认证信息"中。执行以下命令可以查看该 Secret（default-token-cbxf4）的详细信息。

```
kubectl describe secret default-token-cbxf4 -n sa-demo
```

(4) 编辑 "pod-sa-demo1.yaml" 文件在新创建的命名空间中新建一个 Pod。

```yaml
kind: Pod
apiVersion: v1
metadata:
  name: pod-sa-demo1
  namespace: sa-demo
spec:
  containers:
  - name: nginx
    image: nginx
    imagePullPolicy: IfNotPresent
    ports:
    - containerPort: 80
      name: http
```

(5) 创建 Pod，并查看 Pod 的信息。

```
kubectl apply -f pod-sa-demo1.yaml
kubectl get pod pod-sa-demo1 -o yaml -n sa-demo
```

输出的信息如下：

```
    ...
      volumeMounts:
      - mountPath: /var/run/secrets/kubernetes.io/serviceaccount
        name: default-token-cbxf4
        readOnly: true
    ...
    serviceAccount: default
    serviceAccountName: default
    ...
```

可以看出，对于每一个新创建的 Pod 来说，如果没有指定 Service Account，则它会自动挂载默认的 Service Acount，即其值为"default"。

（6）在"sa-demo"命名空间中创建一个名称为"mysa"的 Service Account，并查看当前命名空间中存在的 Service Account。

```
kubectl create serviceaccount mysa -n sa-demo
kubectl get sa -n sa-demo
```

输出的信息如下：

```
[root@master yaml]# kubectl create serviceaccount mysa -n sa-demo
serviceaccount/mysa created
[root@master yaml]# kubectl get sa -n sa-demo
NAME      SECRETS   AGE
default   1         14m
mysa      1         11s
```

在 Service Account 创建成功后，可以使用 RoleBinding 将 Service Account 与 Role 进行绑定，使用 ClusterRoleBinding 将 Service Account 与 ClusterRole 进行绑定。

（7）编辑"sa-demo-role.yaml"文件创建 Role。

```
kind: Role
apiVersion: rbac.authorization.k8s.io/v1
metadata:
  #限定可访问的命名空间为"sa-demo"
  namespace: sa-demo
  #角色名称
  name: sa-demo-role
rules:
  #空字符串表示使用 core API group
- apiGroups: [""]
```

```
    resources: ["namespaces","pods","pods/log"]
    verbs: ["get", "watch", "list", "create", "update", "patch", "delete"]
  - apiGroups: [ "apps" ]
    resources: ["deployments", "daemonsets"]
    verbs: ["get", "list", "watch"]
```

（8）编辑"sa-demo-rolebinding.yaml"文件创建 RoleBinding。

```
kind: RoleBinding
apiVersion: rbac.authorization.k8s.io/v1
metadata:
  namespace: sa-demo
  name: sa-demo-rolebinding
subjects:
- kind: ServiceAccount
  namespace: sa-demo
  # Service Account 的名称
  name: mysa
roleRef:
  kind: Role
  # 角色名称
  name: sa-demo-role
  apiGroup: rbac.authorization.k8s.io
```

 这里将命名空间"sa-demo"中名为"mysa"的 Service Account 绑定到"sa-demo-role"角色上了。

（9）创建 Role 和 RoleBinding。

```
kubectl apply -f sa-demo-role.yaml
kubectl apply -f sa-demo-rolebinding.yaml
```

（10）查看"sa-demo"命名空间中的 Role 和 RoleBinding。

```
kubectl get role,rolebinding -n sa-demo -o wide
```

输出的信息如下：

```
NAME                                            AGE
role.rbac.authorization.k8s.io/sa-demo-role     72s

NAME                                                ROLE                 SERVICEACCOUNTS
rolebinding.rbac.authorization.k8s.io               Role/sa-demo-role    sa-demo/mysa
/sa-demo-rolebinding
```

从输出信息可以看出，命名空间中的"mysa"服务账号通过"sa-demo-rolebinding"绑定到"sa-demo-role"了。

（11）编辑"sa-demo-clusterrole.yaml"文件创建 ClusterRole。

```
kind: ClusterRole
apiVersion: rbac.authorization.k8s.io/v1
metadata:
  # 由于ClusterRole 针对的是集群范围对象，因此不需要定义namespace 字段
  name: sa-demo-clusterrole
rules:
  # 空字符串""表明使用的是core API group
- apiGroups: ["rbac.authorization.k8s.io",""]
  resources: ["pods","pods/log"]
  verbs: ["get", "watch", "list"]
- apiGroups: ["apps"]
  resources: ["namespaces"]
  verbs: ["get", "list", "watch"]
```

（12）编辑"sa-demo-clusterrolebinding.yaml"文件创建 ClusterRoleBinding。

```
apiVersion: rbac.authorization.k8s.io/v1beta1
kind: ClusterRoleBinding
metadata:
  name: sa-demo-clusterrolebinding
roleRef:
  apiGroup: rbac.authorization.k8s.io
  kind: ClusterRole
  name: sa-demo-clusterrole
subjects:
- apiGroup: ""
  kind: ServiceAccount
  namespace: sa-demo
  name: mysa
```

（13）创建 ClusterRole 和 ClusterRoleBinding。

```
kubectl apply -f sa-demo-clusterrole.yaml
kubectl apply -f sa-demo-clusterrolebinding.yaml
```

这里将命名空间"sa-demo"中名为"mysa"的 Service Account 绑定到"sa-demo-clusterrole"角色上了。

（14）查看 ClusterRole 的信息。

```
kubectl get clusterrole
```

输出的信息如下：

```
NAME                            AGE
admin                           11d
cluster-admin                   11d
edit                            11d
flannel                         11d
sa-demo-clusterrole             5m49s
system:aggregate-to-admin       11d
system:aggregate-to-edit        11d
```

（15）查看 ClusterRoleBinding 的信息。

```
kubectl get clusterrolebinding -o \
custom-columns=NAME:.metadata.name,ROLE:.roleRef.name,SERVICEACCOUNTS:.s
ubjects[0].name \
 | grep sa-demo;
```

这里使用 Linux 的管道过滤输出信息，只显示与"sa-demo"相关的信息。

输出的信息如下：

```
NAME                            ROLE                    SERVICEACCOUNTS
sa-demo-clusterrolebinding      sa-demo-clusterrole     mysa
```

18.4.3　服务账号的工作机制

Kubernetes 的 API Service 默认开启了 Service Account 的准入控制功能。因此，在创建 Service Account 后，服务账号将按照以下方式运行。

如果 API Server 没有启用 Service Account 的准入控制功能，则可以在启动 API Server 时加入以下参数：

```
--admission_control=ServiceAccount
```

（1）在 Pod 创建时，通过"spec.serviceAccount"字段挂载指定的 Service Account。如果没有指定该字段，则挂载命名空间中的默认 Service Account，即"default"的 Service Account。

（2）API Server 验证挂载的 Service Account 是否存在。如果存在，则创建 Pod；否则拒绝创建 Pod。

（3）如果在 Pod 中没有指定 ImagePullSecrets，则把 Service Account 中的"ImagePullSecrets"字段加载到 Pod 中。

> 在拉取私有镜像仓库中的镜像时，往往需要先进行用户认证。"ImagePullSecrets"字段用于指定拉取镜像时用户的 Secret 信息。

（4）在 Pod 创建成功并启动容器后，会将 Service Account 的 Token 挂载到"/var/run/secrets/kubernetes.io/serviceaccount/"目录下。

> 可以通过"kubectl describe pod"命令查看到挂载的信息。例如：
> ```
> kubectl describe pod nginx-7bb7cd8db5-wtwnt
> ```
> 输出的挂载信息如下：
> ```
> ...
>     Restart Count:  0
>     Environment:    <none>
>     Mounts:
>       /var/run/secrets/kubernetes.io/serviceaccount from default-token-gp47w (ro)
> ...
> Volumes:
>   default-token-gp47w:
>     Type:        Secret (a volume populated by a Secret)
>     SecretName:  default-token-gp47w
>     Optional:    false
> ...
> ```

第 19 章
Kubernetes 中的日志收集与监控

通过使用命令行的方式,可以查看 Kubernetes 集群的日志,以及监控 Kubernetes 的运行状态。但是这种方式非常麻烦,需要不断输入命令语句。Kubernetes 集群可以很方便地实现可视化的日志收集和监控,利用这种方式可以方便地进行系统的运维和管理。

19.1 收集哪些日志

Kubernetes 的日志收集主要收集以下内容。

1. Kubernetes 集群的系统组件日志

Kubernetes 集群的系统组件(如 API Server、kube-scheduler 和 kube-proxy)在容器运行时会产生运行日志。这些日志信息遵循与应用日志相同的记录原则。

如果 Kubernetes 底层的容器引擎使用的是 Docker,且宿主机上的 systemd 服务可用,则 Kubernetes 会在宿主机的 "/var/log/" 目录下产生日志信息。因此,收集 Kubernetes 集群的系统组件日志就是采集宿主机 "/var/log/" 目录下的日志文件。

2. 应用日志

在 Kubernetes 集群中部署的应用(如 Nginx、Tomcat 等)在运行的过程中也会产生的日志。

通常，应用的日志由其自身来记录和维护。例如：Nginx 将其日志写在"/usr/local/nginx/logs"目录下的文件中；而 Tomcat 将其日志写在"/usr/local/tomcat/logs"目录下的文件中。因此，收集应用的日志就是采集各个应用保存的日志文件。

19.2 日志收集方案

19.2.1 初识 ELK

目前业界比较成熟的解决方案是基于 ELK（Elasticsearch + Logstash + Kibana）来收集、存储和展示日志。

1. 收集日志：Logstash

Logstash 是一个数据实时传输管道：能够将数据实时地从输入端传输到输出端，还能够根据实际的需求在传输过程中加入过滤器来筛选数据。在日志系统中，它常被作为日志采集工具使用。

2. 存储日志：Elasticsearch

Elasticsearch 是一个开源的全文搜索和分析引擎：能够快速地存储、搜索和分析数据。Elasticsearch 可以被看成一个非关系型数据库。在日志系统中，它常被作为存储和搜索日志的工具使用。

3. 展示日志：Kibana

Kibana 是一个开源的分析与可视化平台。在日志系统中，它常作为 Elasticsearch 的输出端使用。用户可以使用 Kibana 搜索、查看 Elasticsearch 中的数据，并以不同的方式（如图表、表格、地图等）进行展现。

19.2.2 日志收集的架构

Kubernetes 集群基于 ELK 的日志收集架构如图 19-1 所示。

Kubernetes 在实现日志收集过程中，除需要 ELK 的支持外，还需要 Filebeat 的支持。

> Filebeat 是用于转发和采集日志数据的轻量级传送程序。它可以被部署在容器中，以指定的方式监视日志文件和目录，并将收集的日志转发到 Logstash 中。

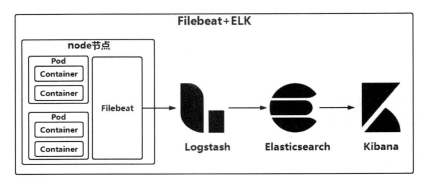

图 19-1

19.2.3 日志收集方案详解

日志收集在具体实现时主要有以下 3 种方式。

方式 1　node 节点上部署专门的日志收集程序

这种方式的核心是：在 node 节点上部署一个日志收集代理来收集该节点的日志信息。为了保证每个 node 节点都能够运行这样一个代理，可以采用 DaemonSet 方式来运行代理，如图 19-2 所示：将宿主机的容器日志文件挂载到 DaemonSet 中，然后由日志收集代理将日志转发到后端的日志收集系统 ELK 中。

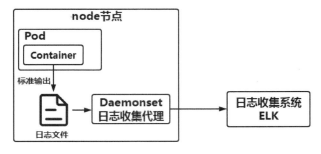

图 19-2

方式 2　Pod 中增加专用的日志收集容器

这种方式的核心是：在每个运行应用的 Pod 中增加一个新的容器，该容器专门用于运行日志收集代理程序。应用可以通过数据卷的方式将日志挂载到这个容器中，最终将日志转发到后端的日志收集系统 ELK 中，如图 19-3 所示。

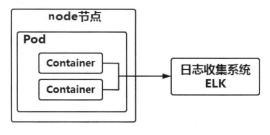

图 19-3

方式 3　通过应用推送日志

这种方式的核心是：在应用中直接将日志信息推送给后端的日志收集系统 ELK 中，如图 19-4 所示。一般在 Kubernetes 集群中不会使用这种方式。

图 19-4

表 19-1 对比了这 3 种方式的区别。

表 19-1

方式	优点	缺点
方式 1	每个 node 节点仅需部署一个日志收集程序即可。资源消耗少，对应用无侵入	将应用日志进行标准输出或者标准错误输出，不支持多行日志
方式 2	能够实现 Pod 之间的低耦合	每个 Pod 需要单独启动一个容器来运行日志收集代理，增加了资源消耗和运维的成本
方式 3	无须部署额外的日志收集工具	侵入应用，增加了应用的复杂度

19.3　实现 Kubernetes 集群的日志收集

本节将具体演示如何使用 ELK 实现 Kubernetes 集群的日志收集。

19.3.1 安装 ELK

下面将 Elasticsearch、Logstash 和 Kibana 都安装在 master 节点上。安装 ELK 需要依赖 JDK，因此在 master 节点上需要先安装 JDK。

在安装与配置 ELK 时，建议不要使用 root 用户，可以创建一个普通用户来进行。

（1）创建"elk"用户，并设置用户密码。

```
useradd elk
passwd elk
```

（2）切换到"elk"用户。

```
su - elk
```

（3）解压缩 JDK 安装包到当前路径下。

```
tar -zxvf jdk-8u181-linux-x64.tar.gz
```

（4）编辑 "~/.bash_profile" 文件设置 JDK 的环境变量，在该文件中输入以下内容：

```
JAVA_HOME=/home/elk/jdk1.8.0_181
export JAVA_HOME
PATH=$JAVA_HOME/bin:$PATH
export PATH
```

（5）生效 JDK 的环境变量。

```
source ~/.bash_profile
```

（6）解压缩 Elasticsearch 安装包到当前路径下。

```
tar -zxvf elasticsearch-6.8.7.tar.gz
```

（7）启动 Elasticsearch。

```
elasticsearch-6.8.7/bin/elasticsearch &
```

在 Elasticsearch 启动成功后，会输出以下日志信息。
```
[z1g-NKv] publish_address {127.0.0.1:9200},
         bound_addresses {[::1]:9200}, {127.0.0.1:9200}
[z1g-NKv] started
```

（8）测试 Elasticsearch 是否正常工作。

```
curl localhost:9200
```

如果输出如下信息，则说明 Elasticsearch 已经正常启动。

```
{
  "name" : "z1g-NKv",
  "cluster_name" : "elasticsearch",
  "cluster_uuid" : "WoMe6IY2TTidcSPkBJrN9Q",
  "version" : {
    "number" : "6.8.7",
    "build_flavor" : "default",
    "build_type" : "tar",
    "build_hash" : "c63e621",
    "build_date" : "2020-02-26T14:38:01.193138Z",
    "build_snapshot" : false,
    "lucene_version" : "7.7.2",
    "minimum_wire_compatibility_version" : "5.6.0",
    "minimum_index_compatibility_version" : "5.0.0"
  },
  "tagline" : "You Know, for Search"
}
```

（9）解压缩 Kibana 安装包到当前路径下。

```
tar -zxvf kibana-6.8.7-linux-x86_64.tar.gz
```

（10）编辑 Kibana 的配置文件"kibana-6.8.7-linux-x86_64/config/kibana.yml"。

```
server.port: 5601
#需要写成master节点的IP地址
server.host: "192.168.79.11"
elasticsearch.hosts: ["http://localhost:9200"]
```

（11）启动 Kibana。

```
kibana-6.8.7-linux-x86_64/bin/kibana &
```

> 在 Kibana 启动成功后，会输出以下日志信息：
> [info][listening] Server running at http://192.168.79.11:5601

（12）访问 Kibana 的 Web 控制台"http://192.168.79.11:5601/"，可以看到如图 19-5 所示界面。

图 19-5

（13）解压缩 Logstash 安装包到当前路径下。

```
tar -zxvf logstash-6.8.7.tar.gz
```

（14）创建 Logstash 的配置文件"logstash-6.8.7/config/logstash.conf"，并输入以下内容：

```
input {
  beats {
    port => 5044
  }
}
output {
  elasticsearch {
    hosts => ["http://localhost:9200"]
    index => "k8s-log-%{+YYYY-MM-dd}"
  }
}
```

 这里在日志的 index 中添加了"k8s-log"便签，是为了方便后续的日志查找。

（15）启动 Logstash。

```
cd logstash-6.8.7
bin/logstash -f config/logstash.conf &
```

启动成功后将输出以下信息：

```
Starting server on port: 5044
Successfully started Logstash API endpoint {:port=>9600}
```

19.3.2 【实战】采集 Kubernetes 系统组件的日志

在 19.1 节中提到，Kubernetes 集群系统组件的日志主要包括 API Server、kube-scheduler 和 kube-proxy 等组件产生的日志。下面来演示如何收集 Kubernetes 系统组件的日志。

（1）创建"k8s-logs.yaml"文件，并输入以下内容：

```
apiVersion: v1
kind: ConfigMap
metadata:
  name: k8s-logs-filebeat-config
  namespace: kube-system
data:
  filebeat.yml: |-
    filebeat.prospectors:
      - type: log
        paths:
          - /messages
        fields:
          app: k8s
          type: module
        fields_under_root: true
    output.logstash:
      hosts: ['192.168.79.11:5044']
---
apiVersion: apps/v1
kind: DaemonSet
metadata:
  name: k8s-logs
  namespace: kube-system
spec:
  selector:
    matchLabels:
      project: k8s
      app: filebeat
  template:
    metadata:
      labels:
        project: k8s
        app: filebeat
    spec:
```

```yaml
containers:
- name: filebeat
  image: collenzhao/filebeat:6.5.4
  args: [
    "-c", "/etc/filebeat.yml",
    "-e",
  ]
  resources:
    requests:
      cpu: 100m
      memory: 100Mi
    limits:
      cpu: 500m
      memory: 500Mi
  securityContext:
    runAsUser: 0
  volumeMounts:
  - name: filebeat-config
    mountPath: /etc/filebeat.yml
    subPath: filebeat.yml
  - name: k8s-logs
    mountPath: /messages
volumes:
- name: k8s-logs
  hostPath:
    path: /var/log/messages
    type: File
- name: filebeat-config
  configMap:
    name: k8s-logs-filebeat-config
```

"k8s-logs.yaml" 文件会挂载宿主机 "/var/log/" 目录下的 message 文件到容器中，并通过 Filebeat 采集 Kubernetes 系统组件的日志。

（2）执行 "kubect apply -f" 命令。

```
kubectl apply -f k8s-logs.yaml
```

（3）查看创建的 ConfigMap、DaemonSet 和 Pod 信息。

```
kubectl get cm,ds,pod -n kube-system | grep k8s-log
```

输出的信息如下：

```
configmap/k8s-logs-filebeat-config
daemonset.extensions/k8s-logs
pod/k8s-logs-pcghr
pod/k8s-logs-v86jt
```

这里只列出了资源的名称。

（4）单击 Kibana 首页中的"Explore on my own"链接进入 Kibana，如图 19-6 所示。

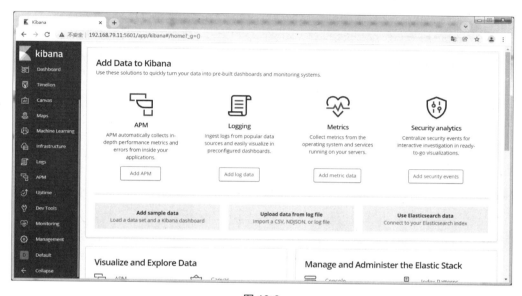

图 19-6

（5）单击左侧的"Management"链接，然后单击"Index Management"链接，在"Index management"的管理页面中便可以看到 Kubernetes 的日志的索引信息，如图 19-7 所示。

图 19-7

（6）单击"Discover"链接，并在"Create index pattern"页面中输入"k8s"。这时 Kibana 将自动对已有的索引信息进行匹配，如图 19-8 所示。

图 19-8

（7）单击"Next step"链接。

（8）在"Configure settings"的"Time Filter field name"下拉框中选择"@timestamp"（表示按照日志的时间戳查看日志信息），然后单击"Create index pattern"按钮，如图 19-9 所示。

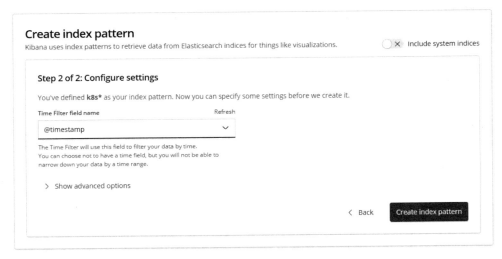

图 19-9

（9）配置完成的界面如图 19-10 所示。

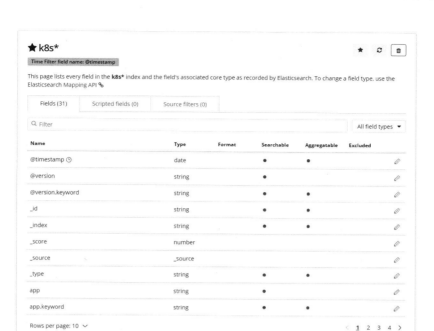

图 19-10

（10）再次单击"Discover"链接就可以看到 Kubernetes 的日志了，如图 19-11 所示。

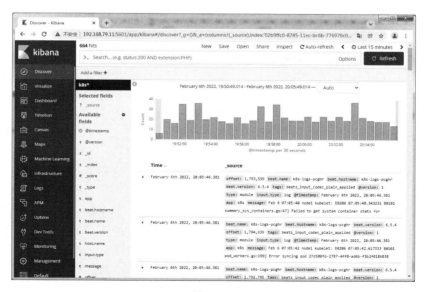

图 19-11

19.3.3 【实战】采集 Nginx 应用的日志

下面演示如何收集部署在 Kubernetes 集群中的 Nginx 应用的日志。

(1)创建 "filebeat-nginx-configmap.yaml" 文件配置 Filebeat 参数。

```yaml
apiVersion: v1
kind: ConfigMap
metadata:
  name: filebeat-nginx-config
data:
  filebeat.yml: |-
    filebeat.prospectors:
      - type: log
        paths:
          - /usr/local/nginx/logs/access.log
        # tags: ["access"]
        fields:
          app: www
          type: nginx-access
        fields_under_root: true
      - type: log
        paths:
          - /usr/local/nginx/logs/error.log
        # tags: ["error"]
        fields:
          app: www
          type: nginx-error
        fields_under_root: true
    output.logstash:
      hosts: ['192.168.79.11:5044']
```

(2)创建 "nginx-deployment.yaml" 文件部署 Nginx 应用。

```yaml
apiVersion: v1
kind: Service
metadata:
  name: nginx-test-service
spec:
  type: NodePort
  ports:
  - name: http
    port: 80
    protocol: TCP
  selector:
    app: www
    type: nginx-demo
```

```yaml
---
apiVersion: apps/v1beta1
kind: Deployment
metadata:
  name: nginx-demo
spec:
  replicas: 2
  selector:
    matchLabels:
      app: www
      type: nginx-demo
  template:
    metadata:
      labels:
        app: www
        type: nginx-demo
    spec:
      containers:
      - name: nginx
        image: nginx
        imagePullPolicy: IfNotPresent
        volumeMounts:
        - name: nginx-logs
          mountPath: /var/log/nginx
      - name: filebeat
        image: collenzhao/filebeat:6.5.4
        args: [
          "-c", "/etc/filebeat.yml",
          "-e",
        ]
        volumeMounts:
        - name: filebeat-config
          mountPath: /etc/filebeat.yml
          subPath: filebeat.yml
        - name: nginx-logs
          mountPath: /usr/local/nginx/logs
      volumes:
      - name: nginx-logs
        emptyDir: {}
      - name: filebeat-config
        configMap:
          name: filebeat-nginx-config
```

在"nginx-deployment.yaml"文件中添加了一个"filebeat"容器,并将其以临时数据卷的形式挂载到 Nginx 日志目录"/usr/local/nginx/logs"下。

(3)执行"kubectl apply -f"命令。

```
kubectl apply -f filebeat-nginx-configmap.yaml
kubectl apply -f nginx-deployment.yaml
```

(4)修改 Logstash 的配置文件"logstash-6.8.7/config/logstash.conf"如下:

```
input {
  beats {
    port => 5044
  }
}
output {
   if [app] == "www" {
      if [type] == "nginx-demo" {
         elasticsearch {
            hosts => ["http://localhost:9200"]
            index => "nginx-demo-%{+YYYY.MM.dd}"
         }
      }
      else if [type] == "nginx-error" {
         elasticsearch {
            hosts => ["http://localhost:9200"]
            index => "nginx-error-%{+YYYY.MM.dd}"
         }
      }
      else if [type] == "javawebdemo" {
         elasticsearch {
            hosts => ["http://localhost:9200"]
            index => "tomcat-catalina-javawebdemo-%{+YYYY.MM.dd}"
         }
      }
   } else if [app] == "k8s" {
      if [type] == "module" {
         elasticsearch {
            hosts => ["http://localhost:9200"]
            index => "k8s-log-%{+YYYY.MM.dd}"
         }
      }
   }
}
```

> 通过 ELK 采集的应用日志类型有很多，因此需要在 Logstash 的配置文件中针对不同应用设置不同的 index 索引。例如：在"nginx-deployment.yaml"文件中定义的 Label 标签：
> ```
> app: www
> type: nginx-demo
> ```
> 这里的 Label 标签将对应 Logstash 输出信息中的以下内容：
> ```
> if [app] == "www" {
>     if [type] == "nginx-demo" {
>         elasticsearch {
>             hosts => ["http://localhost:9200"]
>             index => "nginx-demo-%{+YYYY.MM.dd}"
>         }
>     }
> …
> ```

（5）重启 Logstash 并访问部署的 Nginx 应用。下面是 Logstash 输出的系统组件的日志信息：

```
[k8s-log-2022.02.06] creating index
[k8s-log-2022.02.06/0vz4KGrFSCWBt7MKDQGy2Q] create_mapping [doc]
[nginx-error-2022.02.06] creating index
[nginx-error-2022.02.06/i_ewQy2cRvG5I1riYeNX6g] create_mapping [doc]
[nginx-access-2022.02.06] creating index
[nginx-access-2022.02.06/E_Yqop44TQuwQ5GpnXdREw] create_mapping [doc]
```

（6）按照 19.3.2 节的步骤在 Kibana 中查看 Nginx 的日志信息，如图 19-12 所示。

图 19-12

19.3.4 【实战】采集 Tomcat 应用的日志

采集部署在 Kubernetes 集群中的 Tomcat 应用的日志，与采集 Nginx 应用的日志类似。下面是完整的步骤。

（1）创建 "filebeat-tomcat-configmap.yaml" 文件配置 Filebeat 参数，用于采集 Tomcat 应用的日志。

```yaml
apiVersion: v1
kind: ConfigMap
metadata:
  name: filebeat-tomcat-configmap
data:
  filebeat.yml: |-
    filebeat.prospectors:
    - type: log
      paths:
        - /usr/local/tomcat/logs/localhost_access_log*
      # tags: ["access"]
      fields:
        app: www
        type: javawebdemo
      fields_under_root: true
    output.logstash:
      hosts: ['192.168.79.11:5044']
```

（2）新建 "javawebdemo-deployment.yaml" 文件部署 Java Web 应用。

> 这里还部署了一个 "filebeat" 容器。部署在 Tomcat 中的 Java Web 应用日志会从该容器进行采集。

```yaml
apiVersion: v1
kind: Service
metadata:
  name: javawebdemo-service
spec:
  type: NodePort
  ports:
  - name: http
    port: 8080
  selector:
    app: www
    type: javawebdemo
```

```yaml
---
apiVersion: apps/v1
kind: Deployment
metadata:
  labels:
    app: www
    type: javawebdemo
  name: javawebdemo
spec:
  replicas: 1
  selector:
    matchLabels:
      app: www
      type: javawebdemo
  template:
    metadata:
      labels:
        app: www
        type: javawebdemo
    spec:
      containers:
      - name: k8s-javaweb-demo
        image: collenzhao/k8s-javaweb-demo
        imagePullPolicy: IfNotPresent
        volumeMounts:
        - name: tomcat-logs
          mountPath: /usr/local/tomcat/logs
      - name: filebeat
        image: collenzhao/filebeat:6.5.4
        imagePullPolicy: IfNotPresent
        args: [
          "-c", "/etc/filebeat.yml",
          "-e",
        ]
        securityContext:
          runAsUser: 0
        volumeMounts:
        - name: filebeat-config
          mountPath: /etc/filebeat.yml
          subPath: filebeat.yml
        - name: tomcat-logs
          mountPath: /usr/local/tomcat/logs
      volumes:
      - name: tomcat-logs
        emptyDir: {}
```

```
    - name: filebeat-config
      configMap:
        name: filebeat-tomcat-configmap
```

(3)执行"kubectl apply –f"命令。

```
kubectl apply -f filebeat-tomcat-configmap.yaml
kubectl apply -f javawebdemo-deployment.yaml
```

(4)修改 Logstash 的配置文件"logstash-6.8.7/config/logstash.conf"。

> 这里在 Logstash 的配置中增加了以下匹配规则。
> ```
> else if [type] == "javawebdemo" {
>   elasticsearch {
>     hosts => ["http://localhost:9200"]
>     index => "tomcat-catalina-javawebdemo-%{+YYYY.MM.dd}"
>   }
> }
> ```

完整的配置文件如下:

```
input {
  beats {
    port => 5044
  }
}
output {
   if [app] == "www" {
      if [type] == "nginx-demo" {
         elasticsearch {
            hosts => ["http://localhost:9200"]
            index => "nginx-demo-%{+YYYY.MM.dd}"
         }
      }
      else if [type] == "nginx-error" {
         elasticsearch {
            hosts => ["http://localhost:9200"]
            index => "nginx-error-%{+YYYY.MM.dd}"
         }
      }
      else if [type] == "javawebdemo" {
         elasticsearch {
            hosts => ["http://localhost:9200"]
            index => "tomcat-catalina-javawebdemo-%{+YYYY.MM.dd}"
         }
```

```
      }
   } else if [app] == "k8s" {
      if [type] == "module" {
         elasticsearch {
            hosts => ["http://localhost:9200"]
            index => "k8s-log-%{+YYYY.MM.dd}"
         }
      }
   }
}
```

（5）重启 Logstash，并访问部署的 Java Web 应用。

（6）按照 19.3.2 节的步骤在 Kibana 上查看 Tomcat 的日志信息，如图 19-13 和图 19-14 所示。

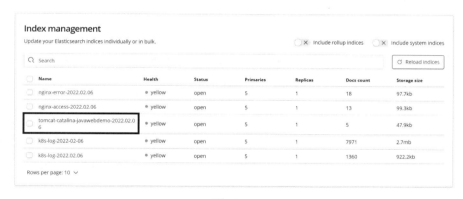

图 19-13

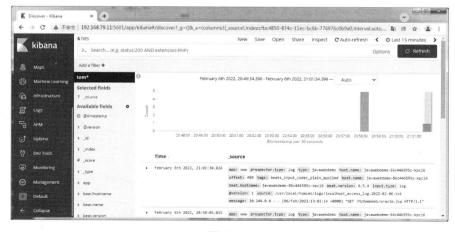

图 19-14

19.4 监控 Kubernetes

在实际的生产环境中，除收集日志外，还需要对 Kubernetes 集群进行有效的监控。

19.4.1 Kubernetes 监控方案

监控 Kubernetes 包含监控基础架构平台和监控正在运行的工作负载。利用 Kubernetes 本身的命令可能无法满足实际的要求，因此在实际的生产环境中，可以将 Kubernetes 与外部的监控系统进行集成。

Kubernetes 监控主要体现为对集群的监控和对 Pod 的监控，包括节点资源利用率、节点数、运行 Pod 的容器指标等。

图 19-15 展示了 Kubernetes 的监控的一种实现的方式。

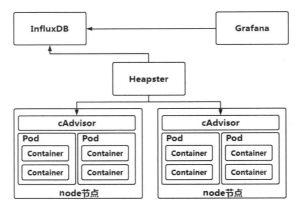

图 19-15

（1）Heapster。

它是集群监控和性能收集的工具。它会与 Kubernetes 的监控进程 cAdvisor 进行通信，将每个 node 节点上的监控进程 cAdvisor 的监控数据进行汇总，然后导到第三方工具（如 InfluxDB）中。目前，Kubernetes 已经将监控进程 cAdvisor 的功能集成到 kubelet 组件中了。

（2）InfluxDB。

它是一个时间序列数据库，用于保存和查询包含大量时间戳的数据。在 Kubernetes 集群监控中，InfluxDB 可用于 Heapster 的后端存储。

（3）Grafana。

它是一款用 Go 语言开发的开源数据可视化工具。在 Kubernetes 集群监控系统中，Grafana 用于监控信息的可视化展示。

19.4.2 【实战】部署 Kubernetes 监控系统

下面基于图 19-15 的架构来部署 Kubernetes 监控系统。

（1）创建"influxdb.yaml"文件部署 influxDB，在文件中输入以下内容：

```yaml
apiVersion: extensions/v1beta1
kind: Deployment
metadata:
  name: monitoring-influxdb
  namespace: kube-system
spec:
  replicas: 1
  template:
    metadata:
      labels:
        task: monitoring
        k8s-app: influxdb
    spec:
      containers:
      - name: influxdb
        image: registry.cn-hangzhou.aliyuncs.com/google-containers/heapster-influxdb-amd64:v1.1.1
        imagePullPolicy: IfNotPresent
        volumeMounts:
        - mountPath: /data
          name: influxdb-storage
      volumes:
      - name: influxdb-storage
        emptyDir: {}
---
apiVersion: v1
kind: Service
metadata:
  labels:
    task: monitoring
    kubernetes.io/cluster-service: 'true'
    kubernetes.io/name: monitoring-influxdb
  name: monitoring-influxdb
  namespace: kube-system
spec:
  ports:
```

```
    - port: 8086
      targetPort: 8086
  selector:
    k8s-app: influxdb
```

（2）创建 "heapster.yaml" 文件部署 Heapster，在文件中输入以下内容：

```
apiVersion: v1
kind: ServiceAccount
metadata:
  name: heapster
  namespace: kube-system
---
kind: ClusterRoleBinding
apiVersion: rbac.authorization.k8s.io/v1beta1
metadata:
  name: heapster
roleRef:
  kind: ClusterRole
  name: cluster-admin
  apiGroup: rbac.authorization.k8s.io
subjects:
  - kind: ServiceAccount
    name: heapster
    namespace: kube-system
---
apiVersion: extensions/v1beta1
kind: Deployment
metadata:
  name: heapster
  namespace: kube-system
spec:
  replicas: 1
  template:
    metadata:
      labels:
        task: monitoring
        k8s-app: heapster
    spec:
      serviceAccountName: heapster
      containers:
      - name: heapster
        image: registry.cn-hangzhou.aliyuncs.com/google-containers/heapster-amd64:v1.4.2
        imagePullPolicy: IfNotPresent
        command:
        - /heapster
```

```
        - --source=kubernetes:https://kubernetes.default
        - --sink=influxdb:http://monitoring-influxdb:8086
---
apiVersion: v1
kind: Service
metadata:
  labels:
    task: monitoring
    kubernetes.io/cluster-service: 'true'
    kubernetes.io/name: Heapster
  name: heapster
  namespace: kube-system
spec:
  ports:
  - port: 80
    targetPort: 8082
  selector:
    k8s-app: heapster
```

由于 Heapster 要连接 API Server 获取每个节点的监控指标，所以，需要 RBAC 授权和每个节点都开启监控指标暴露端口（10255 端口）。

按照以下步骤开启每个节点的 10255 端口。

（1）编辑 "/etc/sysconfig/kubelet" 文件，修改以下参数：
`KUBELET_EXTRA_ARGS="--fail-swap-on=false --read-only-port=10255"`
（2）重启 Kubernetes 服务。
`systemctl restart kubelet.service`
（3）测试是否能获取 metrics 信息。
`curl master:10255/metrics | more`
`curl node1:10255/metrics | more`
`curl node2:10255/metrics | more`
（4）输出的 metrics 信息如图 19-16 所示。

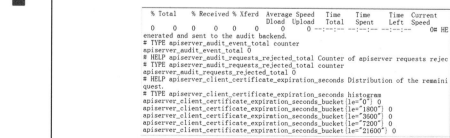

图 19-16

(3)创建"grafana.yaml"文件部署 Grafana,在文件中输入以下内容:

```yaml
apiVersion: extensions/v1beta1
kind: Deployment
metadata:
  name: monitoring-grafana
  namespace: kube-system
spec:
  replicas: 1
  template:
    metadata:
      labels:
        task: monitoring
        k8s-app: grafana
    spec:
      containers:
      - name: grafana
        image: registry.cn-hangzhou.aliyuncs.com/google-containers/heapster-grafana-amd64:v4.4.1
        imagePullPolicy: IfNotPresent
        ports:
          - containerPort: 3000
            protocol: TCP
        volumeMounts:
        - mountPath: /var
          name: grafana-storage
        env:
        - name: INFLUXDB_HOST
          value: monitoring-influxdb
        - name: GF_AUTH_BASIC_ENABLED
          value: "false"
        - name: GF_AUTH_ANONYMOUS_ENABLED
          value: "true"
        - name: GF_AUTH_ANONYMOUS_ORG_ROLE
          value: Admin
        - name: GF_SERVER_ROOT_URL
          value: /
      volumes:
      - name: grafana-storage
        emptyDir: {}
---
apiVersion: v1
kind: Service
metadata:
  labels:
```

```
      kubernetes.io/cluster-service: 'true'
      kubernetes.io/name: monitoring-grafana
    name: monitoring-grafana
    namespace: kube-system
spec:
    type: NodePort
    ports:
    - port : 80
      targetPort: 3000
    selector:
      k8s-app: grafana
```

(4）执行"kubectl apply"命令。

```
kubectl apply -f influxdb.yaml
kubectl apply -f heapster.yaml
kubectl apply -f grafana.yaml
```

(5）查看"kube-system"命名空间，如图 19-17 所示。

```
kubectl get pod,svc -n kube-system
```

```
NAME                                         READY   STATUS            RESTARTS   AGE
pod/coredns-bccdc95cf-b8wg5                  0/1     CrashLoopBackOff  50         3h29m
pod/coredns-bccdc95cf-bk78c                  0/1     CrashLoopBackOff  50         3h29m
pod/etcd-master                              1/1     Running           1          3h28m
pod/heapster-67f9d6df66-8v5d4                1/1     Running           0          10s
pod/kube-apiserver-master                    1/1     Running           1          3h28m
pod/kube-controller-manager-master           1/1     Running           3          3h28m
pod/kube-flannel-ds-amd64-87jh8              1/1     Running           0          3h27m
pod/kube-flannel-ds-amd64-94tbz              1/1     Running           0          3h27m
pod/kube-flannel-ds-amd64-fk4s4              1/1     Running           1          3h27m
pod/kube-proxy-6r9tv                         1/1     Running           0          3h28m
pod/kube-proxy-8gmg8                         1/1     Running           0          3h28m
pod/kube-proxy-r85lw                         1/1     Running           1          3h29m
pod/kube-scheduler-master                    1/1     Running           3          3h28m
pod/monitoring-grafana-5f966856fb-6wz2w      1/1     Running           0          6s
pod/monitoring-influxdb-6cf5cc5c4c-gwq29     1/1     Running           0          14s

NAME                           TYPE        CLUSTER-IP    EXTERNAL-IP   PORT(S)                      AGE
service/heapster               ClusterIP   10.1.92.251   <none>        80/TCP                       10s
service/kube-dns               ClusterIP   10.1.0.10     <none>        53/UDP,53/TCP,9153/TCP       3h29m
service/monitoring-grafana     NodePort    10.1.11.241   <none>        80:32176/TCP                 6s
service/monitoring-influxdb    ClusterIP   10.1.147.7    <none>        8086/TCP                     14s
[root@master yaml]#
```

图 19-17

(6）访问 Grafana 的 Web 控制台"http://192.168.79.11:32176"，如图 19-18 所示。

(7）单击"Create your fist data source"链接，根据表 19-2 的内容添加一个 Influxdb 数据源。

图 19-18

表 19-2

参数	值
name	数据源名称，如：k8s-datasource
type	InfluxDB
url	http://monitoring-influxdb:8086
access	proxy
dataBase	k8s
user	root（可选）
password	root（可选）

（8）单击"Save & Test"按钮创建并测试数据源，如图 19-19 所示。

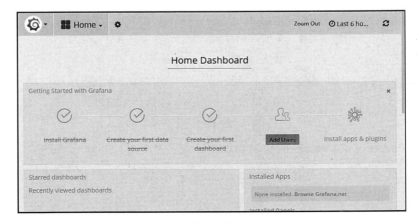

图 19-19

（9）选择"Dashboards"→"Import"（如图 19-20 所示）导入 Grafana 官方提供的监控模板文件。

图 19-20

Grafana 官方提供了很多 Kubernetes 监控的模板文件。这些模板文件可以通过 Grafana Dashboard 导入，从而可视化监控 Kubernetes 集群。例如，以下两个模板文件分别用于展示 Kubernetes 的 node 节点和 Pod 的相关监控信息。
```
kubernetes-node-statistics_rev1.json
kubernetes-pod-statistics_rev1.json
```

（10）导入"kubernetes-node-statistics_rev1.json"模板文件，并选择第（8）步创建的数据源，如图 19-21 所示。

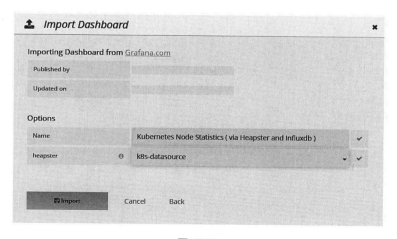

图 19-21

（11）Grafana 监控 Kubernetes 节点信息的页面如图 19-22 所示。

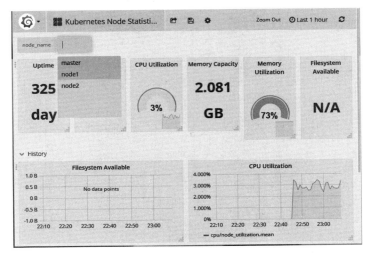

图 19-22

（12）按照第（10）步的方式导入模板文件"kubernetes-pod-statistics_rev1.json"监控 Kubernetes 的 Pod 信息，如图 19-23 所示。

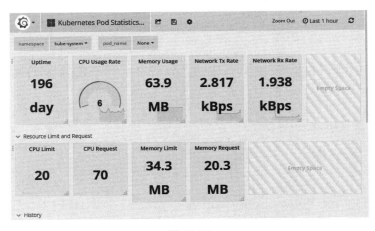

图 19-23

第 20 章

Kubernetes 集成与运维管理

在实际的生产环境中，Kubernetes 需要与外部的系统进行集成以实现更加强大的功能。例如，Kubernetes 与 Jenkins 集成在一起可以实现持续集成与持续部署（即 CI/CD）。另外，在 Kubernetes 的运维管理方面，Helm 作为 Kubernetes 的包管理器，可以用简单的方式在 Kubernetes 上查找、安装、升级、回滚和卸载应用。

20.1 Jenkins 与 Kubernetes 的持续集成与持续部署

在第 8 章中介绍了 CI/CD 的概念，也介绍了如何利用 Jenkins 与 Docker 的集成来实现这样的过程。Kubernetes 作为容器引擎的管理工具，也可以与 Jenkins 进行集成以实现更加强大的 CI/CD。

20.1.1 基于 Kubernetes 的 Jenkins 集群架构

8.2.2 节已经在 master 节点上完成了一个单节点 Jenkins 环境的部署。很明显这种单节点环境是不能满足现实需要的，因此，目前大多公司都采用 Jenkins 集群来实现符合需求的 CI/CD。

Jenkins 集群是主从式架构，会存在一个 Jenkins master 节点和多个 Jenkins slave 节点。这样的主从式架构却存在单点故障问题：

- 如果 Jenkins master 节点发生故障或宕机，则会造成整个 Jenkins 集群无法正常工作，从而影响业务流程的执行。
- 所有 Jenkins slave 节点部署在不同的主机上，若配置环境不一致则会导致管理非常不方便，并可能由于资源分配不均衡而造成维护非常麻烦。

而基于 Kubernetes 来部署 Jenkins 集群（利用虚拟化容器技术），能够很好地解决这样的问题，从而更好地实现 CI/CD。

在基于 Kubernetes 的 Jenkins 集群中，Jenkins master 节点和 Jenkins slave 节点都以 Pod 的形式运行在 Kubernetes 的 node 节点上；并且 Jenkins master 节点上的 Pod 会通过持久化存储的数据卷，将 Jenkins 的配置信息保存到外部的存储介质上以防止数据丢失；Jenkins slave 节点也会根据需求动态地创建、扩容和缩容。

图 20-1 展示了基于 Kubernetes 的 Jenkins 集群架构。

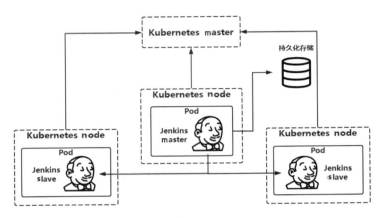

图 20-1

Jenkins master 节点在接收到构建请求时，会动态创建一个 Jenkins slave 节点来执行构建任务。在构建执行完成后，Jenkins master 节点会自动回收该 Jenkins slave 节点。

 在图 20-1 的架构中，Jenkins 集群依然是主从式架构，也存在单点故障的问题。但是，当 Jenkins master 节点出现故障时，Kubernetes 会自动创建一个新的 Pod 来运行 Jenkins master 节点，从而保证了 Jenkins 集群的高可用性。

20.1.2 【实战】Jenkins 与 Kubernetes 的集成

在集成 Jenkins 与 Kubernetes 时，首先应确保 Kubernetes 集群正常运行。由于是由 Jenkins master 节点来接收构建请求，并创建 Jenkins slave 节点来执行构建任务的，因此只需要在 Kubernetes 中部署 Jenkins master 节点即可。

1. 在 Kubernetes 中部署 Jenkins master 节点

下面演示如何在 Kubernetes 中部署一个 Jenkins master 节点。

（1）创建所需的命名空间与上下文环境。

```
kubectl create namespace kubernetes-plugin
kubectl config set-context $(kubectl config current-context) \
--namespace=kubernetes-plugin
```

（2）创建角色绑定以允许 Jenkins Kubernetes Plugin 插件访问 Kubernetes 集群。

```
kubectl create clusterrolebinding permissive-binding \
--clusterrole=cluster-admin \
--user=admin \
--user=kubelet \
--group=system:serviceaccounts
```

（3）编辑 Jenkins master 节点部署的 YAML 文件 "jenkins-master.yaml"。

```yaml
apiVersion: apps/v1beta1
kind: Deployment
metadata:
  name: jenkins-master
  labels:
    k8s-app: jenkins
spec:
  replicas: 1
  selector:
    matchLabels:
      k8s-app: jenkins
  template:
    metadata:
      labels:
        k8s-app: jenkins
    spec:
      containers:
      - name: jenkins-master
        image: jenkins/jenkins:lts-alpine
        imagePullPolicy: IfNotPresent
        volumeMounts:
        - name: jenkins-home
          mountPath: /var/jenkins_home
        ports:
        - containerPort: 8080
          name: web
        - containerPort: 50000
          name: agent
      volumes:
      - name: jenkins-home
        emptyDir: {}
```

```yaml
---
kind: Service
apiVersion: v1
metadata:
  labels:
    k8s-app: jenkins
  name: jenkins-master
spec:
  type: NodePort
  ports:
    - port: 8080
      name: web
      targetPort: 8080
    - port: 50000
      name: agent
      targetPort: 50000
  selector:
    k8s-app: jenkins
```

（4）执行"kubectl apply -f"命令。

```
kubectl apply -f jenkins-master.yaml
```

（5）查看创建的 Pod 信息，如图 20-2 所示。

```
kubectl get pod,deploy,svc
```

```
NAME                                         READY   STATUS    RESTARTS   AGE
pod/jenkins-master-65bff4c56-nckf6           1/1     Running   0          103s

NAME                                         READY   UP-TO-DATE   AVAILABLE   AGE
deployment.extensions/jenkins-master         1/1     1            1           103s

NAME                          TYPE       CLUSTER-IP    EXTERNAL-IP   PORT(S)                          AGE
service/jenkins-master        NodePort   10.1.16.41    <none>        8080:32286/TCP,50000:30713/TCP   103s
[root@master jenkins]#
```

图 20-2

从图 20-2 中可以看到 Jenkins master 节点已经成功运行在 Pod 中。

（6）查看 Jenkins 登录的初始密码。

```
kubectl logs pod/jenkins-master-65bff4c56-nckf6
```

输出的信息如下：

```
*************************************************************
Jenkins initial setup is required. An admin user has been
created and a password generated.
Please use the following password to proceed to installation:
```

```
343d681514ff4e4ab12feef2673d72af

This may also be found at:
/var/jenkins_home/secrets/initialAdminPassword
*************************************************************
```

（7）访问 Jenkins 的首页"http://192.168.79.11:32286"，并按照 8.2.2 节中的步骤配置 Jenkins。配置完成后 Jenkins 的首页如图 20-3 所示。

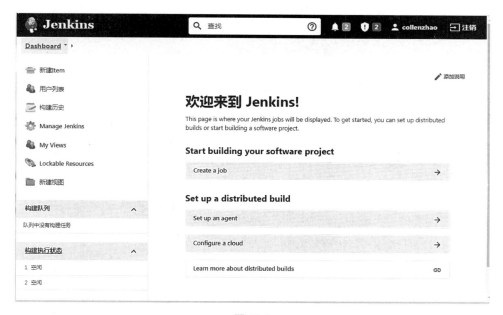

图 20-3

2. 在 Jenkins 上安装与配置 Kubernetes 的插件

由于 Jenkins master 节点在收到构建请求后需要访问 Kubernetes 集群创建 Pod 来运行 Jenkins slave 节点，因此需要在 Jenkins 上安装与配置 Kubernetes 的插件。

（1）单击左侧的"Manage Jenkins"，并选择"System Configuration"中的"Manage Plugin"，进入插件管理页面。

（2）在"可选插件"中搜索"Kubernetes"，如图 20-4 所示。勾选"Kubernetes"复选框后单击"Install without restart"按钮，插件安装完成后返回首页。

（3）为了在 Jenkins 中访问 Kubernetes，需要创建一个 Cloud 环境。单击左侧的"管理 Jenkins"，并选择"Configure Clouds"，如图 20-5 所示。

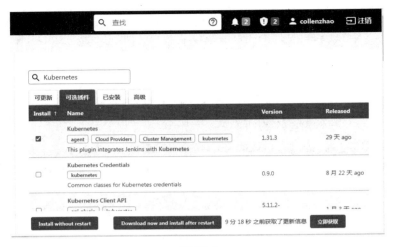

图 20-4

图 20-5

（4）在"Add a new cloud"下拉框中选项"Kubernetes"，如图 20-6 所示。

图 20-6

（5）选项右侧的"Kubernetes Cloud details..."配置 Kubernetes 的插件。按照表 20-1 的内容进行配置。

表 20-1

配置项	值
名称	kubernetes
Kubernetes 地址	https://kubernetes.default
	也可以填写：https://kubernetes.default.svc.cluster.local
Jenkins 地址	http://jenkins-master.kubernetes-plugin:8080

（6）单击"连接测试"按钮。如果 Kubernetes 的插件可以成功访问集群，则出现以下信息：

```
Connected to Kubernetes v1.15.0
```

（7）单击"Save"按钮完成 Kubernetes 插件的配置。

3. 执行 Jenkins Pipeline 任务

 Jenkins Pipeline（流水线）是 Jenkins 提供的工作流执行框架。通过使用 Jenkins Pipeline 可以将若干个独立运行的任务连接起来，从而实现复杂的流程编排。Jenkins Pipeline 是 Jenkins 的核心功能，帮助 Jenkins 实现从 CI/CD 到 DevOps 的转变。

（1）新建一个 Jenkins Pipeline 的流水线任务，如图 20-7 所示。

图 20-7

（2）在流水线定义页面中的"Pipeline script"文本框中输入以下脚本：

```
def label = "jenkins-slave-${UUID.randomUUID().toString()}"
podTemplate(label: label, cloud: 'kubernetes') {
```

```
node(label) {
    stage('Run shell') {
        sh 'sleep 60s'
        sh 'echo hello world.'
    }
}
```

(3)单击"保存"按钮，并单击"立即构建"链接执行任务。

(4)查看创建的 Pod 信息，如图 20-8 所示。

```
kubectl get pod,deploy,svc
```

```
NAME                                                                  READY   STATUS    RESTARTS   AGE
pod/jenkins-master-85444c98d7-nmch7                                   1/1     Running   0          53m
pod/jenkins-slave-e0202453-56a2-43c9-ade4-72a447e231ff-mrb8q-5nmmw    1/1     Running   0          9s

NAME                                         READY   UP-TO-DATE   AVAILABLE   AGE
deployment.extensions/jenkins-master         1/1     1            1           53m

NAME                      TYPE       CLUSTER-IP    EXTERNAL-IP   PORT(S)                          AGE
service/jenkins-master    NodePort   10.1.225.59   <none>        8080:30990/TCP, 50000:30471/TCP  53m
```

图 20-8

从图 20-8 中可以看到，在 Pod 中成功运行了一个 Jenkins slave 节点。当流水线任务执行完成后，Kubernetes 集群会自动回收这个 Pod。

(5)在 Jenkins 任务的 "Console Output" 页面中将看到以下输出信息，如图 20-9 所示。

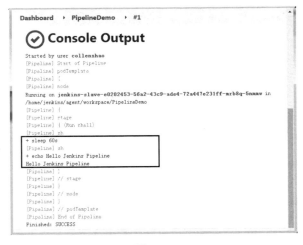

图 20-9

20.2 使用 Helm 简化 Kubernetes 应用的部署和管理

在 Kubernetes 中部署应用，需要创建 Pod、Deployment 和 Service 等资源，并且创建的步骤也比较烦琐。当遇到复杂系统时，Kubernetes 应用的部署和管理会变得更复杂。好在可以使用 Helm 来简化 Kubernetes 应用的部署和管理。

20.2.1 什么是 Helm

Helm 通过打包的方式动态创建 Kubernetes 应用的配置信息，然后生成应用的 YAML 文件，并最终由 kubectl 进行调用完成应用的部署。因此从使用方式上看，Helm 类似于 Linux YUM 的包管理。图 20-10 展示了 Helm 的体系架构。

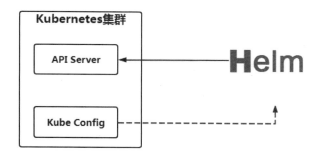

图 20-10

从 Helm 3 开始，Helm 将所有的配置信息存储在 Kubernetes 集群的配置中。在 Helm 中有 3 个非常重要的概念。

- Chart：应用信息的集合，包括应用中对 Kubernetes 资源的定义和依赖关系的说明等。
- Repository：存放 Chart 的仓库。
- Release：Chart 的运行的实例，代表一个正在运行的应用。在 Chart 在 Kubernetes 集群中部署成功后会生成一个 Release。

20.2.2 部署 Helm

在 GitHub 上提供了 Helm 多种操作系统的二进制版本。

下面在 master 节点上部署 Helm，使用的是 helm-v3.5.4-linux-amd64.tar.gz。

（1）解压缩 Helm 安装包，并将 Helm 的可执行命令复制到 "/usr/bin/" 目录下。

```
tar -zxvf helm-v3.5.4-linux-amd64.tar.gz
cd linux-amd64/
mv helm /usr/bin/
```

（2）添加 Helm 的 Repository 仓库。

```
#添加 Helm 官方的 Repository 仓库
helm repo add stable https://charts.helm.sh/stable
```

在这里可以添加多个 Repository 仓库地址，例如：

helm repo add azure http://mirror.azure.cn/kubernetes/charts/

helm repo add aliyun https://kubernetes.oss-cn-hangzhou.aliyuncs.com/charts

（3）查看 Repository 仓库信息。

```
helm repo list
```

输出的信息如下：

```
NAME      URL
stable    https://charts.helm.sh/stable
azure     http://mirror.azure.cn/kubernetes/charts/
aliyun    https://kubernetes.oss-cn-hangzhou.aliyuncs.com/charts
```

（4）Helm 仓库的其他操作。

```
#更新仓库
helm repo update
#删除仓库
helm repo remove aliyun
```

（5）查看 Helm 的配置信息。

```
helm env
```

输出的信息如下：

```
HELM_BIN="helm"
HELM_CACHE_HOME="/root/.cache/helm"
HELM_CONFIG_HOME="/root/.config/helm"
HELM_DATA_HOME="/root/.local/share/helm"
HELM_DEBUG="false"
HELM_KUBEAPISERVER=""
HELM_KUBEASGROUPS=""
HELM_KUBEASUSER=""
HELM_KUBECAFILE=""
HELM_KUBECONTEXT=""
HELM_KUBETOKEN=""
```

```
HELM_MAX_HISTORY="10"
HELM_NAMESPACE="kubernetes-plugin"
HELM_PLUGINS="/root/.local/share/helm/plugins"
HELM_REGISTRY_CONFIG="/root/.config/helm/registry.json"
HELM_REPOSITORY_CACHE="/root/.cache/helm/repository"
HELM_REPOSITORY_CONFIG="/root/.config/helm/repositories.yaml"
```

(6) 在 Repository 仓库中搜索可用的 Charts, 搜索结果如图 20-11 所示。

```
helm search repo
```

图 20-11

默认情况下会搜索所有添加的 Helm Repository 仓库, 也可以指定搜索某一个 Repository 仓库。例如, 以下搜索命令只会搜索阿里云的 Repository 仓库。

helm search repo aliyun

20.2.3 【实战】使用 Helm 管理 Kubernetes

要使用 Helm 管理 Kubernetes, 则需要设置 Helm 管理的 Kubernetes 的环境变量, 执行以下命令即可:

```
export KUBECONFIG=/root/.kube/config
```

这一步非常重要, 在"/root/.kube/config"文件中保存了 Kubernetes 集群的信息, 该信息可以保证 Helm 与 Kubernetes 进行通信。为了方便, 可以将这一步写到"/etc/profile"文件中。

1. 使用 Helm 部署应用

下面使用 Helm 在 Kubernetes 中部署一个 MySQL 数据库服务。

(1) 在 Repository 仓库中搜索可用的 MySQL Charts, 搜索的结果如图 20-12 所示。

```
helm search repo mysql
```

```
NAME                                  CHART VERSION    APP VERSION    DESCRIPTION
aliyun/mysql                          0.3.5                           Fast, reliable, scalable, and easy to use open-...
azure/mysql                           1.6.9            5.7.30         DEPRECATED - Fast, reliable, scalable, and easy...
azure/mysqldump                       2.6.2            2.4.1          DEPRECATED! - A Helm chart to help backup MySQL...
azure/prometheus-mysql-exporter       0.7.1            v0.11.0        DEPRECATED A Helm chart for prometheus mysql ex...
stable/mysql                          1.6.9            5.7.30         DEPRECATED - Fast, reliable, scalable, and easy...
stable/mysqldump                      2.6.2            2.4.1          DEPRECATED! - A Helm chart to help backup MySQL...
stable/prometheus-mysql-exporter      0.7.1            v0.11.0        DEPRECATED A Helm chart for prometheus mysql ex...
aliyun/percona                        0.3.0                           free, fully compatible, enhanced, open source d...
aliyun/percona-xtradb-cluster         0.0.2                           free, fully compatible, enhanced, open source d...
azure/percona                         1.2.3            5.7.26         DEPRECATED - free, fully compatible, enhanced,...
azure/percona-xtradb-cluster          1.0.8            5.7.19         DEPRECATED - free, fully compatible, enhanced,...
azure/phpmyadmin                      4.3.5            5.0.1          DEPRECATED phpMyAdmin is an mysql administratio...
stable/percona                        1.2.3            5.7.26         DEPRECATED - free, fully compatible, enhanced,...
stable/percona-xtradb-cluster         1.0.8            5.7.19         DEPRECATED - free, fully compatible, enhanced,...
stable/phpmyadmin                     4.3.5            5.0.1          DEPRECATED phpMyAdmin is an mysql administratio...
aliyun/gcloud-sqlproxy                0.2.3                           Google Cloud SQL Proxy
aliyun/mariadb                        2.1.6            10.1.31        Fast, reliable, scalable, and easy to use open-...
azure/gcloud-sqlproxy                 0.6.1            1.11           DEPRECATED Google Cloud SQL Proxy
azure/mariadb                         7.3.14           10.3.22        DEPRECATED Fast, reliable, scalable, and easy t...
stable/gcloud-sqlproxy                0.6.1            1.11           DEPRECATED Google Cloud SQL Proxy
stable/mariadb                        7.3.14           10.3.22        DEPRECATED Fast, reliable, scalable, and easy t...
[root@master ~]#
```

图 20-12

（2）执行以下命令部署一个 MySQL 数据库的应用。

```
helm install mysql-demo stable/mysql
```

输出的信息如下：

```
NAME: mysql-demo
LAST DEPLOYED: Thu Feb 10 06:33:49 2022
NAMESPACE: kubernetes-plugin
STATUS: deployed
REVISION: 1
NOTES:
MySQL can be accessed via port 3306 on the following DNS name from within your cluster:
  mysql-demo.kubernetes-plugin.svc.cluster.local
...
```

使用"helm install"命令至少需要两个参数：Release 的名称和 Charts 名称。以这里的命令为例："mysql-demo"是 Release 的名称，而"stable/mysql"是 Charts 的名称。

另外，可以使用"helm list"和"helm status mysql-demo"命令查询 Release 的状态信息。

（3）查看部署的 Pod、Deployment 和 Service 信息，如图 20-13 所示。

```
kubectl get all
```

```
[root@master ~]# kubectl get all
NAME                                    READY   STATUS    RESTARTS   AGE
pod/mysql-demo-5d85fc7bd7-cwpk4         0/1     Pending   0          16m

NAME                   TYPE        CLUSTER-IP   EXTERNAL-IP   PORT(S)    AGE
service/mysql-demo     ClusterIP   10.1.46.71   <none>        3306/TCP   16m

NAME                           READY   UP-TO-DATE   AVAILABLE   AGE
deployment.apps/mysql-demo     0/1     1            0           16m

NAME                                      DESIRED   CURRENT   READY   AGE
replicaset.apps/mysql-demo-5d85fc7bd7     1         1         0       16m
[root@master ~]#
```

图 20-13

这时会发现 Pod 的状态一直是"Pending"状态。

（4）查看 Pod 的详细信息。

```
kubectl describe pod/mysql-demo-5d85fc7bd7-cwpk4
```

输出的信息如下：

```
Events:
... ...  Message
... ...  -------
... ...  pod has unbound immediate PersistentVolumeClaims
```

从 Message 信息中可以看到 Pod 缺少 PVC 资源。

（5）查看 PVC 资源。

```
kubectl get pvc
```

输出的信息如下：

```
NAME          STATUS    VOLUME   CAPACITY
mysql-demo    Pending
```

（6）查看 Charts 的详细信息。

```
helm show all stable/mysql
```

通过输出的信息可以确定这里需要一个 8GB 的 PV 资源。

```
...
# Persist data to a persistent volume
persistence:
  enabled: true
  # database data Persistent Volume Storage Class
  # If defined, storageClassName: <storageClass>
  # If set to "-", storageClassName: "", which disables dynamic provisioning
  # If undefined (the default) or set to null, no storageClassName spec is
  #   set, choosing the default provisioner. (gp2 on AWS, standard on
  #   GKE, AWS & OpenStack)
  #
  # storageClass: "-"
  accessMode: ReadWriteOnce
  size: 8Gi
  annotations: {}
...
```

（7）创建 MySQL 的数据存储目录。

```
mkdir -p /mnt/mysql/data
```

（8）创建"mysql-pv-volume.yaml"文件并在其中输入以下内容：

```
kind: PersistentVolume
apiVersion: v1
metadata:
  name: pv-volume-mysql
  namespace: kubernetes-plugin
  labels:
    type: local
spec:
  capacity:
    storage: 8Gi
  accessModes:
    - ReadWriteOnce
  hostPath:
    path: "/mnt/mysql/data"
```

（9）创建 PV 资源。

```
kubectl apply -f mysql-pv-volume.yaml
```

（10）查看 PV 和 PVC 资源。

```
kubectl get pv,pvc
```

输出的信息如下：

```
NAME                             CAPACITY   ACCESS MODES
```

```
persistentvolume/pv-volume-mysql      8Gi         RWO

NAME                                  STATUS      VOLUME
persistentvolumeclaim/mysql-demo      Bound       pv-volume-mysql
```

 这时 PVC 已经与 PV 成功绑定。

（11）再次查看部署的 Pod、Deployment 和 Service，如图 20-14 所示。

```
kubectl get all
```

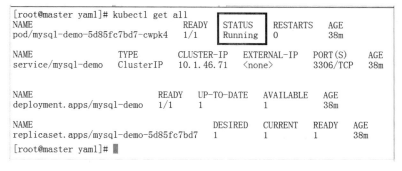

图 20-14

（12）卸载部署的 MySQL 应用。

```
helm uninstall mysql-demo
```

2. 使用 Helm 创建自己的 Charts

用户可以使用 Helm 提供的 Charts 模板创建自己应用的 Charts。下面使用 Helm 创建一个自己的 Nginx Charts，并部署到 Kubernetes 集群中。

（1）生成 Nginx Charts 的模板。

```
helm create my-nginx
```

（2）查看生成的 Charts 模板。

```
tree my-nginx/
```

输出的信息如下：

```
my-nginx/                   Charts 包目录的名称
├── charts                  依赖的子包目录，其中可以包含多个依赖的 Chart 包
├── Chart.yaml              Charts 的描述信息，如 Charts 的名称、版本信息等
```

```
│   ├── templates                          Kubernetes 应用的配置模版目录
│   │   ├── deployment.yaml                Deployment 的部署描述文件
│   │   ├── _helpers.tpl                   公有库定义文件
│   │   ├── hpa.yaml
│   │   ├── ingress.yaml                   Ingress 的部署描述文件
│   │   ├── NOTES.txt
│   │   ├── serviceaccount.yaml            ServiceAccount 的部署描述文件
│   │   ├── service.yaml                   Service 的部署描述文件
│   │   └── tests
│   │       └── test-connection.yaml
│   └── values.yaml
```

用户可以使用这里生成的模板，并编辑其中的 YAML 文件来完成相应的配置。重点是编辑 Deployment、Service 和 Ingress 的描述文件。

（3）下面是一个最简单的 Charts 模板，这里只保留了必要的文件。

```
my-nginx/
├── Chart.yaml
├── templates
│   ├── deployment.yaml
│   └── service.yaml
└── values.yaml
```

（4）编辑 "values.yaml" 文件，输入以下内容：

```yaml
deployname: my-nginx
replicaCount: 2
image:
  repository: nginx
  pullPolicy: IfNotPresent
```

这里定义了 Deployment 的名称、副本数及镜像的相关信息。

（5）编辑 "deployment.yaml" 文件。

```yaml
apiVersion: apps/v1
kind: Deployment
metadata:
  name: {{ .Values.deployname }}
  labels:
    app: my-nginx
```

```
spec:
  replicas: {{ .Values.replicaCount }}
  selector:
    matchLabels:
      app: my-nginx
  template:
    metadata:
      labels:
        app: my-nginx
    spec:
      containers:
      - name: my-nginx
        image: {{ .Values.images.repository }}
        imagePullPolicy: {{ .Values.images.pullPolicy }}
        ports:
        - containerPort: 80
```

在"deployment.yaml"文件中引用了在"values.yaml"文件中定义的变量值。

（6）编辑"service.yaml"文件，输入以下内容：

```
apiVersion: v1
kind: Service
metadata:
  name: my-nginx
spec:
  type: NodePort
  ports:
  - name: http
    port: 80
    protocol: TCP
    targetPort: 80
  selector:
    app: my-nginx
```

"service.yaml"文件也可以使用在"values.yaml"文件中定义的变量值。

（7）验证 Charts 中的各个文件格式是否正确：

```
helm lint my-nginx/
```

输出的信息如下：

```
==> Linting my-nginx/
[INFO] Chart.yaml: icon is recommended
1 chart(s) linted, 0 chart(s) failed
```

（8）打包应用。

```
helm package my-nginx/
```

输出的信息如下：

```
Successfully packaged chart and saved it to: /root/my-nginx-0.1.0.tgz
```

（9）试运行应用。

```
helm install --dry-run my-nginx my-nginx-0.1.0.tgz
```

输出的信息如下：

```
...
---
# Source: my-nginx/templates/service.yaml
apiVersion: v1
kind: Service
metadata:
  name: my-nginx
spec:
  type: NodePort
...
---
# Source: my-nginx/templates/deployment.yaml
apiVersion: apps/v1
kind: Deployment
metadata:
  name: my-nginx
  labels:
    app: my-nginx
spec:
  replicas: 2
  selector:
    matchLabels:
      app: my-nginx
...
```

（10）在 Kubernetes 集群中部署应用。

```
helm install my-nginx my-nginx-0.1.0.tgz
```

(11)查看创建的资源信息,如图 20-15 所示。

```
kubectl get all
```

```
[root@master ~]# kubectl get all
NAME                                READY   STATUS    RESTARTS   AGE
pod/my-nginx-5579d644f6-78x7t       1/1     Running   0          61s
pod/my-nginx-5579d644f6-dq2tx       1/1     Running   0          61s

NAME               TYPE       CLUSTER-IP    EXTERNAL-IP   PORT(S)        AGE
service/my-nginx   NodePort   10.1.22.58    <none>        80:31019/TCP   61s

NAME                       READY   UP-TO-DATE   AVAILABLE   AGE
deployment.apps/my-nginx   2/2     2            2           61s

NAME                                  DESIRED   CURRENT   READY   AGE
replicaset.apps/my-nginx-5579d644f6   2         2         2       61s
[root@master ~]#
```

图 20-15